AF743633

BIOCHEMISTRY RESEARCH TRENDS

HEMAGGLUTININS

STRUCTURES, FUNCTIONS AND MECHANISMS

Biochemistry Research Trends

Additional books and e-books in this series can be found on Nova's website under the Series tab.

BIOCHEMISTRY RESEARCH TRENDS

HEMAGGLUTININS

STRUCTURES, FUNCTIONS AND MECHANISMS

TZI BUN NG
JACK WONG
RYAN TSE
TAK FU TSE
AND
HELEN CHAN
EDITORS

nova science publishers
New York

Copyright © 2019 by Nova Science Publishers, Inc.

All rights reserved. No part of this book may be reproduced, stored in a retrieval system or transmitted in any form or by any means: electronic, electrostatic, magnetic, tape, mechanical photocopying, recording or otherwise without the written permission of the Publisher.

We have partnered with Copyright Clearance Center to make it easy for you to obtain permissions to reuse content from this publication. Simply navigate to this publication's page on Nova's website and locate the "Get Permission" button below the title description. This button is linked directly to the title's permission page on copyright.com. Alternatively, you can visit copyright.com and search by title, ISBN, or ISSN.

For further questions about using the service on copyright.com, please contact:
Copyright Clearance Center
Phone: +1-(978) 750-8400 Fax: +1-(978) 750-4470 E-mail: info@copyright.com.

NOTICE TO THE READER

The Publisher has taken reasonable care in the preparation of this book, but makes no expressed or implied warranty of any kind and assumes no responsibility for any errors or omissions. No liability is assumed for incidental or consequential damages in connection with or arising out of information contained in this book. The Publisher shall not be liable for any special, consequential, or exemplary damages resulting, in whole or in part, from the readers' use of, or reliance upon, this material. Any parts of this book based on government reports are so indicated and copyright is claimed for those parts to the extent applicable to compilations of such works.

Independent verification should be sought for any data, advice or recommendations contained in this book. In addition, no responsibility is assumed by the Publisher for any injury and/or damage to persons or property arising from any methods, products, instructions, ideas or otherwise contained in this publication.

This publication is designed to provide accurate and authoritative information with regard to the subject matter covered herein. It is sold with the clear understanding that the Publisher is not engaged in rendering legal or any other professional services. If legal or any other expert assistance is required, the services of a competent person should be sought. FROM A DECLARATION OF PARTICIPANTS JOINTLY ADOPTED BY A COMMITTEE OF THE AMERICAN BAR ASSOCIATION AND A COMMITTEE OF PUBLISHERS.

Additional color graphics may be available in the e-book version of this book.

Library of Congress Cataloging-in-Publication Data

ISBN: 978-1-53615-708-6
Library of Congress Control Number: 2019943534

Published by Nova Science Publishers, Inc. † New York

CONTENTS

PREFACE

Hemagglutinins refers to glycoproteins which bring about agglutination of erythrocytes or hemagglutination. Hemagglutination can be used to identify surface antigens on erythrocytes (with known antibodies) and hence the blood type of an individual. Hemagglutinins consist of lectins and antibodies. Lectins (from the Latin *legere,* "to select") are non-immune glycoproteins that exhibit reversible binding to specific carbohydrate structures, glycans of glycoproteins, glycolipids and polysaccharides. Lectins own at least one non-catalytic domain, and in some cases a few carbohydrate binding domains, which enable them to effect agglutination of erythrocytes and other cells or precipitation of glycoconjugates. Lectins are found in a constellation of organisms and display an array of activities. Topics discussed in this book encompass algal lectins, plant type 2 ribosome inactivating proteins, edible legume lectins, jacalin, jacalin-related lectin, wheat lectins, rice lectins, banana lectins and potato lectin, immunomodulatory action of plant lectins, piezoelectric assay using the galactose-binding *Bauhinia monandra* leaf lectin and its antibody as a potent tool for detection of antigen-antibody recognition, assays using hemagglutinins of different origins to investigate their mechanisms of action, lectins from medicinal herbs and medicinal mushrooms, C-type lectins from a diversity of invertebrates and vertebrates, fish lectins and amphibian lectins.

Chapter 1 - Viruses, bacteria, fungi, plant and animals exhibit a range of cell-agglutinating and sugar-binding proteins called lectins, or hemagglutinins. These proteins mediate the binding to sugar receptors in cells, promoting cell-cell communication and can initiate an infection, in the case of pathogenic microorganisms. Hemagglutinins identified in the influenza viruses have an important role in the viral infection. In this context, many assays have been developed to measure hemagglutinins in samples as vaccines, and to analyze the sugar-binding properties of hemagglutinins from different influenza virus subtypes. In this chapter, we will comment on the hemagglutination assays of lectins and their mechanisms of action. The interaction of lectins with viruses and bacterial cell wall components as well as the mechanism of action of antimicrobial lectins are demonstrated through microbial assays. The extracellular matrix of microbial biofilms represents a continuous challenge in terms of characterization; lectins are molecules with high sugar-binding specificity used to analyze glycoconjugates within biofilm communities. Lectins can be obtained in milligram quantities from plants and these proteins can harbor different biological properties. Cytotoxic action in tumor cell lines and ability to act against pathogens, such as parasites and insects, are alternatives for evaluations of these plant biomolecules. Interaction of human lectins with human cells, how the diverse mechanisms of action of these proteins can be valuable tools to study human microbiomes, diagnosis and treatment of many diseases will also be approached.

Chapter 2 - Highly purified galactose-binding lectin from *Bauhinia monandra* leaves (BmoLL) was intradermally injected into rabbit to evaluate the immune response. Antiserum developed against BmoLL (anti-BmoLL serum) revealed the immunogenicity of the lectin. Anti-BmoLL IgG was purified from the antiserum on a Protein A-Sepharose CL-4B column. BmoLL and extracts from different tissues of *B. monandra*, with previously detected lectin activity, interacted with anti-BmoLL serum and anti-BmoLL IgG; protein precipitation lines were visualized by double immunodiffusion assay. Anti-BmoLL IgG was immobilized on gold surface for electrochemical and piezoelectric characterization of BmoLL antibody-antigen interaction. An electrochemical immunosensor performed with anti-BmoLL IgG and gold electrode surface estimated BmoLL binding and its electrochemical potential. A piezoelectric immunosensor was also carried out based on a quartz crystal microbalance (QCM) with a gold surface where anti-BmoLL IgG was immobilized to characterize BmoLL recognition. The electrochemical system was sensitive for evaluating anti-BmoLL IgG interactions at different concentrations of BmoLL. The piezoelectric immunosensor with flow injection analysis (FIA) revealed a frequency decrease (15 kHz) after the interaction of BmoLL (454 μg/ml) with anti-BmoLL IgG (1.1 mg/mL) indicating antigen-antibody specificity. Then the piezoelectric assay was a potent tool for detection of antigen-antibody recognition.

Chapter 3 - Lectins are proteins with carbohydrate-binding ability. In plants, lectins have been obtained from leaves, weeds, roots, rhizomes, bulbs, pods, seeds, fruits and flowers. The interaction of lectins with carbohydrates from cell surfaces can trigger diverse intracellular responses, resulting in a range of biological activities that have been exploited for various biotechnological purposes. Among the biological activities of plant lectins, it is possible to mention immunomodulatory action that can be produced through the interaction between lectin to surface glycoproteins of immune cells. Immunomodulators may exhibit positive (immunostimulation) or negative (immunosuppression) activities over a specific immunological function. It is known that lectins can stimulate pro-inflammatory or anti-inflammatory responses. In this paper, we review plant lectins able to induce both profiles, *in vitro* and/or *in vivo*, and their possible mechanisms of action. Plant lectins are capable of altering the release of cytokines, chemokines and nitric oxide by immune cells. One of the mechanisms that may be involved in the pro-inflammatory action of lectins is the activation of Toll-like receptors (TLR). Also, it was observed that the lectin-induced profile of cytokines may vary according to the model, the type of immune cell and the route of administration. Moreover, plant lectins can stimulate the differentiation, activation and proliferation of immune cells. More studies are needed to determine in more details the pathways and mechanisms involved in the immunomodulatory action presented by these biomolecules as well as on their effective application in immunotherapy.

Chapter 4 - Algal lectins manifest an array of exploitable biological activities encompassing anticancer activity, cytokine inducing activity, neutrophil migration inducing activity, anti-influenza virus activity, anti-porcine reproductive and respiratory syndrome virus activity, anti- hepatitis C virus activity, anti-HIV-1 activity, antibacterial activity, antinociceptive and anti-inflammatory activities, and wound healing activity.

Chapter 5 - Jacalin, jacalin-related lectin from *Artocarpus nitidus* subsp. Lingnanensis, wheat germ lectin and rice bran lectins, banana lectin and potato lectin have anticancer activity. Jacalin, jacalin-related lectin from *Artocarpus nitidus* subsp. Lingnanensis, and banana lectin show immunomodulatory activity. Banana lectin has anti-HIV-1 reverse transcriptase activity and jacalin has antiviral and antiparasite activities. The recombinant wheat jacalin-like lectin His(6)-HFR1 protein and rice jacalin-like lectin orysata have anti-insect activity. Potato lectin has antifungal and antibacterial activities.

Chapter 6 - Legume seeds contain carbohydrate binding proteins of non-immune origin that are termed as lectins. Chickpea belongs to the family Leguminosae and is a food legume. The seed of chickpea provide an exceptional source of dietary proteins and is one of the important legumes in both developed and developing countries over the world. Desi and kabuli chickpea seed types are the most ubiquitous. Significant amounts of lectins have been reported to be present in them. Some of the biological roles of chickpea lectin have yet to be fully elucidated. Antifungal, antibacterial, anticancer, anti- HIV reverse transcriptase activities have recently been reported. Owing to its aforesaid characteristics, there is an urgent need to develop chickpea lectin as a therapeutic candidate in the pharmacological industry. Up-to-date research findings on different aspects of chickpea lectins vis-à-vis purification strategies and health impacts are presented systematically herein.

Chapter 7 - Hemagglutinins/lectins are produced by the following medicinal herbs: *Astragalus membranaceus* roots, *Astragalus mongholicus* roots, *Dendrobium findleyanum* pseudobulbs, different tissues of *Dendrobium officinale*, *Dendrobium pendulum* pseudobulbs, *Dioscorea batatas* roots, *Euphorbia characias* latex, *Euphorbia heterophylla* seeds, *Euphorbia marginata* latex, *Euphorbia milii* latex, *Euphorbia neriifolia* latex, *Euphorbia tirucalli* latex, *Hura crepitans* latex, *Momordica charantia* seeds, *Trichoanthes anguina* seeds, *Trichosanthes cordata* seeds, *Trichosanthes dioica* seeds, *Trichosanthes cucumerina* seeds, *Trichosanthes japonica* seeds, *Trichosanthes kirilowii* roots, *Trichosanthes kirilowii* seeds and *Withania somnifera* leaves. They display various biological activities including anticancer, immunomodulatory, antimicrobial, anti-insect and insulin-like activities.

Chapter 8 - Ribosome inactivating proteins (RIPs) are proteins capable of interrupting protein synthesis by virtue of their RNA N-glycosidase activity. These proteins are produced mostly by plants, but are now known to be produced also by fungi, bacteria and insects. Type 1 RIPs are composed of a single chain with RNA N-glycosidase activity. Type 2 RIPs, the focus of this review, have an A chain corresponding to the single chain in type 1 RIPs and in addition, a B chain with hemagglutinating activity. Type 2 RIPs are further divided into toxic and nontoxic type 2 RIPs. The very toxic ones have been used as bioweapons. Immunotoxins based on the type 2 RIPs are potential anticancer agents. Some type 2 RIPs exhibit antiviral, anti-nematode and anti-insect activities. Mutants of the bacterial type 2 RIP Shiga toxin exhibit anticancer and antiangiogenic actions.

Chapter 9 - This review reports the various potentially exploitable biological activities observed in recently reported mushroom hemagglutinins comprising anticancer, immunostimulatory, antifungal, antibacterial, anti-HIV reverse transcriptase, antinematode and anti-insect activities

Chapter 10 - C-type lectins are Ca++-dependent carbohydrate-binding proteins that resemble each other in primary and secondary structures of their carbohydrate- recognition domains, alternatively called C-type lectin domain. A diversity of invertebrates including arthropods, mollusks and echinoderms, and vertebrates produce C-type lectins. C-type lectins protect these organisms against pathogens including viruses and bacteria by interacting with the pathogen-associated molecular patterns on the surfaces of infectious microbes. C-type lectins may have other functions such as playing a role in regeneration in Hydra, food sorting in oysters, and as a receptor for vitellogenin transport in the hemolymph during yolk formation in crustaceans. The C type lectin in venom of the snake *Bothrops pauloensis* may have therapeutic value due to its action on inhibition of collagen deposition and secretion of inflammatory cytokines. C type lectins in venoms of snakes manifested inhibitory activity on breast cancer cells.

Chapter 11 - The fish lectins discussed in this review were reported from many different species and isolated from tissues as diverse as serum, gills, liver, eggs and skin mucus. These lectins can be classified under different categories consisting of galectins, intelectins, lily-type lectin, mannose binding lectins and C-type lectins. Most of the aforementioned fish lectins manifest antibacterial or bacterial cell agglutinating activity. Lectins from bighead carp, black rockfish, cobia, pearl spot, striped murrel, tongue sole and zebrafish exhibit inhibitory activity toward bacteria. Black rockfish, blunt snout bream, Qihe crucian carp, rock bream and turbot lectins induce bacterial cell agglutination.Very few fish lectins have been demonstrated to have antiviral or antifungal activity. Pearl spot and rock bream lectins elicit antiviral activity. Pearl spot lectin has fungal cell agglutinating activity. Several fish lectins show anticancer, immunoenhancing or antiparasite/parasite cell agglutinating activity. The blunt snout bream, catfish, chinook salmon and pearl spot lectins exhibit inhibitory activity toward cancer cells. Black rockfish, chinook salmon, cobia, tongue sole and zebrafish lectins exert an immunoenhancing action. Turbot lectin demonstrates antiparasite activity. Rock bream and turbot lectins have parasite cell agglutinating activity. The antibacterial, bacterial cell agglutinating, immunoenhancing, antiviral, antiparasite and parasite cell agglutinating activities serve to protect the fish against invading pathogenic organisms. An upsurge in the mRNA transcripts of the fish lectins is discerned to provide additional protection in response to bacterial or viral infection.

Chapter 12 - Bullfrog galectin is abundant in the ovary, but is also present in the kidney, liver, sciatic nerve, skeletal muscle, and skin. Bullfrog galectin exerts a suppressive action on all human leukemia cells tested including HL-60, K562, and U937 cells but not human colon cancer Colo 201cells or murine mammary tumor FM3A cells. A sialic acid binding lectin produced from bullfrog eggs demonstrates RNase activity toward poly U and poly C. It acts on the mitochondrial pathway and the endoplasmic reticulum stress-mediated pathway to produce its antitumor action. It inhibits the growth of tumor cells such as P388 and L1210 leukemia cells, human leukemia Jurkat cells, mesothelioma cells and human breast cancer cells. An alpha-N-acetyl galactosamine –binding lectin isolated from bullfrog eggs demonstrated hemagglutinating activity toward blood group A-erythrocytes but not blood group B-or O-erythrocytes. A soluble beta-galactoside- binding S type lectin from unfertilized bullfrog eggs does not depend on Ca++ or SH-reagent for hemagglutinating activity. Japanese egg lectin agglutinates tumor cells. *Xenopus laevis* oocyte cortical granule lectin (XL35) plays a role in fertilization and embryonic development. Salamander (*Andrias davidianus*) skin lectin has antifungal and antibacterial activities

In: Hemagglutinins
Editors: T. Bun Ng, J. Wong, R. Tse et al.
ISBN: 978-1-53615-708-6
© 2019 Nova Science Publishers, Inc.

Chapter 1

ASSAYS INVOLVING HEMAGGLUTININS: USE IN STUDYING THE MECHANISMS OF ACTION OF LECTINS

Luana C. B. B. Coelho[1,*], Priscila M. S. Silva[1,†], Weslley F. Oliveira[1,‡], Maiara C. Moura[1,§], Francis S. Gomes[2,‖] and Maria T. S. Correia[1,¶]
[1]Departamento de Bioquímica, Centro de Biociências,
Universidade Federal de Pernambuco, Recife, Brazil
[2]Instituto de Química e Biotecnologia,
Universidade Federal de Alagoas, Maceió, Brazil

ABSTRACT

Viruses, bacteria, fungi, plants and animals exhibit a range of cell-agglutinating and sugar-binding proteins called lectins, or hemagglutinins (HA). These proteins mediate the binding to sugar receptors in cells, promoting cell-cell communication and can initiate an infection, in the case of pathogenic microorganisms. HA identified in the influenza viruses have an important role in viral infection. In this context, many assays have been developed to measure HA in samples as vaccines, and to analyze the sugar-binding properties of HA

[*] Corresponding Author's Email: lcbbcoelho@gmail.com, Tel: +55 81 999455897.
[†] Author's Email: priscilamssilva@gmail.com.
[‡] Author's Email: weslley.felix93@gmail.com.
[§] Author's Email: maiaraceline@yahoo.com.br.
[‖] Author's Email: francissg85@yahoo.com.br.
[¶] Author's Email: mtscorreia@gmail.com.

from different influenza virus subtypes. In this chapter, we will comment on the hemagglutination assays of lectins and their mechanisms of action. The interaction of lectins with viruses and bacterial cell wall components as well as the mechanism of action of antimicrobial lectins are demonstrated through microbial assays. The extracellular matrix of microbial biofilms represents a continuous challenge in terms of characterization; lectins are molecules with high sugar-binding specificity used to analyze glycoconjugates within biofilm communities. Lectins can be obtained in milligram quantities from plants and these proteins can harbor different biological properties. Cytotoxic action on tumor cell lines and ability to act against pathogens, such as parasites and insects, are alternatives for evaluations of these plant biomolecules. Interaction of human lectins with human cells, and how the diverse mechanisms of action of these proteins can be valuable tools to study human microbiomes, diagnosis and treatment of many diseases will also be approached.

Keywords: hemagglutinin assays, lectins, mechanisms of action

1. Introduction

Animals, plants and bacteria, as well as viruses exhibit proteins known as lectins, also called hemagglutinins (HA), able to bind carbohydrates and to agglutinate erythrocytes, mediating a range of biological processes. HA, among lectins, are sialic acid-binding glycoproteins found on the surfaces of influenza viruses. These molecules affect preferentially cells of the upper respiratory tract, and are related to flu epidemics in human populations (Lewis et al., 2016). HA, essential for virus entry into the target cells, are the main targets for antibodies produced against influenza viruses, besides being useful for definition of influenza viral subtypes by biological assays (Tzarum et al., 2015; Nizet, 2017). Methods based on single-radial immunodifusion (SRID), real-time polymerase chain reaction (*PCR*), glycan-based microarrays and fluorescent-based assays are used for quantifying and characterizing HA binding properties.

The interactions between lectins and carbohydrate structures are suggested to be involved in many functions such as gene transcription, protein folding, protein transport, cell adhesion, signalling, as well as defense responses against pathogens (Van Damme and Lannoo, 2015). Lectin interaction with bacterial glycoconjugates may result in agglutination, inhibition of cell growth and cell death. The mechanisms of action may involve inhibition of respiration, adherence/invasion of bacteria in host cells, and changes in membrane permeability (Gomes et al., 2013; Mukherjee et al., 2014; Qu et al., 2015; Silva et al., 2016). Furthermore, lectins can induce the immune system to produce cytokines and cells that kill microorganisms by opsonization, phagocytosis and oxidative burst (Zhou and Sun, 2015).

Microbial biofilms are defined as interface-associated microorganisms embedded in a self-produced matrix (Neu and Kuhlicke, 2017). In biofilm systems, polymeric carbohydrate structures are located on cell surfaces and throughout the extracellular matrix (Neu et al., 2001). Polysaccharides, major components of this community, can provide diverse benefits to the cells in the biofilm, including adhesion, protection, and structure (Limoli et al., 2015). Lectins specifically recognize carbohydrate moieties on biofilm systems (Dias et al., 2015) and, in this way, can be used as antimicrobial molecules, interfering cell-to-cell or cell-to-surface interactions (Moura et al., 2017), or to characterize biofilms qualitatively and quantitatively (Mlouka et al., 2016). This last method uses fluorescently labeled lectins to bind carbohydrates, allowing their visualization and quantification (Tawakoli et al., 2017a). Additionally, lectin-mediated drug targeting has been applicable as one of the most promising approaches for drug targeting due to its great specificity (Juan et al., 2017).

The plant kingdom is a rich source of lectins, which can be purified by relatively simple procedures (Karnchanatat, 2012). These lectins have distinct biotechnological properties; some plant lectins have antiproliferative action on neoplastic cells and exert a mitogenic stimulus on normal cells. In addition, these lectins can combat parasites and insects that are harmful to plants, animals or humans.

Lectins can bind to glycoconjugates present in human cell membranes. The first human cell line in which this activity was observed was on erythrocytes, leading to agglutination. However, this process has been shown to be related with the lectin specificity because of the different patterns of glycoconjugates. In group A of the ABO system, the terminal monosaccharide of the oligosaccharide chain is N-acetylgalactosamine and those of group B is galactose. When the terminal monosaccharide is fucose, erythrocytes belong to blood group O. However, the ability of lectins to bind selectively to carbohydrate residues is not exclusive to erythrocytes since other cell lines have glycoconjugates. Thus, many applications of lectins are currently being investigated, making these proteins useful as healing and inflammatory inhibitor/activator agents, as differentiating markers to study cancer cells and other diseases, as well as anti-proliferative and mitogenic stimulation for specific cell types. In this scenario, many human lectins have been highlighted as tools with potent biotechnological applications (Lodish et al., 2016; Zubcevic et al., 2016).

2. Virus Hemagglutinin Assay Mechanisms

HA, an antigenic glycoprotein found on the membrane envelope of influenza viruses, has the ability to agglutinate erythrocytes due to the presence of carbohydrate-binding sites (Nizet et al., 2017). It is through HA that the influenza virus binds to host cells and causes infection, playing an important role in the viral life cycle and in the induction of immune responses (Nachbagauer et al., 2016). HA is the main target of neutralizing antibodies produced against influenza viruses, being used as the influenza vaccine potency marker (Wohlbold and Krammer, 2014).

Influenza virus types have been identified by their surface glycoproteins, HA and neuraminidase (NA), based on the antigenic reactivity to polyclonal sera and sequence data. There are four types of influenza viruses: A, B, C and D (WHO, 1980; Lewis et al., 2016). Types A and B are related to seasonal epidemics of influenza in human populations that can reach pandemic proportions. Influenza C infections can cause a mild respiratory illness, and influenza D viruses infect cattle, but do not cause infections in people.

Influenza A viruses are divided into subtypes according to the HA and NA exposed on the viral surface. Eighteen different HA subtypes (H1 to H18) and eleven NA subtypes (N1 to N11) have been identified. Some influenza A subtypes have been found causing seasonal illness in humans, such as the pandemic influenza A in 1918 and 2009 (H1N1) as well as in 1968 (H3N2) (Tscherne and García-Sastre, 2011).

The HA is a homotrimeric protein composed of a fibrous stalk portion and a globular head containing a site for receptors (carbohydrates) recognition in host cells, present in each monomer (Michalek et al., 2015). HA monomer is synthesized as an inactive single polypeptide chain precursor called HA0. This precursor is cleaved, generating an N-terminal HA1 chain and a C-terminal HA2 chain, which are maintained together by disulfide bonds. The glycosylation of HA is by N-linkage, in the endoplasmic reticulum and Golgi apparatus of host cells. The N-terminal region of HA2 is known as the fusion region, a conserved sequence in HA form of all strains from influenza A and B, involved in the influenza virus entry into the target cell and in the development of infection, also used as a potential target for the development of vaccines (Tate et al., 2014).

The active HA mediates the attachment of the influenza virus to its natural receptors, which are terminal monosaccharide sialic acids (N-acethylneuraminic acid or Neu5Ac) residues on the surfaces of target cells, and viral membrane fusion with the endosomal membranes in the cells (Tzarum et al., 2015). The HA specificity for sialic acid depends on the linkage type between sialic acid and the penultimate galactose, as represented in Figure 1. Avian viral HA recognizes sialic acid α 2, 3-linked to galactose, and human viral HA has affinity for sialic acid with an α 2, 6 linkage type (McBride et al., 2016; Byrd-Leotis et al., 2017). This difference of specificity between avian and human viral HA can block the infection of human population by avian influenza viruses.

Influenza virus infection initiates with the recognition and HA binding to the sialic acid linked to galactose, exposed on target cells, showing preference to infect cells in the human upper respiratory tract (Sriwilaijaroen and Suzuki, 2012). The virus is enveloped by the cell membrane, forming an endosome. The endosome content is digested after the pH decreases to 6.0 in the endosomal interior, and the HA unfolds and releases a hydrophobic portion, the fusion peptide. This peptide attaches to endosomal membrane, and the rest of the HA completes fusion with the viral membrane, and the viral genome is transferred to into the cell cytoplasm (Michalek et al., 2015).

Currently, HA measured by single radial immunodiffusion (SRID) is able to detect the subtypes of influenza virus. SRID was recommended by the World Health Organization (WHO) in 1978 and has since been the accepted method for evaluating the influenza vaccine potency. This assay measures the quantity of HA antigen present in a vaccine or other sample, using antibodies for quantitative binding to HA. These antibodies are incorporated in agarose gels and the HA samples are added to wells in the gel. HA diffuses radially from the wells and reacts with the antibody incorporated in the gel, producing a zone of precipitation around the well, generated by the immunological reaction. The area of the zone is calculated to estimate the quantity of HA (Schild et al., 1975).

Alternative methods have been developed to quantify HA, generally based on simple and faster techniques and at the same time with good quality. A real-time reverse transcriptase PCR assay was also developed based on influenza virus genes for detection of influenza virus and HA subtypes (Spackman et al., 2002). A simple slot blot protocol was developed for detection of all subtypes of the influenza A viral HA based on the conserved fusion peptide target by universal antibodies. In this assay, HA is pre-treated with a denaturant agent to expose the fusion peptide from the inside to the HA surface, followed by the fusion peptide detection using a single antibody. Moreover, this assay is easy, inexpensive and the procedures require some hours to detect and quantify subtypes of influenza viral HA (Li et al., 2010).

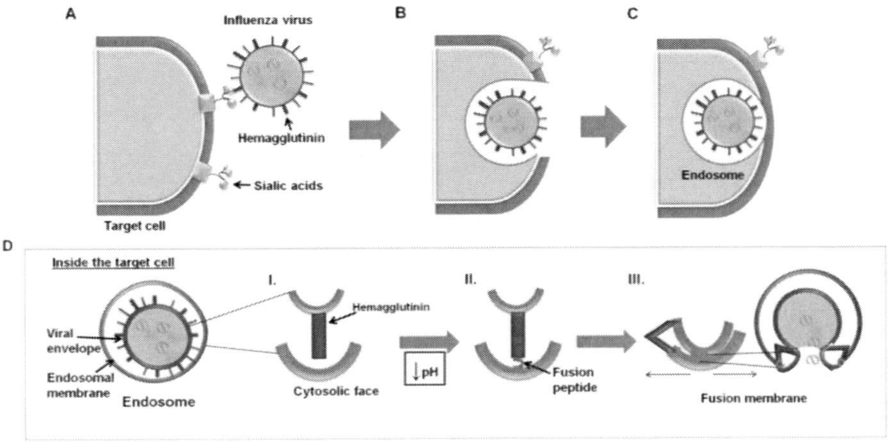

Figure 1. Representation of virus HA binding to sialic acid residues on target cell (A), followed by the entry into the cell (B and C) and the viral membrane fusion with the endosomal membrane (D). Prior to formation of the fusion membrane, the virus is engulfed by the cell membrane (B), resulting in an endosome (C). Inside the target cell as shown in D; the HA (I) is involved in the fusion membrane process. When the pH decreases to 6.0 in the endosome, HA unfolds and the fusion peptide is exposed and binds to endosomal membrane (II), joining the membranes to form a pore, with the viral genome transferred to the cytosol (III).

Assays have also been developed to evaluate avidity and specificity of influenza virus HA, interesting to monitor the adaptation of avian influenza viruses for human receptors. In a miniaturized glycan microarray developed by McBride et al., (2016), glycans with α2,3 or α2,6 sialic acids were coupled to polyacrylamide and printed in a microwell glycan array containing 48 arrays, allowing the evaluation of 6 glycan-binding HAs with respective dilutions and non-sialylated controls at the same time.

Another study reports the elaboration of a homogeneous fluorescence quenching based assay for detection of influenza viral HA antigen (Chen and Neethirajan, 2015), using fluorescent nanoprobes composed by N-glycan-conjugated quantum dots (Gly-QDs), and other with the antibody specific for HA linked to gold nanoparticles (Ab-AuNPs). The HA is recognized by the glycan and antibody, resulting in a binding of HA to the two nanoprobes and the formation of a complex sandwich. The fluorescence of QDs is decreased after the binding, correlated with the HA concentration, with sensing to distinguish the viral subtypes H1 (human) and H5 (avian).

3. Bacterial Lectin Assay Mechanisms

Lectins are a highly diverse class of carbohydrate-recognizing proteins that bind to cells promoting hemagglutination and antimicrobial effects. Antibacterial lectins can recognize and reversibly bind to carbohydrates present on the bacterial surface, such as lipopolysaccharides, found in Gram-negative bacteria, and peptidoglycans, present in cell walls of Gram-positive and Gram-negative bacteria. Lipopolysaccharides are composed of a lipid, a core oligosaccharide and an O-antigen polysaccharide chain, which is composed of oligorhamnans in many phytopathogens, and peptidoglycans are composed by alternating β(1–4) linked N-acetylmuramic acid (MurNAc) and N-acetylglucosamine (GlcNAc) residues, with a short peptide chain attached to MurNAc (Van Damme and Lannoo, 2015).

The interaction between lectins and bacterial glycoconjugates can lead to inhibition and killing of these microorganisms (Balouri et al., 2016). Lectins should have only inhibitory activity, demonstrated by determination of the lowest concentration of the lectin that inhibits the visible growth of an organism after overnight incubation, referred to as the minimum inhibitory concentration (MIC). However, lectins can cause metabolic and/or structural changes that lead to death of the bacteria even after a subculture on a lectin-free media, demonstrated by minimum bactericidal concentration (MBC). Silva et al., (2016) isolated a lectin from the juicy sarcotesta of *Punica granatum* (PgTeL) that showed MIC values against *Aeromonas* sp., *Enterococcus faecalis*, *Escherichia coli*, *Klebsiella* sp., and *Salmonella enterica* serovar Enteritidis, *Staphylococcus aureus*, *S. epidermidis* and *S. saprophyticus*. However, PgTeL was able to inhibit and kill *Micrococcus luteus*, *Serratia marcescens*, and *Streptococcus mutans*. The relation between MBC and MIC can be defined if bacteria are susceptible (MBC/MIC less than or equal to 4) or resistant (MBC/MIC higher than 4) to lectin (Levison, 2004). Based on the MBC/MIC ratio, Gomes et al., (2013) showed that *E. coli, S. aureus* and *Proteus mirabilis* were susceptible whereas *K. pneumoniae, Pseudomonas aeruginosa* and *S. enteritidis* were resistant to *Schinus terebinthifolius* leaf lectin (SteLL).

The difference in glycoconjugate composition between Gram-negative and Gram-positive bacteria can affect the antibacterial activity. A lectin from *Bothrops leucurus* snake venom (BlL) exhibited antibacterial effect against human pathogenic Gram-positive bacteria and was not effective against Gram-negative bacteria (Nunes et al., 2011). The authors suggested that the outer cell wall and the inner membrane of Gram-negative bacteria may be difficult for BlL to access and that the lectin specifically recognizes peptidoglycans of Gram-positive bacteria.

The ability of lectins to mediate cross-linking interactions with carbohydrates of microbial membrane is confirmed by the agglutination assay (Figure 2A), as demonstrated by the lectin from *Aplysia dactylomela* eggs (ADEL) that is able to agglutinate *S. aureus* (Carneiro et al., 2017). The agglutinating phenomenon is due to the linkage of particles or cells suspended in a liquid, resulting in formation of visible aggregates. The agglutination assays can be performed in tubes or plates (Santos et al., 2014). The inhibition of this property by free carbohydrates (Figure 2B) or by a chelating agent specific to the mode of this interaction is dependent on ionic interrelation promoted by metallic ions. The lectin from the Pacific white shrimp *Litopenaeus vannamei* (LvCTL3) lost the property of agglutinating Gram-positive and Gram-negative bacteria when Ca^{2+} ions were chelated with EDTA, suggesting that the lectin is calcium-dependent (Li et al., 2014). Confocal imaging, flow cytometry, circular dichroism, colorimetry, fluorescence, and western blotting were also used to study and confirm this interaction (Zhou and Sun, 2015; Chandrasekaran et al., 2016; Chikalovets et al., 2016; Sun et al., 2016; Arasu et al., 2017; Wongpanya et al., 2017).

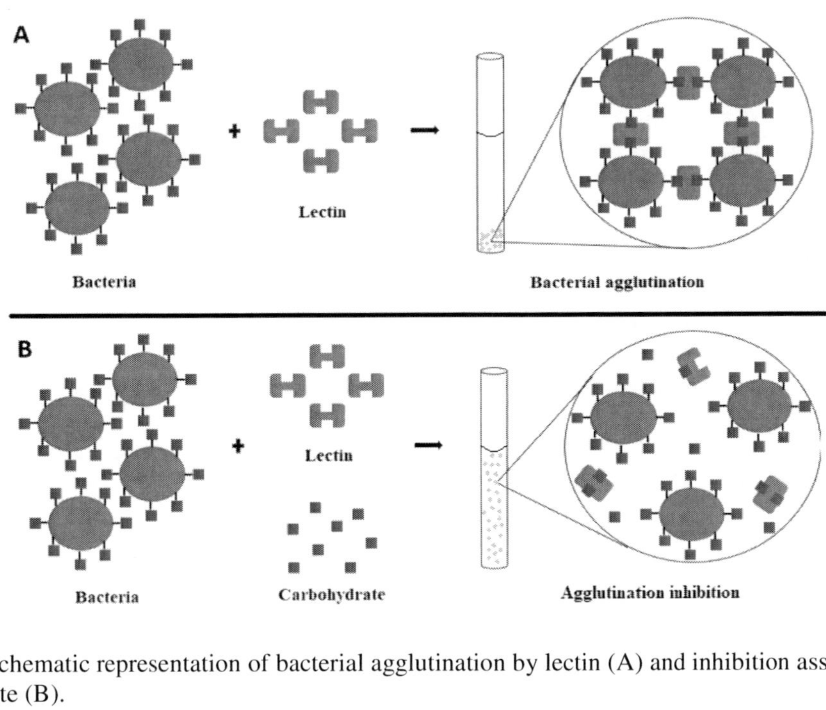

Figure 2. Schematic representation of bacterial agglutination by lectin (A) and inhibition assay with free carbohydrate (B).

The lectin-bacteria interaction can increase the permeability of bacterial membranes as part of their killing mechanism. Mukherjee et al., (2014) demonstrated that the lectin REGIIIα, localized at the human intestinal epithelium, recognizes the bacterial peptidoglycan and then increases cell permeability and leakage of cellular contents through the formation of oligomeric pores on bacterial membrane. The authors also showed that this activity is exclusively against Gram-positive bacteria, since lipopolysaccharide, a major constituent of the Gram-negative outer membrane, inhibited RegIIIα. Similarly, a water soluble *Moringa oleifera* lectin (WSMoL) is able to induce the leakage of proteins from bacterial cells as well as to damage the integrity of the cell wall. This effect on membrane permeability can be observed even in those bacteria in which the lectin only showed bacteriostatic activity (Moura et al., 2015).

The killing mechanisms of antibacterial lectins should involve effects on the respiratory chain and energy production of bacterial cells. A lectin secreted from *Andrias davidianus* skin (ADL) inhibited the respiration of *E. coli, E. aerogenes, S. aureus, B. subtilis* and *Shewanella* sp., with the percent inhibition rates between 13.9 and 59.2%. The mechanism of action suggested by the authors was through inhibition of respiratory metabolism that affects the tricarboxylic acid cycle and hexose monophosphate pathway, since ADL inhibited the same pathway as malonic acid and sodium phosphate, typical inhibitors of this pathways (Qu et al., 2015).

Lectins can inhibit the adhesion and invasion of bacteria in the host cells, besides their ability to cause metabolic and membrane permeability changes. Initial adherence and invasion of epithelial cells are two important factors, which lead to pathogenesis. Infection by pathogenic strains usually involves interactions with multiple components of the cell wall, including carbohydrate-containing molecules that are directly linked to bacterial adherence (Wagner et al., 2014). Therefore, lectins should prevent the interaction between these molecules of the microorganism and human cells. The lectin from the sarcotesta of *Punica granatum* (PgTeL) reduced the adherence and invasion of *Aeromonas* sp., *S. aureus*, *S. marcescens* and *S. enterica* serovar Enteritidis to HeLa cells. The authors noted that the mechanism of action is different from that shown on the growth inhibition, since no correlation was detected between the degree of adherence and invasion inhibition as well as the MIC values (Silva et al., 2016). The mechanisms of adherence and invasion inhibition promoted by a lectin is not clearly understood since it could be diverse, such as in the production of extracellular substances, engulfment of the bacterium, cytoskeletal rearrangements and even a direct interaction with the cellular machinery that regulates the actin cytoskeleton dynamics (Cossart and Sansonetti, 2004).

The mechanism of defense to pathogens promoted by lectins can be through activation of the innate immune system and inflammation that lead to opsonization of the pathogen. The lectin from teleost C-type lectin (CsCTL1) was able to activate monocytes leading to phagocytic activity and increase of the respiratory burst and acid phosphatase activities of phagocytes. A mutant version of this lectin with alteration at a mannose-type carbohydrate motif did not exhibit this activity, confirming the importance of carbohydrate binding domain of CsCTL1 in the immune stimulation against *Vibrio anguillarum* (Zhou and Sun, 2015). The mechanism of action has not been totally elucidated, but the start of immunomodulatory activity is when lectins bind to glycan targets on the cell surface, which have the role of lectin-receptors, and then induce the immune response through mediators such as reactive nitrogen species and cytokines, second messengers released from the membrane (Coelho et al., 2017). TML-1 and TML-2, lectins from the mushroom *Tricholoma mongolicum*, induced T helper type 1 and enhanced the production of nitrite ions and the cytokine tumor necrosis factor by macrophages (Wang et al., 1996). Research over the past decade indicates that mannose-binding lectin (MBL) is essential to activate the complement system and plays an important role in the first hours/days of any primary immune response (Turner, 2003). Mannose-binding lectins from plants also have an immunomodulatory activity. The lectin purified from *Cratylia mollis* (Cramoll 1,4) and its recombinant version (rCramoll) stimulate the production of NO, reactive oxygen species (ROS) and proinflammatory cytokines (interleukin-1β and IFNγ) and phagocytic activity in a cell culture of mouse peritoneal exudates infected and noninfected with *S. aureus* (da Silva et al., 2015).

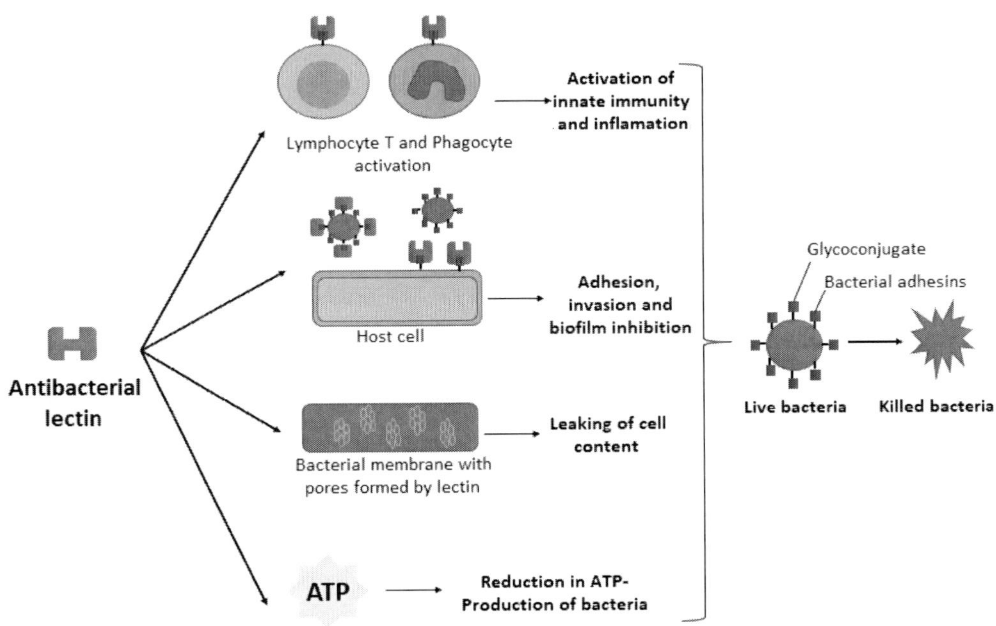

Figure 3. Schematic representation of the modes of action of antibacterial lectins.

The antimicrobial effect of lectins could be studied in an *in vivo* model. Infection of the shrimp *Penaeus monodon* with *Vibrio harveyi* associated with RNAi-mediated silencing of the lectin from this shrimp species (PmCLec) leading to higher cumulative mortality (Wongpanya et al., 2017). Li et al., (2014) also used cumulative mortality to show that lectin from the shrimp *Litopenaeus vannamei* (LvCTL3) promoted reduction of the mortalities of *V. parahaemolyticus*. The authors indicated a possible role of this lectin in shrimp innate immunity defense against bacterial infection. Sun et al., (2016) confirmed that lectins could promote reduction of bacterial recovery from infected tissue. C-type lectins from *Cynoglossus semilaevis* inhibited dissemination of *V. harveyi* and *V. anguillarum* in fish tissues. These effects may result from lectin binding to glycan determinants on the surface of animal defense cells, modulating the immune response. Therefore, the modes of action of antibacterial lectins are diverse (Figure 2) and observed through *in vitro* and *in vivo* assays. Probably the effects of antibacterial lectins detected by *in vitro* assays, such as pore forming, metabolic changes and inhibition of adhesion, invasion and biofilm formation should contribute to kill bacteria and enhance survival in *in vivo* assays.

4. Lectins Interacting with Biofilms

Biofilms consist of hydrated three-dimensional structures of microbial cells embedded in an extracellular matrix (Koo and Yamada, 2016). The extracellular polymeric substance (EPS) matrix has structural and physiological functions, maintaining the mechanical architecture and providing the various chemical and physical signals to the cells (Flemming et al., 2016; Koo and Yamada, 2016; Kumar et al., 2017). This sessile community organization presents significant morphological, physiological and genetic differences from the lifestyles of planktonic fungi and bacteria (Silva et al., 2016; Moura et al., 2017).

The matrix - most of 90% of the biofilm mass - can vary widely between different biofilms and is composed of nucleic acids, proteins, lipids, extracellular enzymes and polysaccharides, mostly produced by the microbial cells themselves (Flemming and Wingender, 2010; Kumar et al., 2017). These biopolymers are responsible for adhesion to biotic or abiotic surfaces, for cohesion in the biofilm and for allowing the intense interactions between close cells (Flemming and Wingender, 2010).

Polysaccharides are often a major fraction of the EPS matrix, varying depending on the microbial species, strain, growth conditions, and the overall environment, forming a complex network that allows the initial adhesion and aggregation of microbial cells (Limoli et al., 2015; Roux et al., 2015). Other important characteristics include the maintenance of the retention of water, the improvement of a protective barrier against host defenses and antimicrobial agents, the retention and stabilization of extracellular enzymes, besides serving as a carbon stock and nutrient source (Flemming and Wingender, 2010; Limoli et al., 2015). Mitchell and collaborators (2015) studied the matrix of the fungus *Candida albicans*, that is mainly formed by α-mannan, β-1,6 glucan and β-1,3 glucan, polysaccharides that together are related to fluconazole resistance of the biofilm. Pel is an exopolysaccharide produced by *P. aeruginosa* strains, which are implicated in biofilm formation. It plays a role in cell adherence, cell-cell interactions via *quorum sensing* molecules and provides a certain level of resistance against aminoglycoside antibiotics (Jennings et al., 2015). The microbial carbohydrate poly-N-acetylglucosamine (PNAG), synthesized by fungi and bacteria, is involved in microbial growth, protection against anti-microbial factors and biofilm formation (Roux et al., 2015).

Lectins possess a wide range of molecular structures and glycol-binding portion that promote specific carbohydrate recognition and important functions in cell signaling, cellular interactions and cellular recognition (Dias et al., 2015). Lectin families comprise a promising class of antimicrobial molecules, since they can access and bind carbohydrates located on the bacterial surface or in biofilm EPS structures, interfering on cell-to-cell or cell-to-surface interactions (Moura et al., 2017). Some studies demonstrated its potential as anti-*quorum sensing* molecules, preventing the expression of genes required for biofilm development and for other virulence factors (Klafke et al., 2013; Hasan et al., 2014). In this way, these molecules have already been used in antivirulence therapies that comprise a new alternative strategy that involves non-lethal mechanisms, which block all or some of the microbial virulence without killing the cells, unlike conventional antibiotics (Silva et al., 2016).

Some research aimed at finding alternatives to combat microbial pathogens has shown the potential of using lectins for this purpose (Cavalcante et al., 2013; Petrova et al., 2016; Moura et al., 2017). The methods used by these studies mostly included determination of the minimum inhibitory concentration, which is defined as the lowest lectin concentration able to promote a reduction of ≥50% in the optical density (600 nm), in comparison with the 100% growth of microbial control. The crystal violet assay is used to evaluate the antibiofilm action of lectins and its ability to destroy preformed bacterial or fungal biofilms. The prevention of biofilm establishment is due to the blocking of microbial adhesion on surface, mainly by the carbohydrate-binding site at the bacterial surface, as recently demonstrated by Moura et al., (2017). It uses a coating containing WSMoL isolated from *M. oleifera* seeds which suppressed adhesion of *Bacillus* cells without interfering their viability. Jayanthi and collaborators (2017) also worked with a lectin isolated from the haemolymph of the blue swimmer crab *Portunus pelagicus* alone and also in combination with a silver nanoparticle AgNPs. The lectin and the synthesized lectin-coated AgNPs, at 100 μg/mL, caused a reduction of the hydrophobicity index, especially that in combination, leading to a decrease in the rate of adhesion of *P. vulgaris*, *P. aeruginosa*, *E. faecalis* and *B. pumilus* cells. Additionally, the lectin-coated AgNPs significantly reduced the EPS quantity of these bacterial biofilms after 24h of treatment. An investigation conducted in 2001 indicates that the galactose-specific lectin from *Psidium guava* binds to *E. coli* cells, blocking its adhesion to the intestinal wall (red cells) presenting, thus, antidiarrheic uses. According to the authors, this action may be related with the blocking of sugars, galactose and mannose, involved in the attachment between *E. coli* and red cells (Coutiño-Rodríguez et al., 2001).

Table 1. Applications of lectins to biofilm matrix studies

Lectin	Source	Specific sugar(s)	Application(s)	Reference
AAL	*Aleuria aurantia* (seeds)	α -Fuc	To study and characterization glycoconjugate tracks of diatoms, cyanobacteria, acidophilic bacteria and miscellaneous strains on a surface	Neu and Kuhlickand (2017)
			Identify and characterize dental biofilms	Tawakoli et al., (2017a)
ABA	*Agaricus bisporus*	Galβ1, 3GalNAc	Identify and characterize dental biofilms	Tawakoli et al., (2017a)
WGA	*Triticum vulgaris* (wheat germ)	GlcNAc	Target clarithromycin delivery to methicillin-resistant *Staphylococcus aureus* in liposome formulation *in vivo* and *in vitro*	Meng et al., (2016)
			Detection of biofilms in nanofiltration membranes during drinking water production	Mlouka et al., (2016)
PNA	*Arachis hypogaea* (Peanut agglutinin)	Gal/ GalNAc	Detection of biofilms in nanofiltration membranes during drinking water production	Mlouka et al., (2016)
UEA	*Ulex europaeus*	α -Fuc	Study of environmental biofilm systems (rotating annular reactors)	Neu et al., (2001)
			Study of *Burkholderia thailandensis* mature biofilms	Tseng et al., (2016)
ConA	*Canavalia ensiformis* (seeds)	α-Man, α-Glc, α-GlcNAc	Study and characterization of exopolysaccharide produced by mature biofilm of *Pseudomonas syringae*	Laue et al., (2006)
			Study the effects of endodontic irrigants at their clinical concentration on *Streptococcus mutans*, *Streptococcus oralis* and *Actinomyces oris* matrix polysaccharides	Tawakoli et al., (2017b)
			Study seawater biofilm polysaccharides in reverse osmosis (SWRO) membranes	Nagaraja et al., (2017)

Besides the use of lectins as antimicrobial and antibiofilm compounds, the capacity to recognize small motifs in different carbohydrates makes lectins important for studying the presence and composition of some glycoconjugates in a variety of different biofilm systems, including environmental, industrial and medical biofilm systems (Zhang et al., 2015; Tawakoli et al., 2017a). Some examples of the studies of biofilm systems using lectin-binding carbohydrates analysis are shown in Table 1.

The application of labelled lectins allows researchers to stain glycoconjugates and also demonstrate their spatial heterogeneity within the biofilm matrix, a component poorly studied, and normally could be combined with general nucleic acid stains to obtain both nucleic acid and glycoconjugate signals (Neu and Kuhlicke, 2017; Tawakoli et al., 2017a).

In addition, labelled lectins may also be employed for studying the effect of different chemicals as antimicrobial agents on the glycoconjugate matrix (Neu and Kuhlicke, 2017).

Fluorescence lectin-binding analysis (FLBA) is a powerful tool to reveal the complexity of extracellular glycoconjugates in biofilm matrix and comprises an important method used nowadays (Tawakoli et al., 2017a). A critical evaluation of FLBA includes optimal incubation time, lectin concentration, fluor conjugate, carbohydrate inhibition, order of addition, and lectin interaction (Neu and Kuhlicke, 2017). For this preparation test the biofilm sample is covered with a few droplets of fluorescently labelled lectins and incubated for a certain time duration and, after removing unbound lectins, the samples were visualized using confocal laser scanning microscopy in an epifluorescence mode (Neu and Kuhlicke, 2017). The lectins could interact with high binding efficiency, indicating that these lectins have specificities that seem to closely match the microbial glycoconjugates of the strains tested, and thus a high signal intensity is detected (Neu and Kuhlicke, 2017; Tawakoli et al., 2017a). Tawakoli and collaborators (2017a) evaluated the use of fluorescently labeled lectins of *Aleuria aurantia* (AAL), *Calystega sepiem* (Calsepa), *Helix pomatia* (HPA), *Lycopersicon esculentum* (LEA) and *Morniga-G* (MNA-G) to study dental biofilms that yielded strong fluorescence signals evidencing its applicability to the characterization of these biofilms. The glycoconjugates in EPS structures by biofilms of the three metal-oxidizing archaea species were studied using 75 commercial lectins of different origins by FLBA analysis (Zang et al., 2014). According to the specificity of 22 lectins, moieties with fucose, glucose, N-acetylgalactosamine (GalNAc), galactose and mannose were detected.

In another point of view, there is a great interest to use lectins to mediate antimicrobial drug delivery to bacterial biofilms (Juan et al., 2017). This mechanism involves the conjugation of lectins and other products with known bactericidal action (Cavalcante et al., 2011). Meng and collaborators (2016) demonstrated that liposomes containing wheat germ agglutinin (WGA) and clarithromycin yielded better results against methicillin-resistant *Staphylococcus aureus* than those of free clarithromycin. Some of these results include MIC determination, antibiofilm activity and biofilm eradication capacity.

5. Evaluations of Plant Hemagglutinins

Lectins were termed phytohemagglutinins due to their ability to agglutinate erythrocytes and because they were first found in plants. However, some hemagglutinins agglutinate certain erythrocytes belonging to specific ABO blood groups and therefore the term lectin, derived from Latin word "legere" denoting to select (Gorakshakar and Ghosh, 2016) was used. It is believed that Peter Hermann Stillmark reported the first plant lectin in 1888, isolated from castor bean (*Ricinus communis*) seeds, the so-called ricin (Lannoo and Van Damme, 2014).

In the plant kingdom, lectins are found in a variety of tissues in a large number of different plants. The families whose lectins have been well characterized include Solanaceae, Fabaceae and Poaceae. However, legume seeds generally have an abundance of lectins with high homology. These lectins can also be stored in leaves, fruits, bulbs, tubers, roots, rhizomes, flowers and nectar (Jaber et al., 2010; Lagarda-Diaz et al., 2017). Sometimes, different tissues of the same plant can produce different lectins with specificity to distinct carbohydrates, and, therefore, can present different bioactivities (Karnchanatat, 2012). Plants synthesize lectins to carry out their defense against pathogens and predators as well as to maintain symbiotic relations with some microorganisms, such as commensal bacteria (De Hoff et al., 2009).

Plants are an important source for obtaining lectins due to their relatively easy extraction and possibly producing milligram quantities of these proteins (Karnchanatat, 2012). Isolation and purification of lectins from plant extracts generally involve precipitation steps of these proteins using solutions containing acids, organic solvents or salts; and chromatographic steps, such as affinity chromatography, ion-exchange chromatography, hydrophobic interaction chromatography, and gel filtration chromatography (Lam and Ng, 2011; Nascimento et al., 2012). To obtain high concentrations of a lectin it is necessary that the extract has a high lectin content to be purified and the use of few steps in the purification process (simple protocol) is important (Lam and Ng, 2011). Then pure plant lectins can be evaluated in different biological assays for potential biotechnological applications.

5.1. Cell Assays

Several plant lectins are tested against tumor cell lines to screen for anticancer activity that these proteins can have. Lectins can cause cytotoxicity in tumor cells after penetration into the cells; promoting aggregation and/or agglutination of such cells through their glycoconjugates on cell surfaces; can interfere in apoptosis and autophagy processes by modulating molecules that participate in specific signaling pathways that culminate in induction of cell death (Fu et al., 2011; Giacometti, 2015).

A screening of lectins for anticancer activity is commonly performed by using the MTT 3-(4,5-dimethyl-2-thiazolyl)-2,5-diphenyltetrazolium bromide) assay. This cytotoxicity test was applied to the following plant lectins: Cramoll (from *C. mollis* seeds) killed human prostate adenocarcinoma cells; CasuL (from *Calliandra surinamensis* leaf pinnulae) displayed antiproliferative activity against chronic myelogenous leukemia and breast cancer cells; and ConV (from *Canavalia virosa* seeds) reduced viability of glioma cells (Figueirôa et al., 2017; Procópio et al., 2017; Osterne et al., 2017). In the MTT assay, after incubation with different concentrations of a lectin, the cancer cells are exposed to the MTT solution. Viable cells, through their oxidoreductase enzymes, reduce MTT to the purple formazan product (Bahuguna et al., 2017). Thus, cells incubated with cytotoxic lectins form less formazan compound, since fewer viable cells are found.

Recognition and interaction of lectins with glycans on the tumor cell surface can lead to intracellular signals culminating in activation of cell death pathways (Lichtenstein and Rabinovich, 2013). Two major cell death pathways are apoptosis and necrosis as depicted in Figure 4. In apoptosis, there is reduction of cellular size, integrity of the plasma membrane is maintained and cellular fragments, named apoptotic bodies, are formed, which have cytoplasm and organelles and can have nuclear fragments. Cells undergoing the necrotic process lose the integrity of their plasma membrane and extravasation of intracellular content, including molecules that can stimulate inflammation, occurs (Elmore, 2007; Rock and Kono, 2008; Nikoletopoulou et al., 2013). To know which death pathway an antitumor lectin induces, generally the cells (after treatment with lectin) are labeled using annexin V and propidium iodide (PI) followed by flow cytometry analysis. Cells in the necrotic process internalize PI, a cationic dye that binds to DNA and is not permeable in cells with an integral plasma membrane; while annexin V is able to label phosphatidylserine exposed on the plasma membrane in apoptotic cell death (Wlodkowic et al., 2011). For example, through annexin V-PI assay, it was confirmed that the lectin purified from *Phaseolus vulgaris* seeds (BKBL) induced apoptosis of human breast cancer cells. The lectin obtained from *Momordica charantia* (MCL) also promoted apoptosis in hepatocellular carcinoma cells (Chan et al., 2012; Zhang et al., 2015). Luz et al., (2017) showed that melanoma cells after treatment with cMoL (coagulant *M. oleifera* lectin) stained with PI and annexin V, indicating that cMoL induced both late apoptosis and necrosis in this cell line.

Plant lectins can also recognize altered N- and O-glycosylation patterns in glycoproteins and glycolipids expressed in malignant cells. Such recognition can be important because of the ability of these lectins to guide localized delivery of an antitumor drug (Poiroux et al., 2017). For example, wheat germ agglutinin (WGA) lectin from *Triticum vulgaris* was used as a carrier of conjugate containing WGA, poly-L-glutamic acid (protein capable of coupling hydrophobic drug molecules) and fluorescein cadaverine (fluorescent dye). This conjugate could be internalized into human urothelial carcinoma cells by the ability of WGA to recognize carbohydrates present on the surface of these cells, since cell-associated fluorescence intensity with conjugate decreased in the presence of N,'N"-triacetylchitotriose (Apfelthaler et al., 2017).

An action opposite to antiproliferative response is reported for some lectins capable of inducing cell proliferation. Cramoll was able to stimulate lymphocyte and splenocyte proliferation (Maciel et al., 2004; Melo et al., 2011a). This mitogenic activity can be evaluated by the MTT assay as mentioned above. Moreover, such an assessment can also be done by autoradiography, through the incorporation of tritiated-thymidine ([^{3}H]-T) molecule into DNA during DNA duplication in the cell cycle (Madhavan et al., 2007; Duque and Rakic, 2011). [^{3}H]-T as an indicator of cell proliferation was useful, for instance, to determine mitogenic capacity of tarin, a lectin purified from taro (*Colocasia esculenta*) tubers, on splenocytes (Pereira et al., 2014).

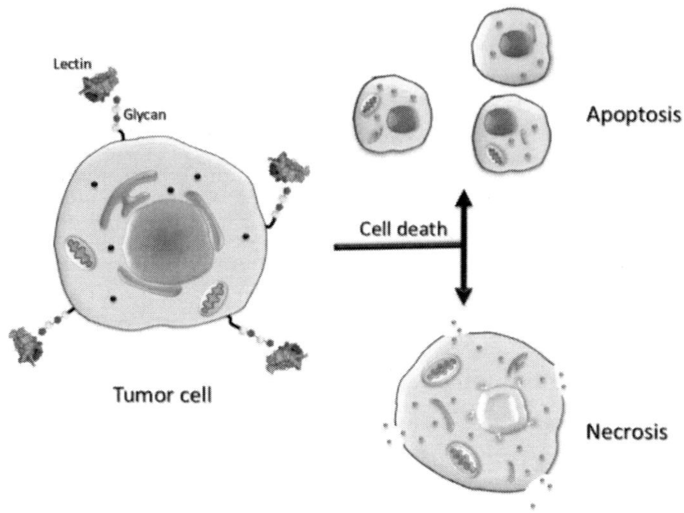

Figure 4. Schematic representation of lectin interaction with surface glycans on tumor cells leading to two main types of cell death: apoptosis and necrosis.

5.2. Response to Pathogens: Antiparasitic and Insecticidal Actions

Anthelmintic activity is found in certain plant lectins. For example, a purified lectin from *Abelmoschus esculentus*, AEL, inhibited juvenile hatching of second-stage of the root-knot nematodes *Meloidogyne incognita* and *M. javanica* (Lacerda et al., 2017). RVL lectin, obtained from *Remusatia vivipara* tubers, also exerted nematicidal activity against *M. incognita* (Bhat et al., 2010). Another lectin from *M. oleifera* seeds, WSMoL, exerted ovicidal and larvicidal effects on gastrointestinal nematodes. These actions were partially inhibited in the presence of fructose, indicating involvement of carbohydrate-binding sites of WSMoL in their nematicidal activity (Medeiros et al., 2017). In these assays, commonly a solution containing a certain amount of eggs or larvae (after hatching eggs) is incubated with the lectin at different concentrations. After the incubation period, larvae are counted with a microscope. In addition, the anthelmintic action of lectins can also be evaluated *in vivo*. In mice infected with *Schistosoma mansoni* and treated with Cramoll, the lectin decreased excretion of *S. mansoni* eggs in feces and reduced the number of these adult worms in the hepatic-portal system of infected animals (Melo et al., 2011b).

The antiparasitic profile of lectins also includes their probable action against protozoa. For instance, macrophages pretreated with ScLL lectin, from *Synadenium carinatum* latex manifested a reduction of intracellular *Leishmania* (*Leishmania*) *amazonensis*, demonstrating an inhibitory action of ScLL on this parasite without altering the viability of macrophages (Afonso-Cardoso et al., 2011). Cramoll lectin was able to stimulate necrotic death of *Trypanosoma cruzi* epimastigotes (Fernandes et al., 2010). To exert toxic effects directly on pathogens, plant lectins can act as adjuvants through their combination with parasitic antigens and improve the host immune action against the pathogen. These lectins can promote immunomodulation by modulating production of certain cytokines and other mediators of immune response (Jandú et al., 2017). ScLL and/or ArtinM (from *Artocarpus heterophyllus* seeds) lectins were used for treatment of mice infected with *Toxoplasma gondii*. It was observed that treatments of animals infected with ScLL alone and ScLL plus ArtinM increased levels of pro- and anti-inflammatory cytokines; such therapies also reduced parasite burden in brain tissues and enabled high survival rates of animals (Souza et al., 2016).

Plant lectins can also wield an insecticidal action due to their ability to interact with glycoproteins and glycolipids of insects affecting their physiological processes; and some lectins cause antinutritional effects and are resistant to proteolytic degradation of digestive enzymes in these organisms (Macedo et al., 2015). cMoL, a chitin-binding lectin, was able to exert an insecticidal effect on *Anagasta kuehniella*, probably by interaction of cMoL with chitinous structures in the midgut of larvae causing nutritional disorders, besides this the lectin resisted enzymatic digestion by proteases of *A. kuehniella* (Oliveira et al., 2011). WSMoL was also able to affect the development of *A. kuehniella*, however this lectin was not able to resist proteolytic digestion after acute exposure to the midgut juice of the organism. WSMoL bound to the peritrophic membrane of *A. kuehniella* causing partial blockage in its pores and, thus, interfered with the action of important digestive enzymes such as α-amylase, α-glucosidase and trypsin (Oliveira et al., 2017). WSMoL and a lectin isolated from *M. oleifera* seed cake (WSMoLc) promoted larvicidal and ovicidal activities on *Aedes aegypti*. WSMoL and WSMoLc had an oviposition-stimulant action, were resistant to enzymatic digestion, and elicited stimulatory effects on some proteases in the larval gut (Santos et al., 2012; Agra-Neto et al., 2014; Oliveira et al., 2016). Termiticidal activity on the termite *Nasutitermes corniger* was observed for *Microgramma vaccinifolia* rhizome lectin (MvRL) and lectins purified from bark (MuBL), heartwood (MuHL) and leaves (MuLL) of *Myracrodruon urundeuva* (Albuquerque et al., 2012; Lima et al., 2017). Therefore, in view of various plant lectins with activities against plant, animal or human pathogens, screening of these different activities in isolated novel lectins is encouraged.

6. Human Lectins and Assay Mechanisms

A large variety of lectins are expressed in human. Many human lectins have immunomodulatory activity. A group of C-type lectins known as macrophage galactose-type lectins (MGL) are glycan-binding receptors on dendritic cells and macrophages from the human immune system that participate in immune response steps, as pathogen recognition, endocytosis, and presentation of antigens to T cells. Human MGL can enhance production of cytokines, such as IL-10, by dendritic cells and induce the proliferation of regulatory T cells and CD8+ T cell responses (Coelho et al., 2017). Previously termed as S-type lectins, galectins are also described as human lectins with ability to induce innate immunity and inflammation. One member of this family, galectin-3, shows the ability to activate neutrophils leading to superoxide production. This activity is also associated with carbohydrate-recognition domains since it is inhibited by lactose. However, the amino-terminal domain is also necessary for the observed activity (Yamaoka et al., 1995).

Glycosylation plays fundamental roles in many biological processes and is known to be organism, tissue, and cell type specific. However, inflammation, infections, immunological disorders, and neoplasia have been associated with glycan changes and this different pattern can be distinguished by lectins (Brooks, 2017). To detect this difference, conjugated lectins have been used as a histochemical marker. Human lectins have been evaluated for this purpose. Saal et al., (2005) biotinylated human galectin-2 and performed immunohistochemistry using an anti-galectin-2 antibody with three pairs of normal tissue and their corresponding malignant counterparts. Galectin-2 showed a tendency for increased staining intensity in malignant tissues of epithelial origin. This encouraged further studies to use lectins as tools for the diagnosis and prognosis in histopathology.

The distinct expression of glycans on the surface of cancer and normal cells allows lectins to specifically recognize malignant cells and exert cytotoxic effects such as inhibition of proliferation and activation of cell death pathways (Coelho et al., 2017). Some human lectins have been evaluated for antitumor activity. Zymogen granule protein 16 (ZG16p), a mannose and heparin/heparan binding lectin, inhibits proliferation of human colorectal cancer cells (Mito et al., 2017). The authors demonstrated that the pathway is through carbohydrate-binding sites since its mutant version lacking mannose- and heparin-binding activities does not inhibit the growth of cancer cells. Human lectins may also affect the adhesion ability of cancer cells. André et al., (1999) showed that the binding of tumor cells to laminin and fibronectin was generally reduced by treatment of cells or matrix with human galectins. The interaction of cancer cells with laminin and fibronectin, proteins of the extracellular matrix, is a key event in tumor invasion and metastasis.

Many researchers have reported the healing effect induced by human lectins. The healing response involves several processes, such as regulation of inflammation, macrophage polarization, angiogenesis, fibroblast to myofibroblast transition and re-epithelialization. The mechanism of action of galectins can be diverse. Galectin-1 show the ability to induce myofibroblast activation, migration, and proliferation by triggering intracellular reactive oxygen species (ROS) production (Lin et al., 2015). Myofibroblasts have a key role in wound healing by secreting growth factors and chemoattractants to create new substrates and proteins in the extracellular matrix. The intestinal epithelial cell restitution promoted by Gal-2 and Gal-4, differently from Gal-1, is achieved via a TGF-beta-independent mechanism and involves increasing cyclin B1 expression and consequently cell cycle progression (Paclik et al., 2008). Galectin 3, expressed in inflammatory cells and fibroblasts, has been demonstrated to influence processes associated with inflammation through its interaction with various cell types including neutrophils, monocytes, and macrophages. In addition, Gal-3 facilitates neutrophil adhesion to laminin, has been implicated in the recruitment of neutrophils and monocytes in the wound healing process and induces angiogenesis by activating focal adhesion kinase-mediated signalling pathways that modulate endothelial cell migration (McLeod et al., 2018). Another lectin that contributes to the healing process is the mannose-binding lectin (MBL). MBL is a component of the complement system of innate immunity and its role in the elimination of microorganisms and modulation of immune and inflammatory response could contribute to wound healing (Coelho et al., 2017).

Glycan markers hold considerable opportunities and challenges for disease diagnosis. Human lectin-based biosensors have been developed to detect and quantify glycans related with some of this disease. The potential of using lectins as a biomarker is that lectins can have characteristics of an ideal biomarker, such as high specificity and sensitivity, and correlation with disease progression and therapy. The ability of lectins to discriminate between different glycan structures has allowed them to be very useful as a biomarker. MBLs and galectins are the human lectins that received more attention in this review. Although MBLs offer protection against invading microorganisms, high levels may confer biological disadvantages in other situations such as leading to aggravate local and systemic inflammation through complement activation and modulation of proinflammatory cytokine production. For example, MBL concentration is higher in patients with diabetes than in healthy controls and can indicate possible poor diabetes control and progression of the disease, with the possibility of the presence or development of diabetic complications. MBLs play a significant role in vascular dysfunction and cardiomyopathy after acute hyperglycemia and are also a biomarker of diabetic retinopathy (Geng et al., 2015; Pavlov et al., 2015). In both cases, MBL was measured by immune-fluorometric assay in serum samples. Galectin-3 also has great potential to be used as a biomarker since various "inflammatory/fibrotic" conditions such as arthritis, asthma, pneumonia, atherosclerosis, and kidney disease have been described as associated with Gal-3. An increased Gal-3 concentration, detected by chemiluminescent microparticle immunoassay, has been shown to be a potent signal to the pathobiology of pulmonary arterial hypertension (Calvier et al., 2016).

The human immune system is able to fight microbial infections through the complement system, which is composed of several proteins acting with an enzymatic cascade until recognition, opsonizing and lysis of the pathogen target cell (Nesargikar et al., 2012). The complement system can be activated by three pathways: the classical pathway, occurring by binding of antibodies to microbial surface; the alternative pathway, which is initiated by spontaneous hydrolysis of the C3 protein, independent of antibody binding; and the lectin pathway, which takes place by binding of lectins, MBL and ficolin, to glycids on pathogenic surface (Ehrnthaller et al., 2011; Nesargikar et al., 2012). MBL can recognize mannose, L-fucose, N-acetylmannosamine and N-acetylglucosamine (GlcNAc); and all existing ficolins (L-, M- and H-ficolin) bind to GlcNAc and N-acetylgalactosamine. Interactions of MBL and ficolin with their respective carbohydrate, as can be observed in Figure 5, lead to activation of mannose-associated serine protease (MASP)-1. MASP-1 activates MASP-2, a serine protease, capable of cleaving the C4 and C2 components of complement, and inducing a cascade that culminates in formation of the membrane attack complex (MAC) molecule leading to pathogen cellular lysis (Chandrasekhar et al., 2013; Mason and Tarr, 2015; Ghebrehiwet, 2016).

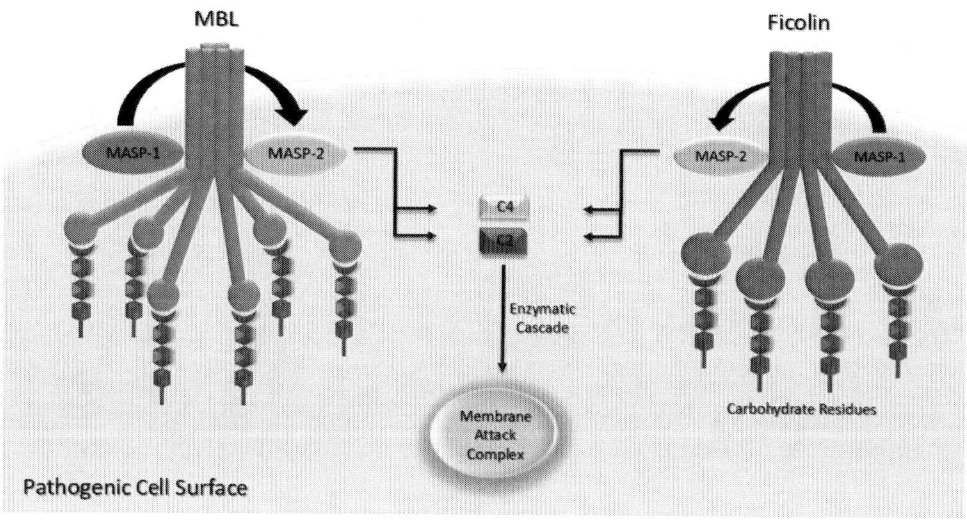

Figure 5. Activation of the complement system through the lectin pathway. MBL and ficolin lectins interact with their respective specific carbohydrate residues on pathogenic cell surface. Such interactions lead to activation of MASP-1, which activates MASP-2, responsible for cleaving C4 and C2 components and leading to an enzymatic cascade; finally, such pathway culminates in the formation of a membrane attack complex, capable of promoting microbial cytolysis.

Conclusion

Lectins of diverse origins show a range of essential roles, functions and applications, from the HA in influenza viruses to complex lectins in plants, or animals. Influenza viral subtype divisions are based on HA types, as well as in HA sialic acid-affinity linkage type. Methods have been developed to characterize HA molecules quickly and with more facility compared to the conventional assay single employing radial immunodiffusion. Lectins from diverse sources can recognize and bind to bacterial glycoconjugates and promote agglutination, growth inhibition and/or death, immune stimulation, metabolic and structural changes as well as inhibition of biofilm formation, adhesion and invasion capacities of bacteria to host, whose mechanisms of action can be visualized by *in vivo* and *in vitro* assays. Lectins are largely used as antimicrobial and antibiofilm molecules, by virtue of their specificity to glycoconjugates, as well as for being a powerful tool to study qualitatively and quantitatively the extracellular matrix of microbial biofilms.

Plant lectins are determined by different biological assays, using cells or pathogens. The MTT assay is widely used to evaluate the cytotoxic action of lectins on tumor cells and mitogenic capacity of these biomolecules. Parasites and insects that cause harmful effects on different organisms can be combated by plant lectins. These proteins can be toxic to eggs and larvae of helminths and produce antinutritional effects on insects. Human lectins show high specificity to glycoconjugates of human cells, and when associated with glycomic studies of diseases can provide a diverse and increasing number of biotechnological applications.

Acknowledgments

The authors acknowledge support of The Conselho Nacional de Desenvolvimento Científico e Tecnológico (CNPq) in providing fellowships (LCBBC and MTSC) and grants. Authors are also grateful to the Coordenação de Aperfeiçoamento de Pessoal de Nível Superior (CAPES), Fundação de Amparo à Pesquisa do Estado de Alagoas (FAPEAL) and the Fundação de Amparo à Ciência e Tecnologia do Estado de Pernambuco (FACEPE).

Competing Interests

The authors have declared that no competing interests exist.

References

Afonso-Cardoso, S. R., Silva, C. V., Ferreira, M. S., Souza, M. A. (2011). Effect of the *Synadenium carinatum* latex lectin (ScLL) on *Leishmania* (*Leishmania*) *amazonensis* infection in murine macrophages. *Experimental Parasitology*, 128(I), 61-67. doi: 10.1016/j.exppara.
2011.02.006.

Agra-Neto, A. C., Napoleão, T. H., Pontual, E. V., Santos, N. D. L., Luz, L. A., Oliveira, C. M. F., Melo-Santos, M. A. V., Coelho, L. C. B. B., Navarro, D. M. A. F., Paiva, P. M. G. (2014). Effect of *Moringa oleifera* lectins on survival and enzyme activities of

Aedes aegypti larvae susceptible and resistant to organophosphate. *Parasitology Research*, 113(I), 175-184. doi: 10.1007/s00436-013-3640-8.

Albuquerque, L. P., Santana, G. M. S., Pontual, E. V., Napoleão, T. H., Coelho, L. C. B. B., Paiva, P. M. G. (2012). Effect of *Microgramma vaccinifolia* rhizome lectin on survival and digestive enzymes of *Nasutitermes corniger* (Isoptera, Termitidae). *International Biodeterioration and Biodegradation*, 75, 158-166. doi: 10.1016/j.ibiod.2012.06.030.

André, S., Kojima, S., Yamazaki, N., Fink, C., Kaltner, H., Kayser, K., Gabius, H. J. (1999). Galectins-1 and-3 and their ligands in tumor biology. *Journal of Cancer Research and Clinical Oncology*, 125(VIII-IX), 461-474. doi: 10.1007/s004320050303.

Apfelthaler, C., Anzengruber, M., Gabor, F., Wirth, M. (2017). Poly – (L) – glutamic acid drug delivery system for the intravesical therapy of bladder cancer using WGA as targeting moiety. *European Journal of Pharmaceutics and Biopharmaceutics*, 115, 131-139. doi: 10.1016/j.ejpb.2017.02.016.

Arasu, A., Kumaresan, V., Palanisamy, R., Arasu, M. V., Al-Dhabi, N. A., Ganesh, M. R., Arockiaraj, J. (2017). Bacterial membrane binding and pore formation abilities of carbohydrate recognition domain of fish lectin. *Developmental and Comparative Immunology*, 67, 202-212. doi: 10.1016/j.dci.2016.10.001.

Bahuguna, A., Khan, I., Bajpai, V. K., Kang, S. C. (2017). MTT assay to evaluate the cytotoxic potential of a drug. *Bangladesh Journal of Pharmacology*, 12(II), 115-118. doi: 10.3329/bjp.v12i2.30892.

Bhat, G. G., Shetty, K. N., Nagre, N. N., Neekhra, V. V., Lingaraju, S., Bhat, R. S., Inamdar, S. R., Suguna, K., Swamy, B. M. (2010). Purification, characterization and molecular cloning of a monocot mannose-binding lectin from *Remusatia vivipara* with nematicidal activity. *Glycoconjugate Journal*, 27(III), 309-320. doi: 10.1007/s10719-010-9279-0.

Balouiri, M., Sadiki, M., Ibnsouda, S. K. (2016). Methods for *in vitro* evaluating antimicrobial activity: A review. *Journal of Pharmaceutical Analysis*, 6(II), 71-79. doi:10.1016/j.jpha.2015.11.005.

Brooks, S. A. (2017). Lectin histochemistry: Historical perspectives, state of the art, and the future. *Histochemistry of Single Molecules: Methods and Protocols*, 93-107. doi: 10.1007/978-1-4939-6788-9_6.

Byrd-Leotis, L., Cummings, R. D., Steinhauer, D. A. (2017). The interplay between the host receptor and influenza virus hemagglutinin and neuraminidase. *International Journal of Molecular Science*, 18, 1541-1563. doi:10.3390/ijms18071541.

Calvier, L., Legchenko, E., Grimm, L., Sallmon, H., Hatch, A., Plouffe, B. D., Schroeder, C., Bauersachs, J., Murthy, S. K., Hansmann, G. (2016). Galectin-3 and aldosterone as potential tandem biomarkers in pulmonary arterial hypertension. *Heart*, 102(V), 390-396. doi: 10.1136/heartjnl-2015-308365.

Carneiro, R. F., Torres, R. C. F., Chaves, R. P., de Vasconcelos, M. A., de Sousa, B. L., Goveia, A. C. R., Arruda, F. V., Matos, M. N. C., Matthews-Cascon, H., Freire, V. N., Teixeira, E. H., Nagano, C. S., Sampaio, A. H. (2017). Purification, biochemical characterization, and amino acid sequence of a novel type of lectin from *Aplysia dactylomela* eggs with antibacterial/antibiofilm potential. *Marine Biotechnology*, 19(I), 49-64. doi: 10.1007/s10126-017-9728-x.

Cavalcante, T. T. A., Carneiro, V. A., Neves, C. C., Duarte, H. S., Martins, M. G. Q., Arruda, F. V. S., Vasconcelos, M. A., Santos, H. S., Cunha, R. M. S., Cavada, B. S., Teixeira, E. H. (2013). A ConA-like lectin isolated from *Canavalia maritima* seeds alters the expression of genes related to virulence and biofilm formation in *Streptococcus mutans*. *Advances in Bioscience and Biotechnology,* 4, 1073-1078. doi: 10.4236/abb.2013.412143.

Cavalcante, T. T. A., Rocha, B. A. M., Carneiro, V. A., Arruda, F. V. S., Nascimento, A. S. F., Sá, N. C., Nascimento, K. S., Cavada, B. S., Teixeira, E. H (2011). Effect of lectins from *Diocleinae* subtribe against oral *Streptococci*. *Molecules,* 16, 3530-3543. doi:10.3390/
molecules16053530.

Chan, Y. S., Wong, J. H., Fang, E. F., Pan, W., Ng, T. B. (2012). Isolation of a glucosamine binding leguminous lectin with mitogenic activity towards splenocytes and anti-proliferative activity towards tumor cells. *PLoS One*, 7(VI), e38961. doi: 10.1371/journal.pone.0038961.

Chandrasekaran, G., Lee, Y., Park, H., Wu, Y., Shin, H. (2016). Antibacterial and antifungal activities of lectin extracted from fruiting bodies of the Korean cauliflower medicinal mushroom, *Sparassis latifolia* (Agaricomycetes). *International Journal of Medicinal Mushrooms*, 18(IV), 291–299. doi: 10.1615/IntJMedMushrooms. v18.i4.20.

Chandrasekhar, A., Dinasarapu, A. R., Thielens, N., Subramaniam, S. (2013). MASP-2. *UCSD Molecule Pages*, 2(II), 1-9. doi: 10.6072/H0.MP.A004275.01.

Chen, L., Neethirajan, S. (2015). A homogenous fluorescence quenching based assay for specific and sensitive detection of influenza virus A hemagglutinin antigen. *Sensors*, 15, 8852-8865. doi:10.3390/s150408852. doi:10.3390/s150408852.

Chikalovets, I. V., Kovalchuk, S. N., Litovchenko, A. P., Molchanova, V. I., Pivkin, M. V., Chernikov, O. V. (2016). A new Gal/GalNAc-specific lectin from the mussel *Mytilus trossulus*: structure, tissue specificity, antimicrobial and antifungal activity. *Fish Shellfish Immunol*, 50, 27-33. doi: 10.1016/j.fsi.2016.01.020.

Coelho, L. C. B. B., Silva, P. M. D. S., Lima, V. L. D. M., Pontual, E. V., Paiva, P. M. G., Napoleão, T. H., Correia, M. T. D. S. (2017). Lectins, interconnecting proteins with biotechnological/pharmacological and therapeutic applications. *Evidence-Based Complementary and Alternative Medicine*, 2017, 1594074. doi: 10.1155/2017/ 1594074.

Cossart, P., Sansonetti, P. J. (2004). Bacterial invasion: The paradigms of enteroinvasive pathogens. *Science*, 304, 242–248. doi: 10.1126/science.1090124.

Coutiño-Rodríguez, R., Hernández-Cruz, P., Giles-Ríos, H. (2001). Lectins in fruits having gastrointestinal activity: their participation in the hemagglutinating property of *Escherichia coli* O157:H7. *Archives of Medical Research,* 32, 251-257. doi: 10.1016/S0188-4409(01)00287-9.

da Silva, L. C. N., Alves, N. M. P., de Castro, M. C. A. B., Pereira, V. R. A., da Paz, N. V. N., Coelho, L. C. B. B., Figueiredo, R. C. B. Q., Correia, M. T. S. (2015). Immunomodulatory effects of pCramoll and rCramoll on peritoneal exudate cells (PECs) infected and non-infected with *Staphylococcus aureus*. *International Journal of Biological Macromolecules*, 72, 848-854. doi: 10.1016/j.ijbiomac.2014.09.045.

De Hoff, P. L., Brill, L. M., A. M. (2009). Plant lectins: the ties that bind in root symbiosis and plant defense. *Molecular Genetics and Genomics*, 282(I), 1-15. doi: 10.1007/s00438-009-0460-8.

Dias, R. O., Machado, L. S., Migliolo, L., Franco, O. L. (2015). Insights into animal and plant lectins with antimicrobial activities. *Molecules,* 20, 519-541. doi:10.3390/ molecules20010519.

Duque, A., Rakic, P. (2011). Different effects of BrdU and ^{3}H-Thymidine incorporation into DNA on cell proliferation, position and fate. *Journal of Neuroscience*, 31(XLII), 15205-15217. doi: 10.1523/JNEUROSCI.3092-11.2011.

Ehrnthaller, C., Ignatius, A., Gebhard, F., Huber-Lang, M. (2011). New insights of an old defense system: structure, function, and clinical relevance of the complement system. *Molecular Medicine*, 17(III-IV), 317-329. doi: 10.2119/molmed.2010.00149.

Elmore, S. (2007). Apoptosis: a review of programmed cell death. *Toxicologic Pathology*, 35(IV), 495-516. doi: 10.1080/01926230701320337.

Fernandes, M. P., Inada, N. M., Chiaratti, M. R., Araújo, F. F. B., Meirelles, F. V., Correia, M. T. S., Coelho, L. C. B. B., Alves, M. J. M., Gadelha, F. R., Vercesi, A. E. (2010). Mechanism of *Trypanosoma cruzi* death induced by *Cratylia mollis* seed lectin. *Journal of Bioenergetics and Biomembranes*, 42(I), 69-78. doi: 10.1007/s10863-010-9268-9.

Flemming, H. C, Wingender, J. (2010). The biofilm matrix. *Nature Reviews Microbiology,* 8, 623-633. doi:10.1038/nrmicro2415.

Flemming, H. C., Wingender, J., Szewzyk, U., Steinberg, P., Rice, S. A., Kjelleberg, S., (2016). Biofilms: an emergent form of bacterial life. *Nature Reviews Microbiology*, 14(IX), 563-575. doi:10.1038/nrmicro.2016.94.

Figueirôa, E. O., Aranda-Souza, M. A., Varejão, N., Rossato, F. A., Costa, R. A. P., Figueira, T. R., Silva, L. C. N., Castilho, R. F., Vercesi, A. E., Correia, M. T. S. (2017). pCramoll and rCramoll lectins induce cell death in human prostate adenocarcinoma (PC-3) cells by impairment of mitochondrial homeostasis. *Toxicology in Vitro*, 43, 40-46. doi: 10.1016/j.tiv.2017.05.016.

Fu, L. L., Zhou, C. C., Yao, S., Yu, J. Y., Liu, B., Bao, J. K. (2011). Plant lectins: targeting programmed cell death pathways as antitumor agents. *International Journal of Biochemistry and Cell Biology*, 43(X), 1442-1449. doi: 10.1016/j.biocel.2011.07.004.

Geng, P., Ding, Y., Qiu, L., Lu, Y. (2015). Serum mannose-binding lectin is a strong biomarker of diabetic retinopathy in chinese patients with diabetes. *Diabetes Care*, 38(V), 868-875. doi: 10.2337/dc14-1873.

Ghebrehiwet, B. (2016). The complement system: an evolution in progress. *F1000Research*, 5, 1-8. doi: 10.12688/f1000research.10065.1.

Giacometti, J. (2015). Plant lectins in cancer prevention and treatment. *Medicina Fluminensis*, 51(II), 211-229.

Gomes, F. S., Procópio, T. F., Napoleão, T. H., Coelho, L. C. B. B., Paiva, P. M. G. (2013). Antimicrobial lectin from *Schinus terebinthifolius* leaf. *Journal of Applied Microbiology*, 114(III), 672-679. doi: 10.1111/jam.12086.

Gorakshakar, A. C., Ghosh, K. (2016). Use of lectins in immunohematology. *Asian Journal of Transfusion Science*, 10(I), 12-21. doi: 10.4103/0973-6247.172180.

Hasan, I., Ozeki, Y., Kabir, R. (2014). Purification of a novel chitin-binding lectin with antimicrobial and antibiofilm activities from a cultivar of potato (*Solanum tuberosum*). *Indian Journal of Biochemistry and Biophysics,* 51, 142-148.

Jaber, K., Haubruge, E., Francis, F. (2010). Development of entomotoxic molecules as control agents: illustration of some protein potential uses and limits of lectins (Review). *Biotechnologie Agronomie Societe Et Environnement*, 14(I), 225-241.

Jandú, J. J. B., Neto, R. N. M., Zagmignan, A., Sousa, E. M., Brelaz-de-Castro, M. C. A., Correia, M. T. S., Silva, L. C. N. (2017). Targeting the immune system with plant lectins to combat microbial infections. *Frontiers in Pharmacology*, 8, 671. doi: 10.3389/fphar.2017.00671.

Jayanthi, S., Ishwarya, R., Anjugam, M., Iswarya, A., Karthikeyan, S., Vaseeharan B. (2017). Purification, characterization and functional analysis of the immune molecule lectin from the haemolymph of blue swimmer crab *Portunus pelagicus* and their antibiofilm properties. *Fish and Shellfish Immunology,* 62, 227-237. doi: 10.1016/j.fsi.2017.01.019.

Jennings, L. K., Storek, K. M., Ledvina, H. E., Coulon, C., Marmont, L. S., Sadovskaya, I., Secor, P. R., Tseng, B. S., Scian, M., Filloux, A., Wozniak, D. J., Howell, P. L., Parsek, M. R (2015). Pel is a cationic exopolysaccharide that cross-links extracellular DNA in the Pseudomonas aeruginosa biofilm matrix. *Proceedings of the National Academy of Sciences of the United States of America,* 112, 11353-1358. doi: 10.1073/pnas.1503058112.

Juan, L. L., Recio, V. G., López, P. J., Juan, T. J., Cordoba-Diaz, M., Cordoba-Diaz, D. (2017). Pharmaceutical applications of lectins. *Journal of Drug Delivery Science and Technology,* 42, 126-133. doi: 10.1016/j.jddst.2017.05.018.

Karnchanatat, A. (2012). Antimicrobial activity of lectins from plants. In: Bobbarala, V. (Ed), *Antimicrobial Agents*, Rijeka, In Tech, pp. 145-178. doi: 10.5772/33456.

Klafke, G. B., Borsuk, S., Gonçales, R. A., Arruda, F. V. S., Carneiro, V. A., Teixeira, E. H., Coelho da Silva, A. L., Cavada, B. S., Dellagostin, O. A., Pinto, L. S. (2013). Inhibition of initial adhesion of oral bacteria through a lectin from *Bauhinia variegata* L. var. variegata expressed in *Escherichia coli*. *Journal of Applied Microbiology,* 115, 1222-1230. doi: 10.1111/jam.12318.

Koo, H., Yamada, K. M (2016). Dynamic cell-matrix interactions modulate microbial biofilm and tissue 3D microenvironments. *Current Opinion in Cell Biology,* 42, 102-112. doi: 10.1016/j.ceb.2016.05.005.

Kumar, P., Mishra, S., Singh, S. (2017). Advanced acuity in microbial biofilm genesis, development, associated clinical infections and control. *Journal des Anti-infectieux,* 19, 20-31. doi: 10.1016/j.antinf.2017.01.002.

Lacerda, J. T. J. G., Lacerda, R. R., Assunção, N. A., Tashima, A. K., Juliano, M. A., Santos, G. A., dos Santos de Souza, M., de Luna Batista, J., Rossi, C. E., de Almeida Gadelha, C. A., Santi-Gadelha, T. (2017). New insights into lectin from *Abelmoschus esculentus* seeds as a Kunitz-type inhibitor and its toxic effects on *Ceratitis capitata* and root-knot nematodes *Meloidogyne* spp. *Process Biochemistry*, 63, 96-104. doi: 10.1016/j.procbio.2017.09.003.

Lagarda-Diaz, I., Guzman-Partida, A. M., Vazquez-Moreno, L. (2017). Legume lectins: proteins with diverse applications. *International Journal of Molecular Sciences*, 18(VI), 1-18. doi:10.3390/ijms18061242.

Lam, S. K., Ng, T. B. (2011). Lectins: production and practical applications. *Applied Microbiology and Biotechnology*, 89(I), 45-55. doi: 10.1007/s00253-010-2892-9.

Lannoo, N., Van Damme, E. J. (2014). Lectin domains at the frontiers of plant defense. *Frontiers Plant in Plant Science*, 5, 1-16. doi: 10.3389/fpls.2014.00397.

Laue, H., Schenk, A., Li, H., Lambertsen, L., Neu, T. R., Molin, S., Ullrich, M. S. (2006). Contribution of alginate and levan production to biofilm formation by *Pseudomonas syringae*. *Microbiology,* 152, 2909-2918. doi: 10.1099/mic.0.28875-0.

Levison, M. E. (2004). Pharmacodynamics of antimicrobial drugs. *Infectious Disease Clinics*, 18(III), 451-465. doi: 10.1016/j.idc.2004.04.012.

Lewis, N. S., Russell, C. A., Langat, P., Anderson, T. K., Berger, K., Bielejec, F., Burke, D. F., Dudas, G., Fonville, J. M., Fouchier, R. A. M., Kellam, P., Koel, B. F., Lemey, P., Nguyen, T., Nuansrichy, B., Peiris, J. S. M., Saito, P., Simon, G., Skepner, E., Takemae, N., ESNIP3 consortium, Webby, Richard J., Reeth, K. V., Brookes, S. M., Larsen, L., Watson, S. J., Brown, I. H., Vincent, A. L. (2016). The global antigenic diversity of swine influenza A viruses. e*Life*, 5:e12217. doi:10.7554/eLife.12217.

Li, C., Jaentschke, B., Song, Y., Wang, J., Cyr, Terry D., Domselaar, G. V., He, R., Li, X. (2010). A simple slot blot for the detection of virtually all subtypes of the influenza A viral hemagglutinins using universal antibodies targeting the fusion peptide. *Nature Protocols*, 5(I), 14-19. doi: 10.1038/nprot.2009.200.

Li, M., Li, C., Ma, C., Li, H., Zuo, H., Weng, S., Chen, X., Zeng, D., He, J., Xu, X. (2014). Identification of a C-type lectin with antiviral and antibacterial activity from pacific white shrimp *Litopenaeus vannamei*. *Developmental and Comparative Immunology*, 46(II), 231-240. doi: 10.1016/j.dci.2014.04.014.

Lichtenstein, R. G., Rabinovich, G. A. (2013). Glycobiology of cell death: when glycans and lectins govern cell fate. *Cell Death and Differentiation*, 20(VIII), 976-986. doi: 10.1038/cdd.2013.50.

Lima, T. A., Fernandes, K. M., Oliveira, A. P. S., Dornelles, L. P., Martins, G. F., Napoleão, T. H., Paiva, P. M. G. (2017). Termiticidal lectins from *Myracrodruon urundeuva* (Anacardiaceae) cause midgut damages when ingested by *Nasutitermes corniger* (Isoptera, Termitidae) workers. *Pest Management Science*, 73(V), 991-998. doi: 10.1002/ps.4415.

Limoli, D. H., Jones, C. J., Wozniak, D. J. (2015). Bacterial extracellular polysaccharides in biofilm formation and function. *Microbiology Spectrum,* 3, 1-30. doi: 10.1128/microbiolspec.MB-0011-2014.

Lin, Y. T., Chen, J. S., Wu, M. H., Hsieh, I. S., Liang, C. H., Hsu, C. L., Hong, T. M., Chen, Y. L. (2015). Galectin-1 accelerates wound healing by regulating the neuropilin-1/Smad3/NOX4 pathway and ROS production in myofibroblasts. *Journal of Investigative Dermatology*, 135(I), 258-268. doi: 10.1038/jid.2014.288.

Lodish, H., Berk, A., Kaiser, C. A., Krieger, M., Bretscher, A., Ploegh, H., Amon, A., Martin, K., Scott, M. P. (2016). *Molecular Cell Biology* (Vol. 8). New York: WH Freeman.

Luz, L. A., Rossato, F. A., Costa, R. A. P., Napoleão, T. H., Paiva, P. M. G., Coelho, L. C. B. B. (2017). Cytotoxicity of the coagulant *Moringa oleifera* lectin (cMoL) to B16-F10 melanoma cells. *Toxicology in Vitro*, 44, 94-99. doi: 10.1016/j.tiv.2017.06.019.

Macedo, M. L. R., Oliveira, C. F. R., Oliveira, C. T. (2015). Insecticidal activity of plant lectins and potential application in crop protection. *Molecules*, 20(II), 2014-2033. doi:10.3390/molecules20022014.

Maciel, E. V. M., Araújo-Filho, V. S., Nakazawa, M., Gomes, Y. M., Coelho, L. C. B. B., Correia, M. T. S. (2004). Mitogenic activity of *Cratylia mollis* lectin on human lymphocytes. *Biologicals*, 32(I), 57-60. doi: 10.1016/j.biologicals.2003.12.001.

Madhavan, H. N. (2007). Simple laboratory methods to measure cell proliferation using DNA synthesis property. *Journal of Stem Cells & Regenerative Medicine*, 3(I), 12-14.

Mason, C. P., Tarr, A. W. (2015). Human lectins and their roles in viral infections. *Molecules*, 20(II), 2229-2271. doi: 10.3390/molecules20022229.

McBride, R., Paulson, J. C., de Vries, R. P. (2016). A miniaturized glycan microarray assay for assessing avidity and specificity of influenza A virus hemagglutinins. *Journal of Visualized Experiments*, 29(CXI), 1-11. doi: 10.3791/53847.

McLeod, K., Walker, J. T., Hamilton, D. W. (2018). Galectin-3 regulation of wound healing and fibrotic processes: insights for chronic skin wound therapeutics. *Journal of Cell Communication and Signaling*, 1-7. doi: 10.1007/s12079-018-0453-7.

Medeiros, M. L. S., Moura, M. C., Napoleão, T. H., Paiva, P. M. G., Coelho, L. C. B. B., Bezerra, A. C. D. S., Silva, M. D. C. (2017). Nematicidal activity of a water soluble lectin from seeds of *Moringa oleifera*. *International Journal of Biological Macromolecules*, 1-24. doi: 10.1016/j.ijbiomac.2017.10.167.

Melo, C. M. L., Melo, H., Correia, M. T. S., Coelho, L. C. B. B., Silva, M. B., Pereira, V. R. A. (2011a). Mitogenic response and cytokine production induced by Cramoll 1,4 lectin in splenocytes of inoculated mice. *Scandinavian Journal of Immunology*, 73(II), 112-121. doi: 10.1111/j.1365-3083.2010.02490.x.

Melo, C. M. L., Lima, A. L. R., Beltrão, E. I. C., Cavalcanti, C. C. B., Melo-Júnior, M. R., Montenegro, S. M. L., Coelho, L. C. B. B., Correia, M. T. S., Carneiro-Leão, A. M. A. (2011b). Potential effects of Cramoll 1,4 lectin on murine *Schistosomiasis mansoni*. *Acta Tropica*, 118(II), 152-158. doi: 10.1016/j.actatropica.2011.01.008.

Meng, Y., Hou, X., Lei, J., Chen, M., Cong, S., Zhang, Y., Ding, W., Li, G., Li, X. (2016). Multi-functional liposomes enhancing target and antibacterial immunity for antimicrobial and anti-biofilm against methicillin-resistant *Staphylococcus aureus*. *Pharmaceutical Research,* 33, 763-775. doi: 10.1007/s11095-015-1825-9.

Michalek, P., Krejcova, L., Adam, V., Kizek, R. (2015). Hemagglutinin structure, membrane fusion and virus entry. *Journal of Metallomics and Nanotechnologies*, 1, 53-56.

Mitchell, K. F., Zarnowski, R., Sanchez, H., Edward, J. A., Reinicke, E. L., Nett, J. E., Mitchell, A. P., Andes, D. R. (2015). Community participation in biofilm matrix assembly and function. *Proceedings of the National Academy of Sciences of the United States of America,* 112, 4092-4097. doi: 10.1073/pnas.1421437112.

Mito, A., Nakano, Y., Saitoh, T., Gouraud, S. S., Yamaguchi, Y., Sato, T., Sasaki, N., Kojima-Aikawa, K. (2017). Lectin ZG16p inhibits proliferation of human colorectal cancer cells via its carbohydrate-binding sites. *Glycobiology*, 28(I), 21-31. doi: 10.1093/glycob/cwx088.

Mlouka, M. A. B., Cousseau, T., Di Martino, P. (2016). Application of fluorescently labelled lectins for the study of polysaccharides in biofilms with a focus on biofouling of nanofiltration membranes. *AIMS Molecular Science,* 3, 338-356. doi: 10.3934/molsci.2016.3.338.

Moura, M. C., Napoleão, T. H., Coriolano, M. C., Paiva, P. M. G., Figueiredo, R. C. B. Q., Coelho, L. C. B. B. (2015). Water-soluble *Moringa oleifera* lectin interferes with growth, survival and cell permeability of corrosive and pathogenic bacteria. *Journal of Applied Microbiology*, 119(III), 666-676. doi: 10.1111/jam.12882.

Moura, M. C., Trentin, D. S., Napoleão, T. H., Barros, M. P., Xavier, A. S., Carneiro, N. P., Paiva, P. M. G., Macedo, A. J., Coelho, L. C. B. B. (2017). Multi-effect of the water-soluble *Moringa oleifera* lectin against *Serratia marcescens* and *Bacillus sp.*: antibacterial, antibiofilm and anti - adhesive properties. *Journal of Applied Microbiology*, 123(IV), 861-874. doi: 10.1111/jam.13556.

Mukherjee, S., Zheng, H., Derebe, M., Callenberg, K., Partch, C. L., Rollins, D., Propheter, D. C., Rizo, J., Grabe, M., Jiang, Q. X., Hooper, L. V. (2014). Antibacterial membrane attack by a pore-forming intestinal C-type lectin. *Nature*, 505(VIICDLXXX), 103-107. doi: 10.1038/nature12729.

Nachbagauer, R., Miller, M. S., Rong, H., Ryder, A. B., Rose, J. K., Palese, P., García-Sastre, A., Krammer, F., Albrecht, R. A. (2016). Hemmagglutinin stalk immunity reduces influenza virus replication and transmission in ferrets. *Journal of Virology*, 90(VI), 3268-3273. doi: 10.1128/JVI.02481-15.

Nagaraja, N., Skillman, L., Xie, Z., Joang, S., Ho, G., Li, D. (2017). Investigation of compounds that degrade biofilm polysaccharides on reverse osmosis membranes from a full scale desalination plant to alleviate biofouling. *Desalination,* 403, 88-96. doi: 10.1016/j.desal.
2016.06.002.

Nascimento, K. S., Cunha, A. I., Nascimento, K. S., Cavada, B. S., Azevedo, A. M., Aires-Barros, M. R. (2012). An overview of lectins purification strategies. *Journal of Molecular Recognition*, 25(XI), 527-541. doi: 10.1002/jmr.2200.

Nesargikar, P. N., Spiller B., Chavez, R. (2012). The complement system: history, pathways, cascade and inhibitors. *European Journal of Microbiology and Immunology*, 2(II), 103–111. doi: 10.1556/EuJMI.2.2012.2.2.

Neu, T. R., Kuhlicke, U. (2017). Fluorescence lectin bar-coding of glycoconjugates in the extracellular matrix of biofilm and bioaggregate forming microorganisms. *Microorganisms,* 5, 1-13. doi:10.3390/microorganisms5010005.

Neu, T. R., Swerhone, G. D. W., Lawrence, J. R. (2001). Assessment of lectin-binding analysis for *in situ* detection of glycoconjugates in biofilm systems. *Microbiology,* 147, 299-313. doi: 10.1099/00221287-147-2-299.

Nikoletopoulou, V., Markaki, M., Palikaras, K., Tavernarakis, N. (2013). Crosstalk between apoptosis, necrosis and autophagy. *Biochimica et Biophysica Acta - Molecular Cell Research*, 1833(XII), 3448-3459. doi: 10.1016/j.bbamcr.2013.06.001.

Nizet, V., Varki, A., Aebi, M. (2017). *Microbial lectins: hemagglutinins, adhesions, and toxins. In Essentials of glycobiology* [Internet], edited by A. Varki, R. D. Cummings, J. D. Esko, et al., Cold Spring Harbor (NY): Cold Spring Harbor Laboratory Press (2015-2017).

Nunes, E. S., De Souza, M. A. A., Vaz, A. F. M., Santana, G. M. S., Gomes, F. S., Coelho, L. C. B. B., Paiva, P. M. G., da Silva, R. M. L., Silva-Lucca, R. A., Oliva, M. L. V., Guarnieri, M. C., Correia, M. T. S. (2011). Purification of a lectin with antibacterial activity from *Bothrops leucurus* snake venom. *Comparative Biochemistry and*

Physiology Part B: Biochemistry and Molecular Biology, 159(1), 57-63. doi: 10.1016/j.cbpb.2011.02.001.

Oliveira, A. P. S., Silva, L. L. S., Lima, T. A., Pontual, E. V., Santos, N. D. L., Coelho, L. C. B. B., Navarro, D. M. A. F., Zingali, R. B., Napoleão, T. H., Paiva, P. M. G. (2016). Biotechnological value of *Moringa oleifera* seed cake as source of insecticidal lectin against *Aedes aegypti*. *Process Biochemistry*, 51(X), 1683-1690. doi: 10.1016/j.procbio.2016.06.026.

Oliveira, C. F. R., Luz, L. A., Paiva, P. M. G., Coelho, L. C. B. B., Marangoni, S., Macedo, M. L. R. (2011). Evaluation of seed coagulant *Moringa oleifera* lectin (cMoL) as a bioinsecticidal tool with potential for the control of insects. *Process Biochemistry*, 46(II), 498-504. doi: 10.1016/j.procbio.2010.09.025.

Oliveira, C. F. R., Moura, M. C., Napoleão, T. H., Paiva, P. M. G., Coelho, L. C. B. B., Macedo, M. L. R. (2017). A chitin-binding lectin from *Moringa oleifera* seeds (WSMoL) impairs the digestive physiology of the Mediterranean flour larvae, *Anagasta kuehniella*. *Pesticide Biochemistry and Physiology*, 142, 67-76. doi: 10.1016/j.pestbp.2017.
01.006.

Osterne, V. J. S., Silva-Filho, J. C., Santiago, M. Q., Pinto-Junior, V. R., Almeida, A. C., Barreto, A. A. G. C., Wolin, I. A. V., Nascimento, A. P. M., Amorim, R. M. F., Rocha, B. A. M., Delatorre, P., Nagano, C. S., Leal, R. B., Assreuy, A. M. S., Nascimento, K. S., Cavada, B. S. (2017). Structural characterization of a lectin from *Canavalia virosa* seeds with inflammatory and cytotoxic activities. *International Journal of Biological Macromolecules*, 94, 271-282. doi: 10.1016/j.ijbiomac.2016.10.020.

Paclik, D., Lohse, K., Wiedenmann, B., Dignass, A. U., Sturm, A. (2008). Galectin-2 and-4, but not galectin-1, promote intestinal epithelial wound healing in vitro through a TGF-beta-independent mechanism. *Inflammatory Bowel Diseases*, 14(X), 1366-1372. doi: 10.1002/ibd.20499.

Pavlov, V. I., Tan, Y. S., McClure, E. E., La Bonte, L. R., Zou, C., Gorsuch, W. B., Stahl, G. L. (2015). Human mannose-binding lectin inhibitor prevents myocardial injury and arterial thrombogenesis in a novel animal model. *The American Journal of Pathology*, 185(II), 347-355. doi: 10.1016/j.ajpath.2014.10.015.

Pereira, P. R., Del Aguila, E. M., Verícimo, M. A., Zingali, R. B., Paschoalin, V. M. F., Silva, J. T. (2014). Purification and characterization of the lectin from taro (*Colocasia esculenta*) and its effect on mouse splenocyte proliferation in vitro and in vivo. *Protein Journal*, 33(I), 92-99. doi: 10.1007/s10930-013-9541-y.

Petrova, M. I., Imholz, N. C. E., Verhoeven, T. L. A., Balzarini, J., Van Damme, E. J. M., Schols, D., Vanderleyden, J., Lebeer, S. (2016). Lectin-Like molecules of *Lactobacillus rhamnosus* GG inhibit pathogenic *Escherichia coli* and *Salmonella* biofilm formation. *PLoS ONE*, 11, e0161337. doi: 10.1371/journal.pone.0161337.

Poiroux, G., Barre, A., van Damme, E. J. M., Benoist, H., Rougé, P. (2017). Plant lectins targeting *O*-glycans at the cell surface as tools for cancer diagnosis, prognosis and therapy. *International Journal of Molecular Sciences*, 18(VI), E1232. doi:10.3390/ijms18061232.

Procópio, T. F., Patriota, L. L. S., Moura, M. C., Silva, P. M., Oliveira, A. P. S., Carvalho, L. V. N., Lima, T. A., Soares, T., Silva, T. D., Coelho, L. C. B. B., Pitta, M. G. R., Rêgo, M. J. B. M., Figueiredo, R. C. B. Q., Paiva, P. M. G., Napoleão, T. H. (2017). CasuL: a new lectin isolated from *Calliandra surinamensis* leaf pinnulae with cytotoxicity to cancer cells, antimicrobial activity and antibiofilm effect. *International Journal of Biological Macromolecules*, 98, 419-429. doi: 10.1016/j.ijbiomac. 2017.02.019.

Qu, M., Tong, C., Kong, L., Yan, X., Chernikov, O. V., Lukyanov, P. A., Jin, Q, Li, W. (2015). Purification of a secreted lectin from *Andrias davidianus* skin and its antibacterial activity. *Comparative Biochemistry and Physiology part C: Toxicology and Pharmacology*, 167, 140-146. doi: 10.1016/j.cbpc.2014.08.008.

Rock, K. L., Kono, H. (2008). The inflammatory response to cell death. *Annual Review of Pathology*, 3, 99-126. doi: 10.1146/annurev.pathmechdis.3.121806.151456.

Roux, D., Cywes-Bentley, C., Zhang, Y-F., Pons, S., Konkol, M., Kearns, D. B., Little, D. J., Howell, P. L., Skurnik, D., Pier, G. B. (2015). Identification of poly-*N*-acetylglucosamine as a major polysaccharide component of the *Bacillus subtilis* biofilm matrix. *The Journal of Biological Chemistry,* 290, 19261-19272. doi: 10.1074/jbc.M115.648709.

Saal, I., Nagy, N., Lensch, M., Lohr, M., Manning, J. C., Decaestecker, C., André, S., Kiss, R., Salmon, I., Gabius, H. J. (2005). Human galectin-2: expression profiling by RT-PCR/immunohistochemistry and its introduction as a histochemical tool for ligand localization. *Histology and histopathology*, 20(IV), 1191-1226. doi: 10.14670/HH-20.1191.

Santos, A. F., Luz, L. A., Napoleão, T. H., Paiva, P. M., Coelho, L. C. B. B. (2014). Coagulation, flocculation, agglutination and hemagglutination: similar properties? *Advances in Chemistry Research*, 20, 51-70.

Santos, N. D. L., Moura, K. S., Napoleão, T. H., Santos, G. K. N., Coelho, L. C. B. B., Navarro, D. M. A. F., Paiva, P. M. G. (2012). Oviposition-stimulant and ovicidal activities of *Moringa oleifera* lectin on *Aedes aegypti*. *PLoS One*, 7(IX), e44840. doi: 10.1371/journal.pone.0044840.

Schild, G. C., Wood, J. M., Newman, R. W. (1975). A single-radial-immunodifusion technique for the assay of influenza hemagglutinin antigen: proposals for an assay method for the hemagglutinin content of influenza vaccines. *Bulletin of the World Health Organization*, 52, 223-231.

Silva, L. N., Zimmer, K. R., Macedo, A. J., Trentin, D. S. (2016). Plant natural products targeting bacterial virulence factors. *Chemical Reviews*, 116, 9162-9236. doi: 10.1021/acs.chemrev.6b00184.

Silva, P. M., Napoleão, T. H., Silva, L. C., Fortes, D. T., Lima, T. A., Zingali, R. B., Pontual, E. V., Araújo, J. M., Medeiros, P. L., Rodrigues, C. G., Gomes, F. S., Paiva, P. M. G. (2016). The juicy sarcotesta of *Punica granatum* contains a lectin that affects growth, survival as well as adherence and invasive capacities of human pathogenic bacteria. *Journal of Functional Foods*, 27, 695-702. doi: 10.1016/j.jff.2016.10.015.

Souza, L. P. F., Ramos, E. L. P., Santana, S. S., Silva, M. V., Santiago, F. M., Mineo, T. W. P., Mineo, J. R. (2016). Lectins from *Synadenium carinatum* (ScLL) and *Artocarpus heterophyllus* (ArtinM) are able to induce beneficial immunomodulatory effects in a murine model for treatment of *Toxoplasma gondii* infection. *Frontiers in Cellular and Infection Microbiology*, 6, 164. doi: 10.3389/fcimb.2016.00164.

Spackman, E., Senne, D. A., Myers, T. J., Bulaga, L. L., Garber, L. P., Perdue, M. L., Lohman, K., Daum, L. T., Suarez, D. L. (2002). Development of a real-time reverse transcriptase PCR assay for type A influenza virus and the avian H5 and H7 hemagglutinin subtypes. *Journal of Clinical Microbiology*, 40(IX), 3256-3260. doi: 10.1128/JCM.40.9.3256-3260.2002.

Sriwilaijaroen, N., Suzuki, Y. (2012). Molecular basis of the structure and function of H1 hemagglutinin of influenza virus. *Proceedings of the Japan Academy, Series B Physical and Biological Sciences*, 88(VI), 226-249. doi: 10.2183/pjab.88.226.

Sun, Y. Y., Liu, L., Li, J., Sun, L. (2016). Three novel B-type mannose-specific lectins of *Cynoglossus semilaevis* possess varied antibacterial activities against Gram-negative and Gram-positive bacteria. *Developmental and Comparative Immunology*, 55, 194-202. doi: 10.1016/j.dci.2015.10.003.

Tate, M. D., Job, E. R., Deng, Y., Gunalan, V., Maurer-Stroh, S., Reading, P. C. (2014). Playing hide and seek: how glycosylation of the influenza virus hemagglutinin can modulate the immune response to infection. *Viruses*, 6, 1294-1316. doi: 10.3390/v6031294.

Tawakoli, P. N., Neu, T. R., Busck, M. M., Kuhlicke, U., Schramm, A., Attin, T., Wiedemeier, D. B., Schlafer, S. (2017a). Visualizing the dental biofilm matrix by means of fluorescence lectin-binding analysis. *Journal of Oral Microbiology,* 9, 1-10. doi: 10.1080/20002297.2017.1345581.

Tawakoli, P. N., Ragnarsson, K. T., Rechenberg, D. K., Mohn, D., Zehnder, M. (2017b). Effects of endodontic irrigants on biofilm matrix polysaccharides. *International Endodontic Journal,* 50, 153-160. doi: 10.1111/iej.12604.

Tscherne, D. M., García-Sastre, A. (2011). Virulence determinants of pandemic influenza viruses. *The Journal of Clinical Investigation*, 121(I),6-13. doi: 10.1172/JCI144947.

Tseng, B. S., Majerczyk, C. D., Passos da Silva, D., Chandler, J. R., Greenberg, E. P., Parsek, M. R. (2016). Quorum sensing influences *Burkholderia thailandensis* biofilm development and matrix production. *Journal of Bacteriology,* 198, 2643-2650. doi: 10.1128/JB.00047-16.

Turner, M. W. (2003). The role of mannose-binding lectin in health and disease. *Molecular Immunology*, 40(VII), 423-429. doi: 10.1016/S0161-5890(03)00155-X.

Tzarum, N., de Vries, R. P., Zhu, X., Yu, X., McBride, R., Paulson, J. C., Wilson, I. A. (2015). Structure and receptor binding of the hemagglutinin from a human H6N1 influenza virus. *Cell Host and Microbe*, 17, 369-376. doi: 10.1016/j.chom.2015.02.005.

Van Damme, E. J., Lannoo, N., Albenne, C., Jamet, E. (Eds.). (2015). *Plant Glycobiology– a sweet world of lectins, glycoproteins, glycolipids and glycans*. Frontiers Media SA.

Wagner, C., Barlag, B., Gerlach, R. G., Deiwick, J., Hensel, M. (2014). The *Salmonella enterica* giant adhesin SiiE binds to polarized epithelial cells in a lectin-like manner. *Cellular Microbiology*, 16(VI), 962-975. doi: 10.1111/cmi.12253.

Wang, H. X., Liu, W. K., Ng, T. B., Ooi, V. E. C., Chang, S. T. (1996). The immunomodulatory and antitumor activities of lectins from the mushroom *Tricholoma mongolicum*. *Immunopharmacology*, 31(II-III), 205-211. doi: 10.1016/0162-3109(95)00049-6.

Wlodkowic, D., Telford, W., Skommer, J., Darzynkiewicz, Z. (2011). Apoptosis and beyond: cytometry in studies of programmed cell death. *Methods in Cell Biology*, 103, 55-98. doi: 10.1016/B978-0-12-385493-3.00004-8.

Wohlbold, T. J., Krammer, F. (2014). In the shadow of hemagglutinin: a growing interest in influenza viral neuraminidase and its role as a vaccine antigen. *Viruses*, 6, 2465-2494. doi:10.3390/v6062465.

Wongpanya, R., Sengprasert, P., Amparyup, P., Tassanakajon, A. (2017). A novel C-type lectin in the black tiger shrimp *Penaeus monodon* functions as a pattern recognition receptor by binding and causing bacterial agglutination. *Fish and Shellfish Immunology*, 60, 103-113. doi: 10.1016/j.fsi.2016.11.042.

World Health Organization (WHO). (1980). A revision of the system of nomenclature for influenza viruses: a WHO memorandum. *Bulletin of the World Health Organization*, 58(IV), 585-591.

Yamaoka, A., Kuwabara, I., Frigeri, L. G., Liu, F. T. (1995). A human lectin, galectin-3 (epsilon bp/Mac-2), stimulates superoxide production by neutrophils. *The Journal of Immunology*, 154(VII), 3479-3487.

Zhang, C. Z., Fang, E. F., Zhang, H. T., Liu, L. L., Yun, J. P. (2015). *Momordica charantia* lectin exhibits antitumor activity towards hepatocellular carcinoma. *Investigational New Drugs*, 33(I), 1-11. doi: 10.1007/s10637-014-0156-8.

Zhang, R. Y., Neu, T. R., Bellenberg, S., Kuhlicke, U., Sand, W., Vera, M. (2015) Use of lectins to in situ visualize glycoconjugates of extracellular polymeric substances in acidophilic archaeal biofilms. *Microbial Biotechnology,* 8 (III), 448-461. doi: 10.1111/1751-7915.12188.

Zhou, Z. J., Sun, L. (2015). CsCTL1, a teleost C-type lectin that promotes antibacterial and antiviral immune defense in a manner that depends on the conserved EPN motif. *Developmental and Comparative Immunology* 50(II), 69-77. doi: 10.1016/j.dci.2015.01.007.

Zubcevic, N., Damir, S., Focak, M., Rukavina, D. (2016). Effects of plant lectins on human erythrocyte agglutination. *Serbian Journal of Experimental and Clinical Research*, 17(III), 207-214. doi: 10.1515/sjecr-2016-0031.

In: Hemagglutinins
Editors: T. Bun Ng, J. Wong, R. Tse et al.
ISBN: 978-1-53615-708-6
© 2019 Nova Science Publishers, Inc.

Chapter 2

BAUHINIA MONANDRA LEAF LECTIN (BMOLL) IS IMMUNOGENIC: ELECTROCHEMICAL AND PIEZOELECTRIC BIOSENSOR RECOGNIZED LECTIN/IGG ANTI-BMOLL INTERACTIONS

***Sandra R. Souza*[1], *Priscila M. S. Silva*[2], *Rosa Amália F. Dutra*[3], *Nicolaas J. Haver*[2], *Sílvio A. O. Ferreira*[2], *José Luiz de Lima Filho*[4], *Maria T. S. Correia*[2] and *Luana C. B. B. Coelho*[2,*]**

[1]Departamento de Educação,
Universidade Federal Rural de Pernambuco, Recife, Brazil
[2]Laboratório de Bioquímica de Proteínas, Departamento de Bioquímica,
Centro de Biociências, Universidade Federal de Pernambuco,
Recife, Brazil
[3]Departamento de Engenharia Biomédica,
Centro de Tecnologia e Geociências,
Universidade Federal de Pernambuco, Recife, Brazil
[4]Laboratório de Imunopatologia Keizo Asami,
Universidade Federal de Pernambuco, Recife, Brazil

ABSTRACT

A highly purified galactose-binding lectin from *Bauhinia monandra* leaves (BmoLL) was intradermally injected into a rabbit to evaluate the protein immune response. The

[*] Corresponding Author's E-mail: lcbbcoelho@gmail.com. Tel: +55 81 999455897.

antiserum developed against BmoLL (anti-BmoLL serum) revealed the immunogenicity of the lectin. Anti-BmoLL IgG was purified from the antiserum on a Protein A-Sepharose CL-4B column. BmoLL and extracts from *B. monandra* different tissues, with previously detected lectin activity, interacted with anti-BmoLL serum and anti-BmoLL IgG; protein precipitation lines were visualized by double immunodiffusion assay. Anti-BmoLL IgG was immobilized on a gold surface for electrochemical and piezoelectric characterization of BmoLL antibody-antigen interaction. An electrochemical immunosensor performed with anti-BmoLL IgG and gold electrode surface was used to estimate BmoLL binding and its electrochemical potential. A piezoelectric immunosensor was also carried out based upon a quartz crystal microbalance (QCM) with a gold surface where anti-BmoLL IgG was immobilized to characterize BmoLL recognition. The electrochemical system was sensitive for evaluating anti-BmoLL IgG interactions at different concentrations of BmoLL. The piezoelectric immunosensor with flow injection analysis (FIA) revealed a frequency decrease (15 kHz) after the interaction of BmoLL (454 μg/ml) with anti-BmoLL IgG (1.1 mg/mL) indicating antigen-antibody specificity. Thus the piezoelectric assay is a potent tool for detection of antigen-antibody recognition.

Keywords: *Bauhinia monandra*, electrochemical biosensor, leaf lectin, piezoelectric biosensor, polyclonal IgG

1. INTRODUCTION

Lectins, which are carbohydrate-binding proteins of non-immune origin and ubiquitous distribution in nature, participate in many cellular events such as cell recognition, cell-cell communication and metastasis (Santos et al., 2014). Hundreds of lectins with different carbohydrate-binding specificities and broad applications in biological research have been isolated from nature.

A galactose-specific lectin named BmoLL has been obtained from the leaf powder of *Bauhinia monandra*, "pata de vaca" (Coelho and Silva, 2000). The genus *Bauhinia* belongs to the family Fabaceae; in Northeastern Brazil the leaves of *B. monandra* are largely used in folk medicine to reduce blood glucose levels. Argolo et al. (2016) conducted a study with a leaf ethanolic extract of *B. monandra*, which induced insulin secretion in rats and produced a hypoglycemic effect. BmoLL was purified with high yield (milligram quantities) from *B. monandra* leaf extract by ammonium sulfate fractionation and guar gel affinity chromatography (Coelho and Silva, 2000). Biological activities have already been reported for BmoLL, such as insecticidal activity against *Callosobruchus maculatus*, *Zabrotes subfasciatus* and *Anagasta kuehniella* larvae, which produced a ca. 50% mortality to *C. maculatus* and *Z. subfasciatus*, and a 40% weight decrease in *A. kuehniella* larvae (Macedo et al., 2007). The negative effects of BmoLL on insect development can be attributed to binding to chitin components in the insect gut, and interaction with glycoconjugates along the digestive tract or with sugar residues linked to digestive enzymes and decrease of α-amylase activity (Macedo et al., 2007).

Other study revealed that BmoLL did not induce a genotoxic or cytotoxic effect in cell-free systems and bacteria, but promoted a decrease in the spontaneous mutation frequency in *Escherichia coli* strains, indicating the nonmutagenic and antioxidant activity of BmoLL (Sisenando et al., 2009). This suggests a protective effect of BmoLL in bacteria against endogenous oxidative damage.

Anti-lectin immunoglobulin (anti-lectin IgG) has been produced based on the *in vivo* immune response of vertebrates, such as mice and rabbits, representing antigen-specific adaptation. Anti-lectin IgG has been used to evaluate the specific antigen-antibody interaction *in vitro* and to search for homology of specific antigen in extracts, protein preparations, purified antigens and unknown solutions (Traini et al., 2014; Denis et al., 2017; Utarabhand et al., 2017).

The antigen-antibody detection used immunological assays, such as the double immunodiffusion assay, through the diffusion of the antigen across a gel, followed by the addition of the antiserum/antibody. The immunological interaction resulted in a precipitin band in the place of antigen-antibody complex formation (Barrs et al. 2015; de Azevedo et al., 2015).

Detection of the antigen-antibody interactions has also been evaluated by biosensing methods, which are rapid, simple to perform and sensitive. Biosensors based on specific antigen-antibody interactions, known as immunosensors, integrate immunological reactions with a transducer device for signal transfer in the biological/electrical interface, providing specific and quantitative detection of an analyte (Zapp et al., 2014; Kokkinos et al., 2016). Potentiometric sensors and immunosensors are considered attractive tools for the study of immunochemical interactions due to their properties as simple instrumentation, diversity of designs and materials, sensitivity, rapid response and low cost (Daems et al., 2014; Moraes and Kubota, 2016). Currently, many electrochemical immunosensors have been developed for specific detection of proteins, biomarkers and pathogens such as bovine serum albumin (Rawat et al., 2017), dengue NS1 protein (Sinawang et al., 2016) myoglobin (Zapp et al., 2014), and pathogenic *Escherichia coli* O157: H7 (Pandey et al., 2017).

A piezoelectric immunosensor developed based on quartz crystal microbalance (QCM) is a mass-sensitive transducer comprised of the quartz crystal and two gold films coating both sides of the crystal pairing to an oscillator circuit. QCM measures mass change through alterations in the resonance oscillation frequency on the surface with the adsorption of biomolecules, since it oscillates in a resonant shear mode in an electric field of high frequency (Ramos-Jesus et al., 2016; Yoshimoto et al., 2016). Antibodies can be adsorbed on the crystal surface to specific binding (Lederer et al., 2011; Della Ventura et al., 2017). The recognition of an antigen results in mass change on the surface, leading to an increase of the dielectric strength of the crystal and reduction of the oscillation of electric frequency (Kasper et al., 2016). The use of piezoelectric QCM-based immunosensors constitutes an alternative to common immunological techniques, as a diagnostic approach of great interest with low cost and real-time analysis. Some QCM-based immunosensors have been developed to evaluate antigen-antibody binding, such as human cardiac troponin T (Mattos et al., 2012), aflatoxin B1 (Chauhan et al., 2015), *Leishmania infantum* antibodies in canine serum (Ramos-Jesus et al., 2016), ATP6AP1 autoantibodies in saliva for detection of breast cancer (Arif et al., 2015), HPV16 DNA (Jearaneukoon et al., 2016) and organophosphorus analysis (Tang et al., 2016). Another example of analytical application of QCM-based immunosensors was demonstrated by Kim et al. (2015) that evidenced the need for the rapid detection of CPV with the objective of reducing morbidity and mortality of infected dogs. They developed a QCM-based immunosensor with an anti-CPV monoclonal antibody immobilized on the surface device, which showed high sensitivity, specificity and rapidity of the diagnosis.

In the present work, BmoLL was used as a potential immunogen to promote the development of an antiserum in a rabbit followed by purification of anti-BmoLL IgG on a Protein A-Sepharose CL-4B column. Double immunodiffusion assay was applied for detection of BmoLL or homologous lectins, and epitope recognition in preparations of *B. monandra* tissues with lectin activity, using anti-BmoLL serum and anti-BmoLL IgG. Electrochemical and piezoelectric QCM-based immunosensors were also developed to evaluate BmoLL-anti-BmoLL IgG interaction, established on signal transference through electron exchange from molecular conformational changes caused by specific analyte binding.

2. Experimental Procedures

2.1. Isolation of BmoLL Lectin

BmoLL lectin was isolated from *B. monandra* leaves by guar gel affinity chromatography (Coelho and Silva, 2000). Briefly, a 10% (w/v) extract of leaf powder in 0.15 M NaCl was submitted to 60% ammonium sulfate fractionation followed by chromatography on a guar gel column which was eluted with 0.33 M galactose in 0.15 M NaCl.

2.2. Development and Purification of Anti-BmoLL IgG

Purified BmoLL (150 μg) was dissolved in 1 mL of 0.01 M citrate-phosphate buffer, pH 6.8, containing 0.15 M NaCl and emulsified with 1 mL of Freund's complete adjuvant for the first intradermal inoculation. One milliliter of Freund's incomplete adjuvant was used in the four following inoculations, and intradermally injected into a male New Zealand breed white rabbit within a fortnight interval (Correia and Coelho, 1995). Ten milliliters of blood were collected from the central artery of the ear, before each inoculation. The collected blood was allowed to clot in glass tubes (45° angle at room temperature, for 1 h) and kept at 4°C, overnight. The serum obtained was centrifuged at room temperature three times at 1300 x g, for 5 min each time. The aliquots were stored at -20°C.

Three milliliters of the antiserum obtained after the third inoculation were applied on a Protein A-Sepharose CL-4B column (6.5 x 1.0 cm, 4.5 mL matrix), from Sigma-Aldrich, for anti-BmoLL IgG purification, and recycled three times. The column was equilibrated with 0.1 M sodium phosphate buffer, pH 8.0, containing 0.15 M NaCl, until the absorbance at 280 nm fell below 0.01. Then, the rabbit antiserum (containing anti-BmoLL IgG) was diluted with 0.1 M phosphate citrate buffer, pH 2.8. Unbound serum molecules were washed off with the same buffer and bound anti-BmoLL IgG was eluted. The fractions were collected manually with a flow rate of 33 mL/h and monitored at absorbance 280 nm >1 and maintained in 1N NaOH between pH 7 and 8. The fractions were stored in aliquots at -20°C (Correia and Coelho, 1995).

2.3. Double Immunodiffusion Assay

Anti-BmoLL serum and purified anti-BmoLL IgG were evaluated by Ouchterlony double-diffusion analysis (Ouchterlony, 1958). A 1% agarose gel (w/v, 2.5 mL) in 0.5 M NaCl was formed on petri plates (2.5 cm x 7.5 cm). The assay aimed to detect the presence of protein(s) homologous to BmoLL in the preparations. Central wells were made for addition of 20 μL of serum/IgG, and four lateral wells were made around each central well for addition of 20 μL of pure BmoLL, commercial lectins, and *B. monandra* tissue preparations (extracts and 0-60% ammonium sulphate fractions) from dried leaves (blades), flowers, stems, secondary roots, petioles, seeds, buds and branches of stems. The 10% (w/v) extracts were obtained by gentle shaking at 4°C overnight in 0.01 M citrate phosphate buffer (pH 6.5) containing 0.15 M sodium chloride. The extracts were submitted to 60% (w/v) ammonium sulphate fractionation (F 0-60%) by addition of solid salt. After 4 h at room temperature, the resuspended precipitate was dialyzed against distilled water, followed by dialysis against the 0.01 M citrate phosphate buffer (pH 6.5) containing 0.15 M sodium chloride, according to Coelho and Silva (2000). The diffusion was performed in humid chambers for 36 h at 4°C. The gels, after diffusion, were washed several times with 0.15 M NaCl for two days and with distilled water in the last wash, followed by drying overnight at room temperature. Dried gels were stained for 10 min in 0.1% (w/v) Coomassie Brilliant Blue R-250 in water and destained in 10% (v/v) acetic acid in ethanol solution.

2.4. Electrochemical Potential Response to BmoLL Lectin

To evaluate the electrochemical response of the immunosensor to BmoLL, the modified gold electrodes were incubated with the lectin at different concentrations (400, 300, 200, 100, 50 and 25 μg/mL) and the electrochemical potentials were measured. The electrochemical potential of BmoLL-anti-BmoLL IgG binding was obtained into an electrochemical cell containing a gold electrode, working electrode, and a silver/silver chloride electrode (Ag/AgCl) used as reference. Both were coupled to a high impedance multimeter to register electrochemical potentials generated by the interaction of BmoLL with anti-BmoLL IgG immobilized on the gold electrode, in the electrolytic medium, 0.15 M NaCl, containing BmoLL. Assays were performed in the absence and presence of 0.2 M galactose.

2.5. Anti-BmoLL IgG Immobilization and Yield Evaluation

The immobilization of anti-BmoLL IgG was performed on a quartz crystal surface for the piezoelectric assay in a Quartz Cristal Microbalance (QCM) to evaluate BmoLL-anti-BmoLL IgG binding.

The quartz crystals (AT-cut, quartz plate ∅ 14 mm, gold electrode ∅ 8 mm on both sides, 10 MHz frequency) were purchased from Universal Sensors (New Orleans, USA) and gold electrodes (10 mm x 10 mm x 0.5 mm) obtained in a goldsmith's workshop. These surfaces were previously cleaned with a solution containing a 7:3 (v/v) mixture of concentrated sulfuric acid (H_2SO_4) and 30% (v/v) hydrogen peroxide (H_2O_2) for 10 min, followed by rinsing with ultrapure water and drying with dry air. Then, 100 μL of a solution containing 30% (w/v) dextran in 0.0001 M NaOH was added to surfaces during 20 h at 25°C. The dextran layer was sensitized with 0.1 M cyanogen bromide (CNBr) and 0.002 M NaOH for 8 h at 25°C, pH 11, after washing three times with 100 μL of PBS (0.01 M, pH 7.5). According to Hage and Caze (2005) CNBr under alkaline conditions induces the formation of cyanate ester radicals, and it is also possible to form imidocarbonate, a compound of lower reactivity. Both compounds allow covalent binding to amide groups. The immobilization of anti-BmoLL IgG on gold surface is presented in Figure 1.

Purified anti-BmoLL IgG was incubated on QCM (25 μL) and gold plates (100 μL) at the concentrations of 0, 1.1 or 2.2 mg/mL for at least 3 h at room temperature. After the immobilization on the abovementioned surfaces, 50 μL of 3% (w/v) casein solution was applied on the surfaces for 3 h to block remaining reactive free groups and minimize nonspecific binding.

In order to evaluate the yield of immobilization on gold plates, 100 μL was collected after each IgG incubation step to measure unbound IgG (U). After the incubation, gold plates were also washed with 100 μL of PBS (0.01 M, pH 7.2) three times to measure unbound IgG after the washes (W1, W2 and W3, respectively). Protein concentrations of U, W1, W2 and W3 were calculated with a standard protein quantification test using Coomassie brilliant blue G250 (Sedmak and Grossberg, 1977). Knowing the initial quantity of IgG (I) incubated on gold plates prior to immobilization at each IgG concentration, the yield of immobilization could be calculated by using the formula [I – (unbound IgG)/ I] × 100%. The relative immobilization/area was also calculated.

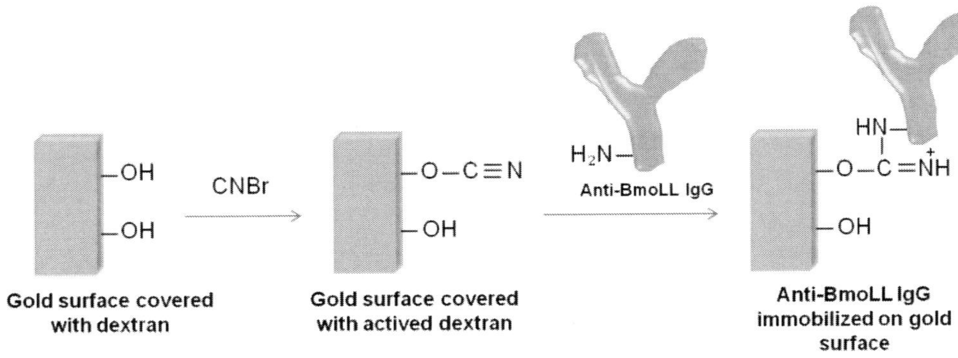

Figure 1. Schematic representation of activation reaction from gold surface covered with dextran and CNBr to establish the covalent immobilization of anti-BmoLL IgG on altered surface.

2.6 Analyses of BmoLL Binding to Immobilized Anti-BmoLL IgG in the Piezoelectric Assay System

A piezoelectric assay based in QCM immunosensor integrated in a flow injection analysis (FIA) system was elaborated to determine the QCM responses to BmoLL. The experimental set-up consisted of one peristaltic pump and an oscillator circuit coupled to a frequency counter (GFC-8131, Good Will), which contained a reference crystal with the same characteristics as the QCM immunosensor crystal. An acrylic flow cell was used to support the QCM immunosensor, maintaining one gold surface in contact with the liquid. BmoLL was introduced into system at a flow rate of 330 μL/min and at a concentration of 454 μg/mL or in 10^{-3}, 10^{-2}, 10^{-1}, 5^{-1} and 2^{-1} dilutions, corresponding to the concentrations of 0.04, 0.4, 40, 200 and 91 μg/mL, respectively.

3. Results and Discussion

BmoLL was isolated in milligram amounts from *B. monandra* leaves following a previously established protocol. Purified BmoLL is in fact a lectin since its hemagglutinating activity was galactose inhibited. BmoLL has been partially characterized by electrophoretic techniques. It is composed of a glycosylated polypeptide subunit of 33 kDa, and a non-glycosylated polypeptide of 26 kDa (Coelho and Silva, 2000).

A range of biological applications has been reported for BmoLL. The lectin was used for composition of lectin-decorated and lectin-loaded nanoparticles of poly(ε-caprolactone) and dextran. BmoLL was incorporated into the nanoparticles, and adsorbed onto nanoparticle surfaces. The biorecognition activity of lectin was maintained. It suggested that BmoLL could recognize and interact with the nanoparticles' polymers, resulting in BmoLL-based nanoparticles (Rodrigues et al., 2003).

BmoLL leaf lectin was able to agglutinate an endophytic active strain of *Pseudomonas aeruginosa* from *B. monandra* leaves, suggesting a defense mechanism against phytopathogenic microorganisms in plants (Ramos et al., 2016). The lectin also showed anti-inflammatory and antinociceptive effects in mice, reducing paw edema, leukocyte migration, and animal contortions (Campos et al., 2016). Similar to other anti-inflammatory lectins, BmoLL can inhibit vascular inflammation by a reduction of rolling and adhesion of immune cells to the endothelium, as well as a decrease in leukocyte migration and cytokines involved. The antinociceptive effect of BmoLL has been associated with its anti-inflammatory activity, and a direct action of BmoLL on specific central nervous system receptors to promote pain relief (Campos et al., 2016).

The production of anti-BmoLL IgG in the rabbit serum (anti-BmoLL serum) was detected after the third inoculation of BmoLL, using double immunodiffusion assay on agarose gel plate. The reactivity of anti-BmoLL serum with BmoLL was confirmed by the appearance of a single immunoprecipitation band in the agarose gel. These results revealed the immunogenic response of BmoLL in a rabbit, showing that the lectin was recognized as an antigen by leucocytes and induced the production of antibodies in the rabbit serum (Denis et al., 2017).

BmoLL antiserum was successively used for purification of anti-BmoLL IgG on a protein A-Sepharose CL-4B column (Figure 2). The immunoprecipitation patterns of anti-BmoLL serum and anti-BmoLL IgG with BmoLL, commercial lectins and *B. monandra* tissue preparations interacting with anti-BmoLL serum and anti-BmoLL IgG in double immunodiffusion assay were displayed in Figure 3. Leaf extract and 0-60% ammonium sulphate fraction exhibited double immunoprecipitation bands after reaction with anti-BmoLL antiserum or anti-BmoLL IgG, indicating the existence of two populations of antibodies, one specific for the lectin, and another that recognized the presence of a lectin isoform. The reactivity antigen-antibody was also confirmed for the extracts from different *B. monandra* tissues; the immunoprecipitation bands were more remarkable in reactions with anti-BmoLL serum (Figure 3).

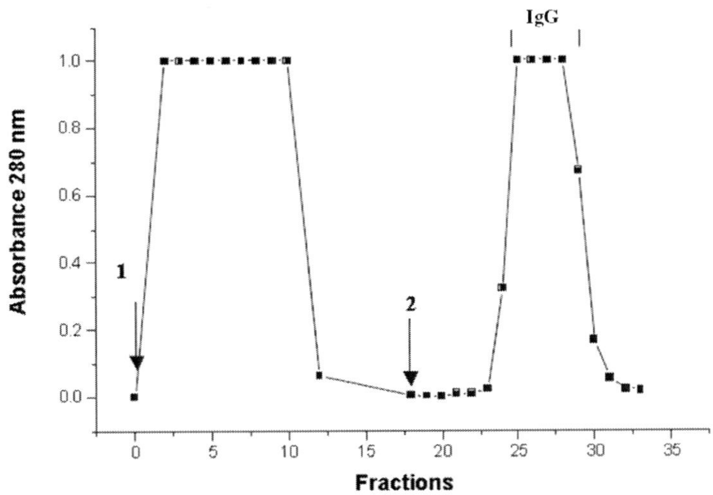

Figure 2. Protein A affinity chromatogram of anti-BmoLL IgG preparation. Antiserum (3 mL) was applied to a 4.5 mL column and fractions (1mL) were collected at 33 mL/h. Unbound proteins were washed off with 0.1M sodium phosphate buffer pH 8.0 (event 1), followed by the elution of the IgG with 0.1M glycine pH 2.8 (event 2). Eluted fractions with $A_{280} \geq 1$ were pooled and stored at -20°C.

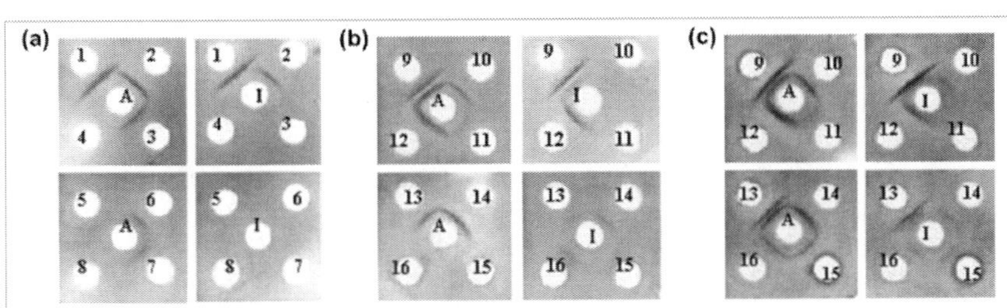

Figure 3. Double immunodifusion assay on agarose gels. Cross-reactions of anti-BmoLL serum (A) or anti-BmoLL IgG (I) with lectins (a), with lectin(s) in tissue extracts from *B. monandra* (b) and in 0-60% ammonium sulphate fractions derived therefrom (c). Lateral wells: 1, BmoLL; 2, *B. purpurea* lectin; 3, *Ulex europaeus* lectin I; 4, *Ulex europaeus* lectin II; 5, *Bandeiraea simplicifolia* lectin II; 6, *Dolichos biflorus* lectin; 7, *Triticum vulgaris* lectin; 8, *Lens culinaris* lectin; 9, leaves; 10, flowers; 11, stems; 12, roots; 13, petioles; 14, seeds; 15, buds; and 16, branches of stem.

The interactions between BmoLL and anti-BmoLL IgG were also evaluated by electrochemical and piezoelectric techniques, immobilizing the anti-lectin antibody on the metal surface and evaluating the signals generated by the sensing interfaces. Despite the fact that electroactive proteins exhibit adsorption on metallic surfaces, factors such as temperature, pH, ionic strength, protein conformation and concentration in the electrolytic medium can affect this process (Smiechowski et al., 2006).

In this context, the anti-BmoLL IgG immobilization and its yield were evaluated on gold plates. Measurement of the immobilization yield is important for estimating the maximal antigen binding capacities of the biosensor. Anti-BmoLL IgG was immobilized in initial quantities of 0, 110 and 220 μg from the solutions with the concentrations of 0,

1.1 and 2.2 mg/mL, respectively. Immobilization of the anti-lectin antibody on the gold plates provided at least two advantages. Firstly, the presence of lectin can be detected in distinct plant tissues; secondly, direct immobilization of the lectin may reduce its activity by severe changes of the conformation during the immobilization process. The concentrations of unbound IgG were measured after immobilization and washes with PBS with a protein quantification method according to Sedmak and Grossberg (1977).

Yields of immobilized anti-BmoLL IgG of 51 μg (46%) and 73 μg (33%) were observed when the initial quantities of 110 and 220 μg were immobilized, respectively. However, the concentration of 1.1 mg/mL (110 μg) was more stable after the PBS wash-steps (Figure 4), according to the anti-BmoLL IgG/area immobilization ratio on gold plate calculated after each PBS wash-steps. A low removal of IgG was observed at a concentration of 1.1 mg/mL, showing an immobilization ratio of 1.4 μg/mm^2 from anti-BmoLL IgG on gold plates after the third PBS wash-step. Considerable and continuous IgG release occurred at a concentration of 2.2 mg/mL along the washes, resulting in an immobilization area of 1.8 μg/mm^2 after the third wash.

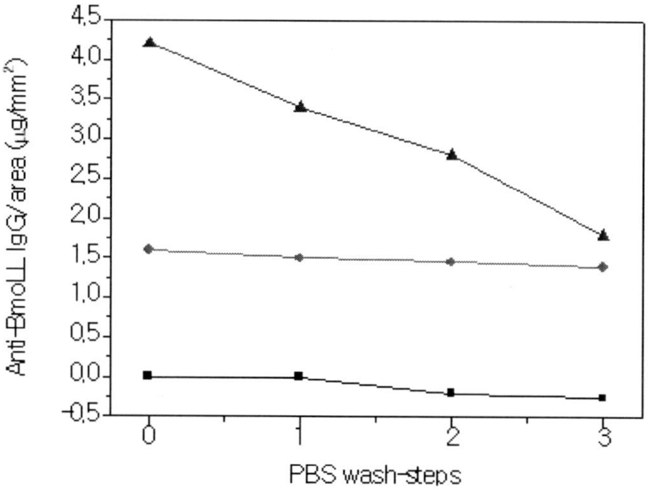

Figure 4. Anti-BmoLL IgG/area ratio of immobilization on gold plates (μg/mm^2) at the anti-BmoLL IgG concentrations of 0 (■), 1.1 (●) and 2.2 (▲) mg/mL. IgG washed-off was measured by protein quantification method and calculated values of immobilized anti-BmoLL IgG are showed.

Electrochemical biosensors have been developed to detect analytes through enzymatic reactions, interaction of antibodies and antigens (immunosensors), the process of recognition of whole cells, as well as different interactions between biomolecules (Liu, 2012). The potentiometric technique has been used for electrochemical analysis to evaluate the maintenance of the biomolecule activity after immobilization through the recognition of specific ligand. Souza et al. (2001) constructed the first potentiometric biosensor with Cramoll 1,4 lectin immobilized to active glass surface with aminopropiletoxisilane (APTES) to define analytical and quantitative forms of lectin interaction with glucose.

According to Souza et al. (2003) the development of techniques to define biomolecule/electrode interface parameters will be able to provide information on charged groups adsorbed or immobilized to the electrode surface revealing specific interactions in biological systems. Cramoll 1,4 was also adsorbed on nafion beads to characterize their interaction with glucose using potential measurements (Souza et al., 2017).

Herein, an electrochemical immunosensor based on anti-BmoLL IgG was developed onto gold electrodes and the maintenance of anti-BmoLL IgG activity was evaluated in the presence of BmoLL lectin by potential measurements. The interaction anti-BmoLL IgG with BmoLL generated electrochemical potentials increasing proportionally with a rise in the concentration of lectin (Figure 5). A potential signal of 48 mV was obtained for BmoLL at 200 μg/mL, increasing significantly to 115 mV at the lectin concentration of 400 μg/mL.

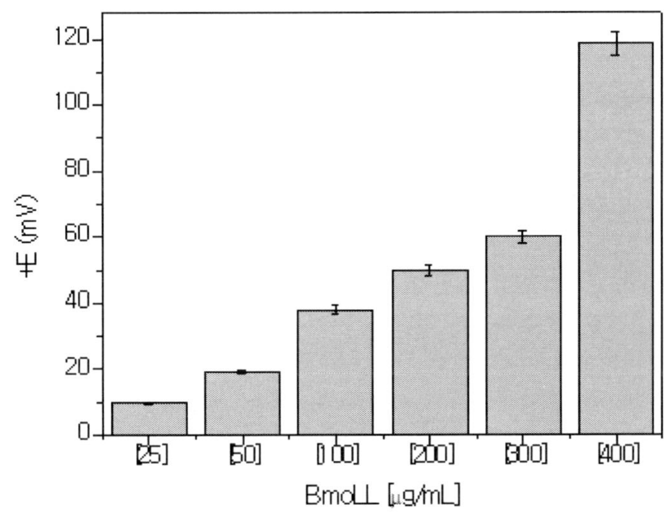

Figure 5. Electrochemical potentials generated by the interaction of anti-BmoLL IgG with different BmoLL concentrations on the gold electrode. Data represent mean+/-standard deviation, sample number = 5.

The electrochemical potentials measured were compared for interactions in the absence or presence of 0.2 M galactose. The responses were similar, suggesting that BmoLL-anti-BmoLL IgG interaction is independent of BmoLL carbohydrate binding site (Figure 6). To determine the specificity of anti-BmoLL IgG, the electrochemical potential was also measured in the presence of Concanavalin A (Con A) under the same conditions for BmoLL. The signal obtained was insignificant, corresponding to approximately 5% of signal from BmoLL interaction (115 mV) (Figure 7). Con A is a lectin from *Canavalia ensiformis* seeds, specific to glucose and mannose and commonly used for recognition of glucose/mannose residues on cell surfaces and biomolecules, being a tool in the development of biosensors for disease detection (Lei and Chang, 2007).

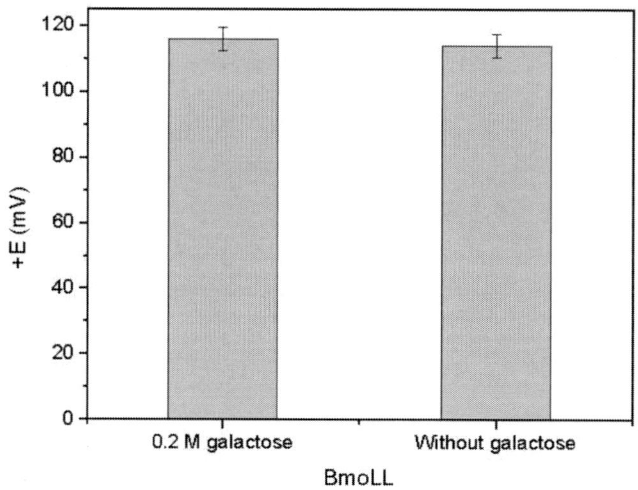

Figura 6. Stable electrochemical potentials generated by the interaction of anti-BmoLL IgG with BmoLL (400 μg/mL) on gold electrodes, in the presence and absence of 0.2 M galactose. Data represent mean+/-standard deviation, sample number = 5.

The variations of the electrochemical potentials resulting from the interaction between biomolecules are intrinsically related to the immobilization process on the metallic surface and the conformational changes occurring on molecule surfaces (Souza et al. 2001; Smiechowski et al., 2006; Bueno et al., 2014). The construction of an electrochemical biosensor must take into consideration the choice of reference and working electrodes as well as immobilization process of the biomolecules and is intrinsically dependent upon the electron-transferring ability (Liu, 2012; Chokkareddy et al., 2017).

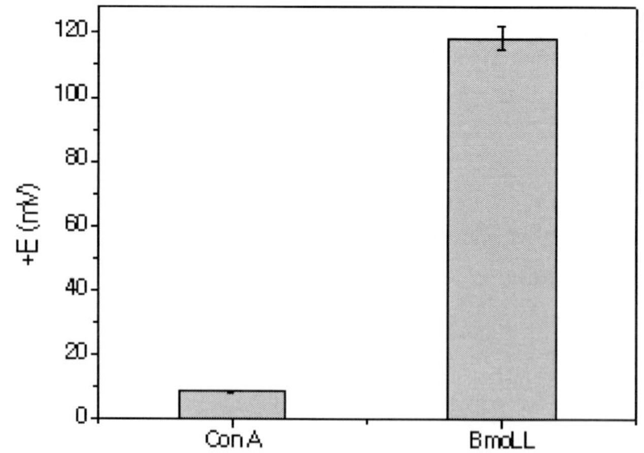

Figura 7. Electrochemical potentials generated by the interaction of Con A and BmoLL with anti-BmoLL IgG immobilized on gold electrode. Data represent mean+/-standard deviation, sample number = 5.

The analyses using the piezoelectric system revealed a reduction of the frequency of 5 KHz (22.5 to 17.5 KHz) observed with 91 μg/mL BmoLL interacting with 1.1 mg/mL anti-BmoLL IgG, showing a strong recognition of the lectin by the immobilized ligand (Figure 8). The change in the piezoelectric assay frequency generated by adsorption of substances on the surface of a quartz crystal exhibits an analytical potential. This ability is disclosed by researchers, detecting organophosphorus (Tang et al., 2016), human papillomavirus (HPV) type58 DNA (Jearanaikoon, 2016) breast cancer (Arif et al., 2015), and antigen-antibody interactions (Kim et al., 2015).

A delay in response of 3.5 min occurred; the time the sample needed to pass the FIA-system before entering the reaction chamber of QCM (Figure 8). The control showed a frequency of 19.5 KHz, without immobilized IgG at different concentrations of BmoLL. There was no reduction of the frequency after incubation with 0.04, 0.4, 40, 200 and 454 μg/mL BmoLL.

These data reveal that due to the use of small amounts of antibodies and antigens and in real response time, the piezoelectric system is a powerful analytical tool for immunological determinations.

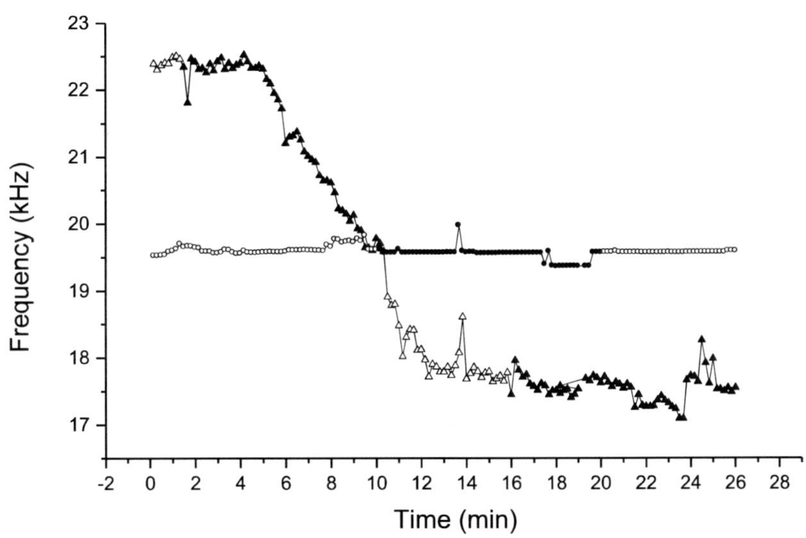

Figure 8. The piezoelectric assay showed BmoLL (91 μg/mL) binding to immobilized anti-BmoLL IgG, 1.1 mg/mL (▲). The control without immobilized IgG showed no reaction after incubation with BmoLL (●). The system was also incubated with PBS (open symbols).

CONCLUSION

BmoLL was highly immunogenic and induced the development of anti-BmoLL IgG in rabbit serum. The appearance of immunoprecipitation bands in the double immunodiffusion assay confirmed the reactivity of anti-BmoLL serum and anti-BmoLL IgG for BmoLL, tissue preparations with lectin activity, and homologous lectins. The assay of immobilization yield was effective in detecting immobilized anti-BmoLL IgG on a gold surface; the concentration of 1.1 mg/mL was more stable after wash-steps. The electrochemical immunosensor based on anti-BmoLL IgG was sufficiently sensitive for revealing the conformational changes occurring at different concentrations of BmoLL after interacting with anti-BmoLL IgG. The system was specific in recognizing BmoLL, showing no significant interaction between anti-BmoLL IgG and Con A lectin. The assay conducted with a piezoelectric biosensor based on anti-BmoLL IgG revealed changes in the frequency signals after the interaction of the antibody with BmoLL, confirming the reactivity of antibody-antigen.

Acknowledgments

Authors are grateful to the *Conselho Nacional de Desenvolvimento Científico e Tecnológico (CNPq)*, for grants and fellowships (JLLF, RAFD, MTSC and LCBBC), the *Coordenação de Aperfeiçoamento de Pessoal de Nível Superior (CAPES)* and the *Fundação de Amparo à Ciência e Tecnologia do Estado de Pernambuco (FACEPE)*. Authors acknowledge Prof. Dr. Kátia Cristina Silva Freitas (UFRPE) for discussions related to Figure 1.

References

Argolo, A. C. C., Dantas, V. C. R., Saturnino, A. C. R. D., Santos, A. F. S., Brandão-Neto, J., Paixão, A. D. O. & Coelho, L. C. B. B. (2016). Leaf ethanolic extract of *Bauhinia monandra* increases insulin secretion in rats subjected to intrauterine malnutrition. *International Journal of Biochemistry Research & Review*, *9*(IV), 1-9. doi: 10.9734/IJBCRR/2016/22434.

Arif, S., Qudsia, S., Urooj, S., Chaudry, N. & Arshad, A. (2015). Blueprint of quartz crystal microbalance biosensor for early detection of breast cancer through salivary autoantibodies against ATP6AP1. *Biosensors and Bioelectronics*, *65*, 62-70. doi: 10.1016/j.bios.2014.09.088.

Barrs, V. R., Ujvari, B., Dhand, N. K., Peters, I. R., Talbot, J., Johnson, L. R., Billen, F., Martin, P., Beatty, J. A. & Belov, K. (2015). Detection of *Aspergillus*-specific antibodies by agar gel double immunodiffusion and IgG ELISA in feline upper respiratory tract aspergillosis. *The Veterinary Journal*, *203*(III), 285-289. doi: 10.1016/j.tvjl.2014.12.020.

Bueno, L., El-Sharif, H. F., Salles, M. O., Boehm, R. D., Narayan, R. J., Paixão, R. L. C. & Reddy, S. M. (2014). MIP-based electrochemical protein profiling. *Sensors and Actuators B: Chemical*, *204*, 88-95. doi: 10.1016/j.snb.2014.07.100.

Campos, J. K. L., Araújo, C. S. F., Araújo, T. F. S., Santos, A. F. S., Teixeira, J. A., LIMA, V. L. M. & Coelho, L. C. B. B. (2016). Anti-inflammatory and antinociceptive activities of *Bauhinia monandra* leaf lectin. *Biochimie Open*, *2*, 62-68. doi: 10.1016/j.biopen.2016.03.001.

Chauhan, R., Solanki, P. R., Singh, J., Mukherjee, I., Basu, T. & Malhotra, B. D. (2015). A novel electrochemical piezoelectric label free immunosensor for aflatoxin B1 detection in groundnut. *Food Control*, *52*, 60-70. doi: 10.1016/j.foodcont.2014.12.009.

Chokkareddy, R., Bhajanthri, N. K. & Redhi, G. G. (2017). An enzyme-induced novel biosensor for the sensitive electrochemical determination of isoniazid. *Biosensors*, *7*(XXI), 2-12. doi: 10.3390/ bios7020021.

Coelho, L. C. B. B. & Silva, M. B. R. (2000). Simple method to purify milligram quantities of the galactose-specific lectin from the leaves of *Bauhinia monandra*. *Phytochemical Analysis*, *11*, 295–300. doi: 10.1002/1099-1565(200009/10)11:53.0.CO;2-S.

Correia, M. T. S. & Coelho, L. C. B. B. (1995). Purification of a glucose/mannose specific lectin, isoform 1, from seeds of *Cratylia mollis* Mart. (Camaratu Bean). *Applied Biochemistry Biotechnology*, *55*, 261-273. doi: 10.1007/BF02786865.

Daems, D., De Waela, K., Vissenbergc, K., Van Campb, G. & Nagelsa, L. (2014). Potentiometric sensors doped with biomolecules as a new approach to small molecule/biomolecule binding kinetics analysis. *Biosensors and Bioelectronics*, *54*, 515-520. doi: 10.1016/j.bios.2013. 11.045.

de Azevedo, P. Z., Sylvestre, T. F., Cavalcante, R. S., de Carvalho, L. R., Moris, D. V., de Oliveira, M. L. C. S. & Mendes, R. P. (2015). Evaluation of the double agar gel immunodiffusion test and of the enzyme-linked immunosorbent assay in the diagnosis and follow-up of patients with chronic pulmonary aspergillosis. *PLOS ONE*, *10*(VIII), e0134841. doi: 10.1371/journal.pone.0134841.

Della Ventura, B., Iannaccone, M., Funari, R., Ciamarra, M. P., Altucci, C., Capparelli, R., Roperto, S. & Velotta, R. (2017). Effective antibodies immobilization and functionalized nanoparticles in a quartzcrystal microbalance-based immunosensor for the detection of parathion. *PLOS ONE*, *12*(II), e0171754. doi: 10.1371/journal.pone. 0171754.

Denis, M., Ramasamy, S. M., Thayappan, K. & Munusamy, A. (2017). Immune response of anti-lectin Pjlec antibody in freshwater crab *Paratelphusa jacquemontii*. *International Journal of Biological Macromolecules*, *104*(Pt A), 1212-1222. doi: 10.1016/j.ijbiomac.2017. 07.034.

Hage, D. S. & Cazes, J. (2005). *Chromatografic Science*. Handebook of affinity chromatography. University of Nebraska. Linoln, Nebraska, U.S.A. Taylor & Francis Group. v. 92.

Jearanaikoon, P., Prakrankamanant, P., Leelayuwat, C., Wanram, S., Limpaiboon, T. & Promptmas, C. (2016). The evaluation of loop-mediated isothermal amplification-quartz crystal microbalance (LAMP-QCM) biosensor as a real-time measurement of HPV16 DNA. *Journal of Virological Methods*, *229*, 8-11. doi: 10.1016/j.viromet. 2015.12.005.

Kasper, M., Traxler, L., Salopek, J., Grabmayr, H., Ebner, A. & Kienberger, F. (2016). Broadband 120 MHz impedance quartz crystal microbalance (QCM) with calibrated resistance and quantitative dissipation for biosensing measurements at higher harmonic frequencies. *Biosensors*, *6*, 23-35. doi: 10.3390/bios6020023.

Kim, Y. K., Lim, Seong-Lin., Choi, S., Cho, I., Park, E. & An, D. (2015). A novel assay for detecting canine parvovirus using a quartz crystal microbalance biosensor. *Journal of Virological Methods*, *219*, 23-24. doi: 10.1016/j.jviromet.2015.03.015.

Kokkinos, C., Economou, A. & Prodromidis, M. I. (2016). Electrochemical immunosensors: critical survey of different architectures and transduction strategies. *Trends in Analytical Chemistry*, *79*, 88-105. doi: 10.1016/j.trac.2015.11.020.

Lederer, T., Steher, B. P., Bauer, S., Jakoby, B. & Hilber, W. (2011). Utilizing a high fundamental frequency quartz crystal resonator as a biosensor in a digital microfluidic platform. *Sensors and Actuators A*, *172*, 161-168. doi: 10.1016/j.sna.2011.04.032.

Lei, H. Y. & Chang, C. P. (2007). Induction of autophagy by concanavalin A and its application in anti-tumor therapy. *Autophagy*, *3*(IV), 402-404. doi: 10.4161/auto.4280.

Liu, C. C. (2012). Electrochemical based biosensors. *Biosensors*, *2*, 269-272. doi: 10.3390/bios2030269.

Macedo, M. R. L., Freire, M. G. M., da Silva, M. B. R. & Coelho, L. C. B. B. (2007). Insecticidal action of *Bauhinia monandra* leaf lectin (BmoLL) against *Anagasta kuehniella* (Lepidoptera: Pyralidae), *Zabrotes subfasciatus* and *Callosobruchus maculatus* (Coleoptera: Bruchidae). *Comparative Biochemistry and Physiology Part A*, *146*, 486-498. doi: 10.1016/j.cbpa.2006.01.020.

Mattos, A. B., Freitas, T. A., Silva, V. L. & Dutra, R. F. (2012). A dual quartz crystal microbalance for human cardiac troponin T in real time detection. *Sensors and Actuators B*, *161*, 439-446. doi: 10.1016/ j.snb.2011.10.058.

Moraes, A. C. M. & Kubota, L. T. (2016). Recent trends in field-effect transistors-based immunosensors. *Chemosensors*, *4*(IV), 20. doi: 10.3390/chemosensors4040020.

Ouchterlony, O. (1958). Diffusion-in-gel methods for immunological analysis. *Progress in Allergy*, *5*, 1-78. doi: 10.1159/000391318.

Pandey, C. M., Tiwari, I., Singh, V. N., Sood, K. N., Sumana, G. & Malhotra, B. D. (2017). Highly sensitive electrochemical immunosensor based on graphene-wrapped copper oxide-cysteine hierarchical structure for detection of pathogenic bacteria. *Sensors and Actuators B Chemical*, *238*, 1060-1069. doi: 10.1016/j.snb.2016.07. 121.

Ramos, S. A. F., Silva, L. C. N., Correia, M. T. S., Araújo, J. M. & Coelho, L. C. B. B. (2016). Endophytic microorganisms from *Bauhinia monandra* leaves: isolation, antimicrobial activities and interaction with galactose-specific lectin BmoLL. *African Journal of Microbiology Research*, *10*(XVII), 600-607. doi: 10.5897/AJMR 2015.7896.

Ramos-Jesus, J., Pontes-de-Carvalho, L. C., Melo, S. M. B., Alcântara-Neves, N. M. & Dutra, R. F. (2016). A gold nanoparticle piezoelectric immunosensor using a recombinant antigen for detecting *Leishmania infantum* antibodies in canine serum. *Biochemical Engineering Journal*, *110*, 43-50. doi: 10.1016/j.bej.2016.01.027.

Rawat, A., Singh, K. P., Vasmatkar, P. & Baral, P. (2017). Highly sensitive gold nanoparticle based electrochemical biosensor for detection of antigen antibody interactions. *Bioscience Biotechnology Research Communications*, *10*(I), 83-90.

Rodrigues, J. S., Santos-Magalhães, N. S., Coelho, L. C. B. B., Couvreur, P., Ponchel, G. & Gref, R. (2003). Novel core(polyester)-shell(polysaccharide) nanoparticles: protein loading and surface modification with lectins. *Journal of Controlled Release*, *92*, 103-112. doi: 10.1016/S0168-3659(03)00296-7.

Santos, A. F. S., da Silva, M. D. C., Napoleão, T. H., Paiva, P. M. G., Correia, M. T. S. & Coelho, L. C. B. B. (2014). Lectins: function, structure, biological properties and potential applications. *Current Topics in Peptide & Protein Research*, *15*, 41-62.

Sedmak, J. J. & Grossberg, S. E. (1977). A rapid, sensitive, and versatile assay for protein using Coomassie brilliant blue G250. *Analytical Biochemistry*, *79*(I-II), 544-552. doi: 10.1016/0003-2697(77)90428-6.

Sinawang, P. D., Rai, V., Ionescu, R. E. & Marks, R. S. (2016). Electrochemical lateral flow immunosensor for detection and quantification of dengue NS1 protein. *Biosensors and Bioelectronics*, *77*, 400-408. doi: 10.1016/j.bios.2015.09.048.

Sisenando, H. A. A. A. C. N., Macedo, M. F. S., Saturnino, A. C. R. D., Coelho, L. C. B. B. & de Medeiros, S. R. B. (2009). Evaluation of the genotoxic potential of *Bauhinia monandra* leaf lectin (BmoLL). *Food and Chemical Toxicology*, *47*, 303-308. doi: 10.1016/j.fct.2008.11.010.

Smiechowski, M. F., Lvovich, V. F., Roy, S., Fleischaman, A., Fissell, W. H. & Riga, A. T. (2006). Electrochemical detection and characterization of proteins. *Biosensors and Bioelectronics*, *22*, 670-677. doi: 10.1016/j.bios.2006.02.008.

Souza, S. R., Correia, M. T. S., Pessoa, M. M. A., Kennedy, J. F., Lima-Filho, J. L. & Coelho, L. C. B. B. (2001). A novel model to characterize the electric double layer of lectins from *Cratylia mollis* (Camaratu bean) and *Canavalia ensiformis* adsorbed on metallic sufarce. *Carbohydrate Polymers*, *46*, 191-193. doi: 10.1016/S0144-8617(00) 00299-X.

Souza, S. R., Dutra, R. F., Correia, M. T. S., Pessoa, M. M. A., Lima-Filho, J. L. & Coelho, L. C. B. B. (2003). Electrochemical potential of free and immobilized *Cratylia mollis* seed lectin. *Bioresource Technology*, *88*, 255-258. doi: 10.1016/S0960-8524(03)00011-7.

Souza, S. R., Santana, E. G., Souza, J. E. G., Silva, P. M. S., Correia, M. T. S. & Coelho, L. C. B. B. (2017). Charge transference detection through cyclic voltammograms in the interaction of *Cratylia mollis* seed lectin-carbohydrate inhibitor. *Biotechnology Journal International*, *18*(III), 1-8. doi: 10.9734/BJI/2017/32672.

Tang, S., Ma, W., Xie, G., Su, Y. & Jiang, Y. (2016). Acetylcholinesterase-reduced graphine oxide hybrid films for organophosphorus neurotoxin sensing via quartz crystal microbalance. *Chemical Physics Letters*, *660*, 199-204. doi: 10.1016/j.cplett. 2016.08.025.

Traini, S., Piccolo, E., Tinari, N., Rossi, C., La Sorda, R., Spinella, F., Bagnato, A., Lattanzio, R., D'Egidio, M., Di Risio, A., Tomao, F., Grassadonia, A., Piantelli, M., Natoli, C. & Iacobelli, S. (2014). Inhibition of tumor growth and angiogenesis by SP-2, an anti–lectin, galactoside binding soluble 3 binding protein (LGALS3BP) antibody. *Molecular Cancer Therapeutics*, *13*(IV), 916-925. doi: 10.1158/1535-7163.MCT-12-1117.

Utarabhand, P., Rittidach, W., Rattanaporn, O., Runsaeng, P. & Hedrick, J. L. (2017). Sialic acid-specific lectin participates in an immune response and ovarian development of the banana shrimp *Fenneropenaeus merguiensis*. *Comparative Biochemistry and Physiology*, *203*, 132-140. doi: 10.1016/j.cbpb.2016.10.005.

Yoshimoto, M., Matsunaga, T., Tanaka, M. & Kurosawa, S. (2016). Determination of thermodynamic parameters for enolization reaction of malonic and metylmalonic acids by using quartz crystal microbalance. *Analytical Chemistry Research*, *8*, 9-15. doi: 10.1016/ j.ancr.2016.03.002.

Zapp, E., Westphal, E., Gallardo, H., de Souza, B. & Vieira, I. C. (2014). Liquid crystal and gold nanoparticles applied to electrochemical immunosensor for cardiac biomarker. *Biosensors and Bioelectronics*, *59*, 127-133. doi: 10.1016/j.bios.2014.03.026.

In: Hemagglutinins
Editors: T. Bun Ng, J. Wong, R. Tse et al.
ISBN: 978-1-53615-708-6
© 2019 Nova Science Publishers, Inc.

Chapter 3

A REVIEW ON THE IMMUNOMODULATORY EFFECTS OF PLANT LECTINS

Leydianne L. S. Patriota*[1]*, Jéssica S. Brito*[1]*,
Bruno R. Barboza*[2]*, Patrícia M. G. Paiva*[1]*,
***Cristiane M. L. Melo*[3] *and Thiago H. Napoleão*[1,*]**
[1]Departamento de Bioquímica, Centro de Biociências,
Universidade Federal de Pernambuco, Recife, Pernambuco, Brazil
[2]Faculdade Estácio do Recife, Estácio-FIR, Recife, Pernambuco, Brazil
[3]Departamento de Antibióticos, Centro de Biociências,
Universidade Federal de Pernambuco, Recife, Pernambuco, Brazil

ABSTRACT

Lectins are proteins with carbohydrate-binding ability. In plants, they have been obtained from leaves, weeds, roots, rhizomes, bulbs, pods, seeds, fruits and flowers. The interaction of lectins with carbohydrates from cell surfaces can trigger diverse intracellular responses, resulting in a range of biological activities that have been exploited for various biotechnological purposes. Among the biological activities of plant lectins, it is possible to mention immunomodulatory action that can be given through the interaction between lectin to surface glycoproteins of immune cells. Immunomodulators may exhibit positive (immune-stimulation) or negative (immunosuppression) activities over a specific immunological function. It is known that lectins can stimulate pro-inflammatory or anti-inflammatory responses. In this paper, we review plant lectins able to induce both profiles, *in vitro* and/or *in vivo*, and their possible mechanisms of action. Plant lectins are capable to alter the release of cytokines, chemokines and nitric oxide by immune cells. One of the mechanisms that may be involved in the pro-inflammatory action of lectins is the activation of Toll-like receptors (TLR). Also, it was observed that the lectin-induced cytokines profile

[*] Corresponding Author's E-mail: thiagohn86@yahoo.com.br.

may vary according to the model, the type of immune cell and the route of administration. Moreover, plant lectins can stimulate the differentiation, activation and proliferation of immune cells. More studies are needed to determine in more details the pathways and mechanisms involved in the immune-modulatory action presented by these biomolecules as well as on their effective application in immunotherapy.

INTRODUCTION

Considering the diversity of diseases and infectious agents that can affect humans, the biotechnological field has searched for molecules from natural sources with both diagnostic and therapeutic potential. Among these molecules, there are the lectins: proteins able to bind specifically and reversibly carbohydrates, including simple sugars (monosaccharides, disaccharides), polysaccharides, or oligosaccharide moieties present in glycoconjugates (glycoproteins and glycolipids) (Coelho et al., 2017).

Lectins are widely found in nature, being isolated from micro-organisms, plants, invertebrates and vertebrates. In plants, lectins have been obtained from leaves, weeds, roots, rhizomes, bulbs, pods, seeds, fruits and flowers. The interaction of lectins with carbohydrates from cell surfaces can trigger diverse intracellular responses, resulting in a range of biological activities that have been exploited for several biotechnological purposes (Paiva et al., 2011, 2013; Dan et al., 2016). Among the various biological properties exerted by the lectins there are antimicrobial, antitumor, insecticidal and immunomodulatory activities (Paiva et al., 2013; Yau et al., 2015; Patriota et al., 2017; Procópio et al., 2017).

The immune system comprises a range of defense mechanisms that recognize and respond to a wide variety of pathogens (Kumar & Bot, 2016). Immunomodulation can be defined as any type of intervention intended to restore the normal function of the immune system or control it for therapeutic purposes; in this way, immunomodulators may exhibit positive (immunostimulation) or negative (immunosuppression) effects (Mahima et al., 2012; Dhama et al., 2015).

Studies have reported that lectins have immunomodulatory activity through interaction with surface glycoproteins of immune cells (Souza et al., 2013; Ditamo et al., 2016). This may trigger different immune-modulatory effects, including cell proliferation and activation, as well as stimulus of cytokines production (Brito et al., 2017; Patriota et al., 2017). The immunological effects promoted by lectins have been pointed as possible strategies for modulation of disturbed immune responses involved in pathological conditions, such as autoimmunity, cancer, allergy, and infections (Souza et al., 2013; Jantan et al., 2015).

It is known that lectins may stimulate both pro-inflammatory (Prasanna & Venkatesh, 2015; Marinković et al., 2016) or anti-inflammatory (Campos et al., 2016) responses. In this study we review lectins with immunomodulatory activity, *in vitro* and/or *in vivo* conditions, their possible mechanisms of action, and potential applications.

IMMUNE SYSTEM

The immune system is responsible for defense of the body against a wide variety of pathogens or lesions. It is divided into innate (or non-specific) and adaptive (or specific) immunities (Figure 1). The innate system is a general response and does not change with repeated exposure. On the other hand, the adaptive system increases its response appropriately in the case of a repeated exposure to the same antigen and exhibits specificity and memory. Each of these two systems consists of cellular and humoral components (mainly proteins). The integration between the two systems is mediated by many molecules like interleukins, immune-globulins, growth factors and chemical mediators produced and released by both immunological and mesenchymal cells (Abbas et al., 2015; Cota & Midwinter, 2015; Kumar & Bot, 2016).

The innate immune response represents the body's first line of defense against pathogens and is an important determinant for directing the adaptive response (Bottazzi et al., 2010; Mantovani et al., 2011). It represents a rapid and stereotyped response, promoted by effector cells like macrophages, neutrophils, dendritic cells and natural killer cells (Figure 1), to a large but limited number of stimuli (Abbas et al., 2015). It is also composed of a system of soluble molecules, called the complement system, present in all individuals, regardless of previous contact with immunogens or aggressors. The mechanisms of innate response are activated by the recognition of structures present on the surfaces of microorganisms (pathogen-associated molecular patterns, PAMPs), structures that are expressed by cells that have suffered some type of aggression (damage-associated molecular patterns, DAMPs) or release of granules present in intracellular vesicles from immune cells (vesicle-associated membrane proteins, VAMPs) (Abbas et al., 2015). Although the ability to recognize PAMPs is a key feature of the innate immune system, this system is unable to remember an invading pathogen and to respond in a more vigorous and rapid way against this antigen (Vazquez et al., 2015).

In adaptive immunity, antigens are recognized by membrane receptors of the system effector cells, the B and T lymphocytes (Figure 1). When they are recognized by membrane receptors on B lymphocytes, these cells differentiate into cells that produce antibodies (plasmocytes). The T lymphocytes, when recognizing antigens through their membrane receptors, formulate different functions within the immune response, being generally subdivided into two cell types: cytotoxic T lymphocytes and T helper (Th) lymphocytes (Figure 1). Cytotoxic T lymphocytes (also called $CD8^+$ T lymphocytes) can destroy cells infected by viruses and intracellular bacteria. T helper cells (also called $CD4^+$) are responsible for secreting cytokines that act on the recruitment of cells of the immune system and their activation mechanisms (Abbas et al. 2015; Parham, 2015).

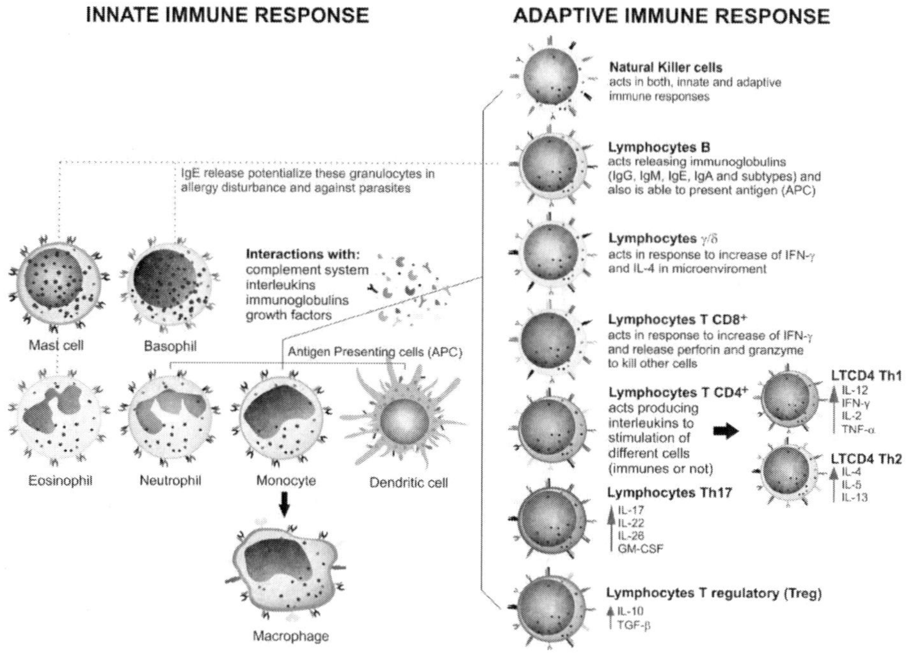

Figure 1. The integration between the innate and adaptive immune systems. Cells belonging to innate immune system are named basal cells and are involved in many effectors mechanisms. Mast cells and basophils act in allergic manifestations. Eosinophils are effectors against parasites. These three cells present high histamine production when stimulated by immunoglobulin E, released by lymphocytes B. Neutrophils, macrophages and dendritic cells are involved with phagocytosis and antigen presenting mechanism through the Major Histocompatibility Complex (MHC) types I and II. Because of this, these cells also are named Antigen Presenting Cells (APC). Lymphocytes are representatives of adaptive immune response. Natural killer cells, involved in immune vigilance to cancer acts in both immunities. The other lymphocytes are stimulated by many signals as well as the presence of antigen (presented by APC, secretions, toxins, virus, DNA, etc.), complement system, interleukins, immunoglobulins, chemical mediators (reactive oxygen species, superoxide, nitric oxide, histamine, among others) and growth factors. Each response and lymphocyte activation are generated inducing fast effector mechanism of removal and immunological memory against the antigen.

B cells can respond directly to antigens, which can induce a rapid antibody response, or may encounter antigen through interactions with other immune cells such as dendritic cells or macrophages (Vazquez et al., 2015). Activation of T lymphocytes may occur by binding of antigens or mitogens on receptors expressed on their surface (Cui et al., 2017). To initiate their effector functions, naïve CD8 T cells must first be activated by specialized members of the innate immune system. T cells act locally upon contact with infected host cells containing pathogen-associated peptides presented by MHC-I (Russell & Ley, 2002). The differentiation of naïve CD8 T cells into effector and memory cells is driven by antigen and/or inflammation exposure and harmonized by the induction of specific transcription factors (Kaech & Cui, 2012).

Studies conducted in the last decade have demonstrated the role of the innate immune system in the control of adaptive immunity (Iwasaki & Medzhitov, 2015). For example, the response of the Th lymphocytes (part of adaptive immunity) to an antigen is driven by the innate immune response, so that the subtype in which a Th lymphocyte will differentiate depends on the environment in which the antigen-presenting cell (APC; Figure 1), initially found the antigen. The presentation of major histocompatibility class (MHC) II-treated antigen, together with co-stimulatory molecules and precise cytokine combinations, boosts the differentiation of Th naïve lymphocytes into specific profile effector cells being Th1, Th2 or Th17 types (Mirchandani et al., 2014; Ciraci et al., 2016).

MEDIATORS OF IMMUNE RESPONSE

The integration between the innate and adaptive immunities is mediated by molecules called cytokines, which are proteins produced by a wide variety of cells such as neutrophils, monocytes, macrophages, B cells and T cells (Figure 1) (Abbas et al., 2015; Cota & Midwinter, 2015). These cytokines act through cell surface receptors to regulate growth, maturation and responsiveness of some populations of cells (Dinarello, 2007; Ramani et al., 2015).

Cytokines include lymphokines, monokines, chemokines, interleukins (ILs), interferons (IFN), colony stimulating factors (CSFs) and growth factors (GFs) and have a very short half-life and act locally at picomolar concentrations. Cytokines form complex interreactive networks with autocrine, paracrine and endocrine characteristics. The number of cytokines identified is increasing, but in general they are classified into four families: alpha-helix family (includes IL-2 subfamily, IFN subfamily and IL-10 subfamily), IL-1 family, IL-17 family and cytokines of cysteine nodules (superfamily of β-transforming growth factor [TGF-β]) (Ramani et al., 2015; Stenken & Poschenrieder, 2015).

Functionally, there are immunomodulatory cytokines of types 1 and 2. Type 1 favors the regulation of cellular immune responses and has a pro-inflammatory function (IFN-γ, tumor necrosis factor α [TNF-α], IL-1, IL-2) or acts in acute phases of some diseases; type 2 controls the production of antibodies and has anti-inflammatory and/or regulatory function, as well as acts in chronic phases of some diseases (TGF-β, IL-4, IL-6, IL-10, IL-13) (Chaplin, 2010; Ramani et al., 2015). The interaction between these different responses promoted by immune cells may be additive, synergistic or antagonistic, or may involve the induction of one cytokine by another (Commins et al., 2010).

Functional pleiotropy and redundancy are characteristics of cytokines, which often show overlapping activities that regulate proliferation or differentiation, depending on the type and stage of development of the target cells involved (Scheller et al., 2011; Stenken & Poschenrieder, 2015). Cytokines can be pleiotropic with respect to their ability to act on multiple targets and physiological effects. Examples of pleiotropic cytokines are TNF-α and IFN-γ.

Chemokines are cytokines of low molecular mass and with high homology to each other which, according to the position of the cysteine residues at the N-terminus, being subdivided into C, CC, CXC and CX3C groups (Salanga & Handel, 2011; Bachelerie et al., 2014). In addition, a fifth subclass of chemokines has been proposed to complement this superfamily, consisting of cytokines in which the specific cysteine residue at the N-terminus is lacking (Asare et al., 2013; Zernecke & Weber, 2014). Chemokines are chemoattractants, acting on the recruitment of leukocytes to sites of inflammation. After immobilization to cell surface glycosaminoglycans, chemokines interact with transmembrane receptor-specific proteins, leading to intracellular signaling and migration of leukocytes to inflammatory sites, a process that is critical to the immune response (Salanga & Handel, 2011).

The ability of Th cells to respond to infection through cytokine secretion remains a paradigm in immunology. Th cells can be divided into at least four distinct subsets based on the cytokine secretion profile: Th1, Th2, Th17, and Treg (Zhu et al., 2010). Th1 cells produce IL-2, IFN-γ and lymphotoxin-a; Th2 cells are characterized by the production of IL-4, IL-5, IL-9, IL-10 and IL-13 and preferentially act on the humoral response of the adaptive immune system (Paul & Zhu, 2010); Th17 cells produce IL-17 and Treg cells produce IL-10, IL-35 and TGF-β. Thus, the ability of naïve T cells to differentiate into distinct Th strains and the roles of Th subsets and individual cytokines in infection, allergy, and autoimmunity have been the subject of several decades of intense research (Spits & Di Santo, 2011; Mirchandani et al., 2012; Walker et al., 2013). Presentation of the antigen by APC, costimulatory molecules and precise combinations of cytokines promote the differentiation of naïve Th cells in Th1, Th2 or Th17 cells (Ciraci et al., 2016). Th cells help B cells to produce antibodies, improve and maintain CD8 cell responses, regulate macrophage function, orchestrate immune responses against a wide variety of pathogenic microorganisms, and regulate immune cells both to control autoimmunity and to adjust the magnitude and persistence of responses (Zhu et al., 2010).

As mentioned before, the activation of B cells leads to clonal expansion and differentiation into plasmocytes producing specific antibodies (immunoglobulins). There are five classes of immunoglobulins produced by humans: IgG, IgD, IgG, IgA, and IgE, which differ in heavy chain structure and have different effector functions. The effector mechanisms are different depending on the class of immunoglobulin involved in the response (Vidarsson et al. 2014).

Nitric oxide (NO) is a free radical that is produced by enzymes called nitric oxide synthases (NOS). In most animal cells, three types of NOS are described: NOS1 (neuronal), NOS2 (inducible) and NOS3 (endothelial) (Lee et al., 2006). NOS2 (or iNOS), which produces higher levels of NO, is expressed in cells only after immunological and inflammatory stimuli (Kapral et al., 2015; Mohsenzadegan et al., 2015). NO is an important cellular signal molecule involved in a variety of physiological and pathological processes in mammals, including humans. Because it is a pluripotent molecule in many different cells, NO can act on many target cells. The physiological consequences of NO production and its cellular targets are being discovered (Fan et al., 2016). NO showed to have pro- and anti-inflammatory properties. The balance between these two effects depends on "where" and "how much" NO is produced. Low amounts of NO appear to be beneficial since, for example, they inhibit the adhesion and migration of inflammatory cells (Laroux et al., 2001; Lo Faro et al., 2014). Another major function of NO during inflammation is microbial death inside phagocytes, through the generation of ONOO- (Lo Faro et al., 2014).

IMMUNOMODULATION IN THE TREATMENT OF DISEASES

Immunomodulation can be defined as any type of intervention intended to restore the normal function of the immune system or control it for therapeutic purposes. Immunostimulation prevents infections through vaccines (which activate humoral immunity), and acts against infections stabilized by displacement of the cell-mediated immune response. In addition, it can act against cancer using cytokines, tumor-specific antibodies and tumor infiltration by lymphocytes. On the other hand, immunosuppression is generally used in allergies, autoimmunity and organ transplantation (Dhama et al., 2015).

Immunomodulators are key components in disease prevention because they strengthen the weak immune system or moderate its functioning when the defense mechanisms are overly active. In recent years, its market share has been increasing due to its wide range of medical applications (El Enshasy & Hatti-Kaul, 2013). The immunomodulation may be either specific or non-specific. The specific is limited to a single antigen, such as vaccination; the non-specific implies in generalized changes in the immune response, both innate and adaptive, leading to a modification in host reactivity for many different antigens (Dhama et al., 2015).

Immune responses are important in pathological conditions such as multiple sclerosis, arthritis, psoriasis, inflammatory bowel disease, ulcers, and fibromyalgia (Rao et al., 2007). On the other hand, the spread of chronic infectious disease, malignant tumors, and organ-specific autoimmune diseases is usually the result of a collapse of the fundamental immune response. Thus, the scientific community has been oriented to identify new products with the intention of developing new therapeutic alternatives against inflammation (Krifa et al., 2013; Mustapha et al., 2016).

Some components of the immune system can be modulated, for example NO, cytokines and chemokines and thus have been used as targets for therapeutic routes in marketed products. Indications include immune-mediated inflammatory diseases, rheumatoid diseases, arthritis, asthma and psoriasis, for example (Ramani et al., 2015). The cytokine TNF-α is known for its antitumor properties and has been used for the treatment of patients with advanced sarcomas and metastatic melanomas (Galluzzi et al., 2009; Bertazza & Mocellin, 2010). Modulation of the immune system by a biotherapeutic cytokine may result in desirable or undesirable effects, depending on the treated disease. Thus, it is important to characterize the potential adverse effects of the cytokine and the drug that induces it at *in vitro* and/or *in vivo* studies (Ramani et al., 2015).

In addition, immune cells can also be modulated. For example, the increase in the amount of CD8 T cells is beneficial in cases of viral infection (Gainey et al., 2012; Reilly et al., 2012). Furthermore, clonal expansion and activation of CD8 T lymphocytes can be positively modulated in case of other diseases such as cancer (Berinstein et al., 2015; Jackson et al., 2014) or negatively, to reduce rejection in transplants, for example (Faria et al., 2008; Gowdy et al., 2015).

Glycosylation of Immune Cells

Glycans have important biological functions in the process of cell adhesion, maturation, renewal of proteins, cellular activation, etc. (Varki, 2006; Caramelo & Parodi, 2007). Specific glycosylation patterns are required for the normal activity of many immune molecules/cells (Johannssen & Lepenies, 2017).

Changes in the glycome can occur in response to environmental and genetic stimuli and are frequently associated with the acquisition of altered cellular phenotypes (Comelli et al., 2006). The immune-cell glycome is altered during cell differentiation, activation and apoptosis, and these alterations may be related to homeostasis or diseases mechanisms, including immune cell trafficking, differentiation and cytokine-receptor activation (Van Kooyk & Rabinovich, 2008). Since the discovery that selectins and their glycan ligands are important for the regulation of leukocyte trafficking, it has been shown that additional features of the vertebrate immune system are also controlled by endogenous cellular glycosylation (Marth & Grewal, 2008).

Glycans can function as ligands for lectins and as steric elements that alter molecular interactions at the cell surface and between extracellular compartments. As an example, during their maturation, the antigen receptors of B and T cells are post-translationally modified with *N*- and *O*-glycan chains, both composed of carbohydrates such as mannose, *N*-acetylglucosamine, *N*-acetylgalactosamine, fucose, sialic acid, and galactose (Marth & Grewal, 2008). In addition, *N*-acetylgluco-saminyltransferase 5, a key enzyme in the *N*-glycosylation pathway, regulates T cell activation and cytokine secretion (Morgan et al., 2004; Demetriou et al., 2001). Also, a glycosylation-dependent mechanism mediated by the C-type lectin MGL, expressed by APCs, has been shown to regulate effector T cells (Van Vliet et al., 2006).

Glycans appear to dictate multiple aspects of adaptive immunity affecting both T and B cells. Antigen receptors on T cells (TCR) and B cells (BCR), as well as MHC molecules are likewise glycoproteins (Wolfert & Boons, 2013), like several other critical molecules involved in immune response (Lyons et al., 2015). Toll-like receptors (TLRs), cytokines, and cytokine receptors also comprise glycoproteins that have multiple *N*-glycosylation sites (Partridge et al., 2004; Weber et al., 2004).

T cell–mediated processes, including activation, differentiation and homing, are accompanied by a programmed remodeling of cell surface glycans, which themselves are the products of a repertoire of glycosyltransferases and glycosidases acting sequentially and dictating the glycosylation 'signature' of each effector cell type (Daniels et al., 2002; Chen et al., 2006). The responsibility of 'decoding' the biological information encrypted by this glycosylation 'signature' is assigned in part to endogenous glycan-binding proteins or lectins, whose expression and function are regulated at sites of T cell activation and inflammation (Daniels et al., 2002).

Lectins

Lectins are proteins able to bind specifically and reversibly carbohydrates, including simple sugars (monosaccharides, disaccharides), polysaccharides, and oligosaccharide moieties present in glycoconjugates (glycoproteins and glycolipids) (Coelho et al., 2017) via non-covalent bonds, such as hydrogen bonds, van der Waals interaction, and hydrophobic interactions (Singh et al., 2015). Lectins have different molecular structures, biological properties, physical and chemical characteristics (Liu et al., 2010). Their molecular structure may have several domains, among them there is at least one non-catalytic domain that is responsible for the recognition of carbohydrate (Van Damme et al., 2011). The non-immune origin of lectins differentiates them from anti-carbohydrate immunoglobulins that can also agglutinate cells. While the antibodies are structurally similar, lectins differ in amino acid composition (Paiva et al., 2010), three-dimensional structure and molecular weight (Coelho et al., 2017).

Lectins are widely found in nature and have been isolated from bacteria, fungi, plants and animals. In plants, lectins have been obtained from leaves, weeds, roots, rhizomes, bulbs, pods, seeds, fruits and flowers (Paiva et al., 2013). Lectin-glycan binding is a mean of molecular recognition, which organisms can use to identify and decode the biological information that is present in their own cellular glycome as well as the glycomes of other organisms (Marth & Grewal, 2008).

Glycans present on cell membranes mediate cell-cell and cell-matrix environment interactions. The interaction of lectins with carbohydrates from cell surfaces can trigger diverse intracellular responses, resulting in a range of biological activities that have been exploited for various biotechnological purposes. Among the biological activities described for lectins, we can mention insecticidal (Albuquerque et al., 2012; Oliveira et al., 2017; Lima et al., 2017), antiviral (Fang et al., 2010; Sato et al., 2012), antifungal (Albuquerque et al., 2014; Wu et al., 2016; Silva et al., 2018), antibacterial (Singh et al., 2015; Al-Mamun et al., 2016; Silva et al., 2016a; Moura et al., 2017), anti-angiogenic (Ambasta et al., 2015), antinociceptive (Campos et al., 2016), antitumor (Shi et al., 2014; Yau et al., 2015; Singh et al., 2016; Procópio et al., 2017), anti-inflammatory (Campos et al., 2016), mitogenic (Dhuna, 2010; Chan et al., 2012; Singh et al., 2015), immunomodulatory (Melo et al., 2010b; Prasanna & Venkatesh, 2015; Ditamo et al., 2016; Brito et al., 2017; Singh et al., 2017), and gastroprotective (Lacerda et al., 2016) effects.

PLANT LECTINS IN IMMUNOMODULATION

The first report of plant lectin was made by Stillmark in 1888, studying the protein called ricin from *Ricinus communis*. There are evidences that lectins act as storage proteins and play a role in plant defense (Hoff et al., 2009; Lannoo & Van Damme, 2014).

Plant materials represent an excellent source of immunomodulators and have been appointed as a new approach for combating infections caused by resistant microorganisms (Hancock et al., 2012). Among the biological activities of plant lectins, there is the immunomodulatory action, that can be given through the interaction between lectin and surface glycoproteins of immune cells. This lead to interesting application of plant lectins in therapy of human diseases (Pervin et al., 2015; Cui et al., 2017).

Lectins can induce lymphoproliferation and myeloproliferation (Gao et al., 2013; Katrlík et al., 2010; Sung et al., 2013; Läubli et al., 2014; Sartim et al., 2014) and activate cells of the immune system, such as macrophages (Dias-Netipanyj et al., 2016), mast cells (Barbosa-Lorenzi et al., 2016) and lymphocytes (Melo et al., 2011). Several plant lectins may act as TLR agonists (Unitt & Hornigold, 2011). Some of these lectins can induce the release of Th1 cytokines upon interaction with glycosylated receptors on macrophages and/or dendritic cells, such as TLR types 2 and 4 (Coltri et al., 2010; Park et al., 2010). In addition, lectins can induce neutrophil migration. Nonetheless, the migration can be augmented by the action of lectins on resident cells, such as mast cells and macrophages (Figueiredo et al., 2009).

These interactions between plant lectins and immune cells trigger the transduction of signals linked to the production of cytokines (Katrlík et al., 2010). The production of some cytokines can induce efficient immune responses against tumors or microbial infections (Silva et al., 2014) for example. The immunomodulatory effect exerted by plant lectins can also protect against infections; as example, we can mention immunization against infections of *Leishmania amazonensis* (Barral-Netto et al., 1996) and bacterial infections (Souza et al., 2013).

Studies have shown that some lectins may induce NO secretion, which may partially contribute to their antitumor abilities (Kitagaki et al., 2009; Wong & Ng, 2006; Nóbrega et al., 2012; Singh et al., 2017). Lectins that induce NO production are important because this molecule is active against microorganisms, parasites, and tumor cells and exerts an important function in immune system modulation, as discussed before.

Table 1. Immunomodulatory effects of some plant lectins

Lectin (plant); specificity	Cell type	Immunomodulatory activities	References
Abrin (*Abrus precatorius*); galactose	Murine splenocytes	Promote IFN-γ and IL-6 release, inducing Th1 response and T cell activation	Bhutia et al. (2009)
ACA (*Allium cepa*); mannose	Murine thymocytes	Induction of Th1 response through IFN-γ and IL-2 production	Prasanna & Venkatesh (2015)
ALL (*Artocarpus lingnanensis*); galactose	T lympho-cytes	Activation and maintenance of homeostasis of T lymphocytes and Th1/Th2 cytokine secretion	Cui et al. (2017)
AntinM (*Artocarpus heterophyllus*); mannose	Macrophages and dendritic cells	Stimulus of IL-12 production (development of Th1 immunity), release of IL-2 and IFN-γ and NO synthesis in activated macrophages	Souza et al. (2013)
	T cells	Mitogenic effect on murine CD4 and CD8 T cells	Silva et al. (2017)
ApuL (*Alpinia purpurata*); complex sugars	Human peripheral blood mononuclear cells (PBMCs)	Induction of a pro-inflammatory release of cytokines belonging to Th1 (IFN-γ, TNF-α, and IL-6) and Th17 (IL-17A) profiles as well as of nitric oxide and regulatory IL-10. Stimulus of differentiation and activation of both T CD8 and CD4 lymphocytes	Brito et al. (2017)
BanLec (*Musa paradisiaca*); glucose/mannose	Murine splenocytes	Induction of a proinflammatory response in murine splenocytes with increased IFN-γ, IL-2 and TNF-α	Cheung et al. (2009)
	Lymphocytes	Stimulation of IL-4 and IL-10 secretion	Stojanović et al. (2010)
CFL (*Cratlylia argentea*); glucose/manose	Peritoneal macrophages	Induction of IL-6, IL-12 and IFN-γ	Batista et al. (2017)
Cramoll (*Cratylia mollis*); glucose/mannose	Spleen lymphocytes	Stimulation of ROS, cytosolic Ca^{2+}, and pro-inflammatory IL-1β expression	Melo et al. (2010a)
	Mice splenocytes	Stimulation of Th1 response through higher IFN-γ and low IL-10 production and anti-inflammatory response through NO suppression	Melo et al. (2010b)
	Mice splenocytes	Th1 induction (high production of IFN-γ), increase in NO production.	Melo et al. (2011)
	Peritoneal exudate cells	Enhancement of nitric oxide, superoxide anion and cytokines (IL-1, IL-6, IFN-γ and TNF-α).	Silva et al. (2015)
	Mouse splenocytes	Induction of IFN-γ, IL-6, IL- 17A, IL-22, and IL-23	Oliveira et al. (2013)
	Human PBMCs	Induction of IL-6, IL-17A, IL-22, and IL-23	Oliveira et al. (2013)

Table 1. (Continued)

Lectin (plant); specificity	Cell type	Immunomodulatory activities	References
ConA (*Canavalia ensiformis*); glucose/mannose	Mice splenocytes	Stimulus of IFN-γ and IL-6 production and NO release	Melo et al. (2011)
	In vivo (mice)	Elevation of serum levels of TNF-α, INF-γ, and IL-4	Li et al. (2016)
ConBr (*Canavalia brasiliensis*); glucose/mannose	Peritoneal macrophages	Enhanced expression of IL-1 β, TNF- α and IL-6	Batista et al. (2017)
DLL (*Dolichos lablab*); glucose/mannose	Human and murine lymphocytes	Mitogenicity and stimulation of IL-2 production	Vigneshwaran et al. (2017)
Jacalin (*Artocarpus integrifolia*); galactose	Human macrophages	Increased proinflammatory cytokines TNF, IL-6, IL-12, and IL-1β, as well as the anti-inflammatory cytokine IL-10	Polli et al. (2016)
MvFL (*Microgramma vacciniifolia*); complex sugars	Human PBMCs	Increases TNF-α, IFN-γ, IL-6, IL-10, and NO production. Stimulates differentiation and activation of CD8 cells.	Patriota et al. (2017)
PHA (*Phaseolus vulgaris*); galactose/*N*-acetylglucosamine/ mannose	Murine spleen cells	Induction of IL-17 production	Silva et al. (2016b)
	Human lymphocytes	Increase of IFN-γ, IFN-α, IL-1α, IL-4, IL-8, and TNF-α	Ponomareva et al. (2012)
PTL (*Pinellia ternata*); mannose	Mouse peritoneal macrophages	Activation of macrophages inducing the production of pro-inflammatory cytokines (TNF-α, IL-1β and IL-6)	Yu et al. (2015)
VAA (*Viscum album*); *N*-acetyl-galactosamine/ galactose	Human dendritic cells	Cell activation and activation of CD4 + T lymphocytes. Stimulated IFN-γ secretion and other CD4 + T cytokines IL-4, IL-13 and IL-17.	Saha et al. (2016)
VCA-B (*Viscum album* coloratum); *N*-acetyl-galactosamine/ galactose	Dendritic cells	Stimulation of the maturation and activation of dendritic cells, increase of IL-1β, IL-6, IL-12 and TNF-α, initiating Th1 response. CD8 T cell activation	Kim et al. (2014, 2017)

Plant lectins may stimulate pro-inflammatory (Marinković et al., 2016; Prasanna & Venkatesh, 2015) or anti-inflammatory responses (Campos et al., 2016) or even induce both Th1 and Th2 profiles at the same time (Singh et al., 2017). The Table 1 summarizes results obtained with respect to the cytokine profile and cell effects induced by some lectins.

LECTINS AND PRO-INFLAMMATORY RESPONSE

ArtinM is a lectin from *Artocarpus heterophyllus*, also known as artocarpin, that has immunomodulatory activity attributed to the recognition of *N*-glycans in the ectodomain of TLRs expressed on the surface of innate-immunity cells (neutrophils, mast cells, dendritic cells, and macrophages), and consequent induction of IL-12 production and development of the Th1 response Souza et al. (2013). Administration of ArtinM or recombinant form (rArtinM) to mice before or after fungal inoculation induced Th1 immunity, as attested by high TNF-α and IL-12 levels as well as low IL-4 concentrations (Coltri et al., 2010).

ConA is a lectin from seeds of *Canavalia ensiformis*. It is a well-known immunomodulatory protein, so that it is used for comparative purposes (Melo et al., 2011; Sang et al., 2017). After ConA administration in mice, the levels of inflammatory cytokines are elevate resulting subsequently in abundant liver necrosis with mass lymphocytes infiltration (Kato et al., 2013). ConA-induced hepatitis appears to be related to the production of chemokines and adhesion molecules that play an important role in the inflammatory process (Luo et al., 2015). Among the inflammatory cytokines induced by ConA, the IFN-γ has been reported to be a critical mediator of liver injury in the pathogenesis of ConA-induced hepatitis (Sang et al., 2017). ConA-induced liver injury is associated with changes in levels of inflammatory cytokines. Two hours after ConA injection, it was noticed the elevation of serum cytokines TNF-α, INF-γ, and IL-4 (Li et al., 2016).

Cramoll 1,4 is a lectin from seeds of *Cratylia mollis* able to promote effector activity of spleen lymphocytes increasing ROS, cytosolic Ca^{2+}, and pro-inflammatory IL-1β expression, without cause apparent cell morphology damage (Melo et al., 2010a). At non-cytotoxic concentrations, Cramoll 1,4 induced a Th1 response in mice splenocytes *in vitro* through higher IFN-γ and low IL-10 production and showed anti- inflammatory action through NO suppression (Melo et al., 2010b). Later, Cramoll 1,4 was shown to induce *in vivo* a Th1 response while inhibiting Th2 response, whereas high production of IFN-γ was observed in mice splenocytes, when treated with lectin, and no change in IL-10 cytokine production. The increase in NO production induced by Cramoll 1,4 was also observed, but its suppression was observed after 6 days of treatment. Likewise, results showed that Cramoll 1,4 lectin induces a proliferative response through cellular and humoral parameters in mice previously inoculated with Con A and Cramoll lectins (Melo et al., 2011). Other study reported the stimulation by Cramoll 1,4 of IL-6, IL-17A, IL-22, and IL-23 production in human peripheral blood mononuclear cells (PBMCs) and mice splenocytes (Oliveira et al., 2013). This lectin could be considered a potent immunomodulatory compound.

MvFL is a lectin isolated from *Micrograma vacciniifolia* fronds, able to induce a pro-inflammatory Th1 response by human PBMCs. There was an increase in the production of IFN-γ, TNF-α and NO. MvFL was also able to induce the production of IL-6, a pleiotropic cytokine (Patriota et al., 2017).

PTL is a lectin from the bulbs of *Pinellia ternata* that are widely used in Chinese traditional medicine (Zuo et al. 2012). PTL was able to activate macrophages inducing the production of TNF-α, IL-1β and IL-6 in a dose and time dependent manner. PTL-stimulated macrophages triggered a pro-inflammation effect on mice peritoneal macrophages, which was possibly associated with the ROS overproduction, activation of the NF-κB pathway and necrosis-like damage. Also, PTL showed pro-inflammatory activity *in vitro* and stimulated chemotactic activity of neutrophils (Yu et al., 2015).

VAA is a lectin obtained from *Viscum album* (known as European mistletoe) leaves. Different *V. album* preparations can induce human dendritic cell activation and activation of CD4 T lymphocytes (Saha et al., 2016). It is believed that the cytotoxicity of VAA on melanoma is due to its effects on immune cells. This lectin induces apoptosis of these cells, which release their immunomodulatory content, which in turn act against the tumor. This view is supported by a recent clinical study in melanoma patients, in which immunostimulation seemed to be the primary mode of action (Schumacher & Pfüller, 2015; Trefzer et al., 2014).

VCA lectins were isolated from seeds of *Viscum album coloratum*, also known as Korean mistletoe. VCA-B is a potent stimulator of the maturation and activation of dendritic cells, inducing the secretion of IL-1β, IL-6, IL-12p70, and TNF-α, initiating a Th1 response. These data suggest that this lectin can exhibit immunoadjuvant action as a tumor therapeutic (Kim et al., 2014). The treatment of mice peritoneal macrophages with VCA-C induced macrophage activation and TNF-α production, which was not observed when TLR-4 was blocked using a neutralizing antibody or TLR-4-deficient macrophages (Park et al., 2010).

ApuL is a lectin from inflorescence of *Alpinia purpurata* and was able to induce a pro-inflammatory Th1 profile (since there was IFN-γ, TNF-α and NO release) by human PBMCs, as well as Th17 response with IL-17A production. At the same time, ApuL also induced production of the regulatory cytokine IL-10 (Brito et al., 2017). Silva et al. (2016b) also demonstrate the stimulus of IL-17 production, in this case by the lectins L-PHA, E-PHA, and ConA on mice splenocytes. The report of induction of Th17 immunity opens a new approach to investigate the potential to confer protection against pathogens, especially against fungal diseases.

Silva et al. (2014) encourage further studies for the characterization of plant lectins as novel agonists and modulators of TLRs. According to these authors, these proteins may act by increasing the immune response of the host against microbial infections, thus overcoming their immunosup-pressive mechanisms and offering an alternative to combat the increasing drug resistance.

PLANT LECTINS AND ANTI-INFLAMMATORY RESPONSE

Cbol is a lectin purified from seeds of *Canavalia boliviana* and study conducted by Bezerra et al. (2014) determined its anti-inflammatory capacity using *in vivo* and *in vitro* models. Cbol was able to inhibit carrageenan-induced paw edema and neutrophil migration, demonstrating an anti-inflammatory potential. *In vitro* experiments with neutrophils demonstrated that IL-8-induced chemotaxis was inhibited by the anti-inflammatory capacity of Cbol, suggesting that lectin acts on resident cells by inhibiting cytokine production. MPLEC is a lectin isolated from seeds of *Mucuna pruriens*. It also showed anti-inflammatory activity, with antiedematogenic properties in the carrageenan-caused intense edema (Lacerda et al., 2015).

LSL is a lectin purified from *Lonchocarpus sericeus* seeds that acts in the reduction of the inflammatory response caused by peritonitis, through the inhibition of neutrophil migration, adenosine deaminase activity and bacterial colonization in the peritoneal fluid (Alencar et al., 2005). The data suggested that LSL acts on the inhibition of neutrophil recruitment through competition with common ligands of selectin. Napimoga et al. (2007) found that treatment of BALB/c mice with LSL, after 15 min or 4 h administration of carrageenan intraperitoneally, reduced neutrophil migration as well as myeloperoxidase levels. In addition, they demonstrated that LSL was able to inhibit cytokines (IL-1 and TNF-) and chemokines (CXCL1 and CCL3) release. The lectin was also able to interfere in the leukocyte iteration in the rolling and endothelial adhesion.

Through studies in an experimental model in mice, Silva et al. (2013) investigated the antinociceptive and anti-inflammatory capacities of the PBL, a lectin purified from seeds of *Parkia biglobosa*. PBL was able to reduce the number of contortions induced by acetic acid, causing a peripheral analgesic probably associated with the inhibition of inflammatory processes. The investigation of the effect of PBL on experimental peritonitis model in mice demonstrated that PBL reduced carrageenan-promoted leukocyte migration, acting specifically on neutrophils. The authors suggested that PBL acts on resident cells modulating the immune response by inducing the release of anti-inflammatory cytokines.

Synadenium carinatum lectin (ScLL) is a latex lectin able to reduce the pathological sequelae associated with the chronic inflammatory disease asthma in experimental animal models (Rogerio et al., 2007). ZOSL, a lectin purified from seeds of *Zizyphus oenoplia*, showed anti-allergic and anti-inflammatory activities *in vivo* as it has prevented both Arthus reaction and anaphylactic shock in Wistar rats (Butle et al., 2016).

Some Plant Lectins Stimulate Both Pro- and Anti-inflammatory Responses

Sansone et al. (2016) have suggested that cytokines profile induced by lectins may vary according to the model (*in vivo/in vitro*), the type of immune cell and the route of administration used. For example, BanLec, a lectin isolated from *Musa paradisiaca* pulp, was showed to induce *in vitro* a pro-inflammatory response in murine splenocytes with increased IFN-γ, IL-2 and TNF-α (Cheung et al., 2009), as well as stimulation of IL-4 and IL-10 secretion in lymphocytes (Stojanović et al., 2010). However, another study revealed a decrease in the level of IFN-γ after exposure *in vivo* to BanLec, without significant differences for IL-2 or IL-4 levels in peripheral blood of rats. Recombinant BanLec also induced the production of IFN-γ, IL-2 and IL-4 (Peumans et al., 2002).

ACA is a lectin obtained from bulbs of *Allium cepa* that induces mitogenicity in murine thymocytes, being mitogenic for T lymphocytes without exerting influence on B lymphocytes. ACA activated murine thymocytes resulting in release of IL-2 and IFN-γ in a dose-dependent manner and, at lower concentration, induced the production of these cytokines more potently than at higher concentration. ACA also induced NO synthesis in activated macrophages (Prasanna & Venkatesh, 2015). This is a mannose-binding lectin whose carbohydrate-binding domain may play a critical role in the induction of pro-inflammatory responses by immune cells. A study in immunosuppressed rats showed that ACA promoted the development/restoration of lymphoid cells and induction of Th2 response, thus possessing a good potential to be used to ameliorate compromised immune status during chemotherapy (Kumar & Bot, 2016).

ALL, lectin from seeds of *Artocarpus lingnanensis*, can regulate various immunological processes, including activation and maintenance of homeostasis of T lymphocytes and Th1/Th2 cytokine secretion by lymphocytes through activation of CD45. Cui et al. (2017) demonstrated that the interaction of ALL with CD45 was through the carbohydrate recognition site.

BBL is a lectin from *Bauhinia bauhinioides* seeds that demonstrated important *in vivo* vasodilator effects. Its edematogenic activity was reversed by its specific sugars, although this occurred only partially, which indicates a close relationship between their pro-inflammatory actions and carbohydrate-binding properties (Silva et al., 2011). Although early studies with BBL showed induction of pro-inflammatory activity, recent studies demonstrated that this lectin reduced carrageenan-induced paw edema in Wistar rats and inhibited leukocyte migration and TNF-α release in experimental model of peritonitis (Girão et al., 2015).

Jacalin, a non-cytotoxic plant lectin extracted from the seeds of *Artocarpus integrifolia,* has already been reported as a general T cell mitogen (Gattass et al., 1988) that modulates Th1/Th2 cytokine secretion (Kulshreshtha et al., 2007) by binding CD45 on T cell surface. K562 cells showed a marked increase in mRNA levels corresponding to proinflammatory cytokines IL-1β and IL-6 after 3 and 6 h of treatment with jacalin; a slight increase was also observed in IFN production after 6 h. In this way, jacalin may contribute to an environment conducive to tumor progression (Lavanya et al., 2016). In their study, Polli et al. (2016), showed that jacalin induces the secretion by human macrophages of TNF, IL-6, IL-1β, IL-12p70 and the Th2 cytokine IL-10, not modifying only the basal production of TGF-β.

Conclusion

In this review, we compiled various plant lectins with both pro-inflammatory and/or anti-inflammatory profiles, *in vitro* and/or *in vivo*, and their possible mechanism of action. Most them are capable to stimulate the release of cytokines (most pro-inflammatory) and NO. Also, plant lectins can stimulate differentiation, activation and proliferation of immune cells. It is already known that some plant lectins may act as TLR agonists, but more studies are needed to determine other pathways and mechanisms involved in the immunomodulatory action presented by these biomolecules.

Acknowledgments

The authors express their gratitude to the *Conselho Nacional de Desenvolvimento Científico e Tecnológico* (CNPq) for research grants and fellowship (PMGP and THN) as well as to the *Fundação de Amparo à Ciência e Tecnologia do Estado de Pernambuco* (FACEPE; APQ-0108-2.08/14) for financial support and graduate scholarship (IBPG-1712-2.08/16).

References

Abbas, A., Lichtman, A., Pillai, S. (2015). Cellular and Molecular Immunology, 8th ed. Elsevier Health Sciences. Philadelphia.

Albuquerque, L. P., Santana, G. M. S., Pontual, E. V., Napoleão, T. H., Coelho, L. C. B. B. and Paiva, P. M. G. (2012). Effect of *Microgramma vaccinifolia* rhizome lectin on survival and digestive enzymes of *Nasutitermes corniger* (Isoptera, Termitidae). *Int. Biodeterior. Biodegrad.* 75: 158–166.

Albuquerque, L. P., Santana, G. M. S., Napoleão, T. H., Coelho, L. C. B. B., Silva, M. V. and Paiva, P. M. G. (2014). Antifungal activity of *Microgramma vacciniifolia* rhizome lectin on genetically distinct *Fusarium oxysporum* f. sp. lycopersici races. *Appl. Biochem. Biotechnol.* 172: 1098–1105.

Alencar, N. M. N., Cavalcante, C. F., Vasconcelos, M. P., Leite, K. B., Aragão, K. S., Assreuy, M. S., Nogueira, N. P., Cavada, B. S. and Vale, M. R. (2005). Anti-inflammatory and antimicrobial effect of lectin from *Lonchocarpus sericeus* seeds in an experimental rat model of infectious peritonitis. *J. Pharm. Pharmacol.* 57: 919–922.

Al-Mamun, M. A., Akter, Z., Uddin, M. J., Ferdaus, K. M. K. B., Hoque, K. M. F., Ferdousi, Z. and Reza., M. A. (2016). Characterization and evaluation of antibacterial and antiproliferative activities of crude protein extracts isolated from the seed of *Ricinus communis* in Bangladesh. *BMC Complement. Altern. Med.* 16: 211–221.

Ambasta, R. K., Jha, S. K., Kumar, D., Sharma, R., Jha, N. K. and Kumar, P. (2015). Comparative study of anti-angiogenic activities of luteolin, lectin and lupeol biomolecules. *J. Transl. Med.* 13: 307–316.

Asare, Y., Schmitt, M. and Bernhagen, J. (2013). The vascular biology of macrophage migration inhibitory factor (MIF): Expression and effects in inflammation, atherogenesis and angiogenesis. *Thromb. Haemost.* 109: 391–398.

Bachelerie, F., Ben-Baruch, A., Burkhardt, A. M., Combadiere, C., Farber, J. M., Graham, G. J., Horuk, R., Sparre-Ulrich, A. H., Locati, M., Luster, A. D., Mantovani, A., Matsushima, K., Murphy, P. M., Nibbs, R., Nomiyama, H., Power, C. A., Proudfoot, A. E. I., Rosenkilde, M. M., Rot, A., Sozzani, S., Thelen, M., Yoshie, O. and Zlotnik, A. (2014). Update on the extended family of chemokine receptors and introducing a new nomenclature for atypical chemokine receptors. *Pharmacol. Rev.* 66: 1–79.

Barbosa-Lorenzi, V. C., Cecilio, N. T., Buranello, N. T. A., Pranchevicius, M. C. X., Goldman, M. H. S., Pereira-da-Silva, G., Roque-Barreira, M. C., Jamur, M. C. and Oliver, C. (2016). Recombinant ArtinM activates mast cells. *BMC Immunol.* 17: 22.

Barral-Netto, M., Von Sohsten, R. L., Teixeira, M., Santos, W. L. C., Pompeu, M. L., Moreira, R. A., Oliveira, J. T. A., Cavada, B. S., Falcoff, E. and Barral, A. (1996). In vivo protective effect of the lectin from *Canavalia brasiliensis* on BALB/c mice infected by *Leishmania amazonensis*. *Acta Trop.* 60: 237–250.

Batista, J., Ralph, M., Vaz, R., Souza, P., Silva, A., Nascimento, D., Souza, L., Ramos, M., Mastroeni, P. and Lima-Filho, J. (2017). Plant lectins ConBr and CFL modulate expression toll-like receptors, pro-inflammatory cytokines and reduce the bacterial burden in macrophages infected with *Salmonella enterica* serovar typhimurium. *Phytomed.* 25: 52–60.

Berinstein, N. L., Karkada, M., Oza, A. M., Odunsi, K., Villella, J. A., Nemunaitis, J. J., Morse, M. A., Pejovic, T., Bentley, J., Buyse, M., Nigam, R., Weir, G. M., MacDonald, L. D., Quinton, T., Rajagopalan, R., Sharp, K., Penwell, A., Sammatur, L., Burzykowski, T., Stanford, M. M. and Mansour, M. (2015). Survivin-targeted immunotherapy drives robust polyfunctional T cell generation and differentiation in advanced ovarian cancer patients. *Oncoimmunol.* 4: e1026529.

Bertazza, L., and Mocellin, S. (2010). The dual role of tumor necrosis factor (TNF) in cancer biology. *Curr. Med. Chem.* 17: 3337–3352.

Bezerra, G. A., Viertlmayr, R., Moura, T. R., Delatorre, P., Rocha, B. A. M., Nascimento, K. S., Figueiredo, J. G., Bezerra, I. G., Teixeira, C. S., Simões, R. C., Nagano, C. S., De Alencar, N. M. N., Gruber, K. and Cavada, B. S. (2014). Structural studies of an anti-inflammatory lectin from *Canavalia boliviana* seeds in complex with dimannosides." *PLoS One* 9: e97015.

Bhutia, S. K., Mallick S. K. and Maiti, T. K. (2009). In vitro immunostimulatory properties of *Abrus* lectins derived peptides in tumor bearing mice. *Phytomed.* 16: 776–782.

Bottazzi, B., Doni, A., Garlanda, C. and Mantovani, A. (2010). An integrated view of humoral innate immunity: pentraxins as a paradigm. *Annu. Rev. Immunol.* 28: 157–183.

Brito, J.S., Ferreira, G. R. S., Klimczak, E., Gryshuk, L., Santos, N. D. L., Patriota, L. L. S., Moreira, L. R., Soares, A. K. A., Barboza, B. R., Paiva, P. M. G., Navarro, D. M. A. F., Lorena, V. M. B., Melo, C. M. L., Coriolano, M. C. and Napoleão, T. H. (2017). Lectin from inflorescences of ornamental crop *Alpinia purpurata* acts on immune cells to promote Th1 and Th17 responses, nitric oxide release, and lymphocyte activation. *Biomed. Pharmacother.* 94: 865–872.

Butle, A., Talmale, S. and Patil, M. B. (2016). Potential in vivo immunomodulatory effects of the most active lectin isolated from seeds of *Zizyphus oenoplia*. *J. Clin. Cell. Immunol.* 7: 1–6.

Campos, J. K. L., Araíjo, C. S. F., Araújo, T. F. S., Santos, A. F. S., Teixeira, J. A., Lima, V. L. M. and Coelho, L. C. B. B. (2016). Anti-inflammatory and antinociceptive activities of *Bauhinia monandra* leaf lectin. *Biochim. Open* 2: 62–68.

Caramelo, J. J. and Parodi, A. J. (2007). How sugars convey information on protein conformation in the endoplasmic reticulum. *Semin. Cell Dev. Biol.* 18: 732–742.

Chan, Y. S., Wong, J. H., Fang, E. F., Pan, W. and Ng, T. B. (2012). Isolation of a glucosamine binding leguminous lectin with mitogenic activity towards splenocytes and anti-proliferative activity towards tumor cells. *PLoS One* 7: e3896.

Chaplin, D. D. (2010). Overview of the immune response. *J. Allergy Clin. Immunol.* 125: S3–S23.

Chen, G. Y., Osada, H., Santamaria-Babi, L. F. and Kannagi, R. (2006). Interaction of GATA-3/T-bet transcription factors regulates expression of sialyl Lewis X homing receptors on Th1/Th2 lymphocytes. *Proc. Natl. Acad. Sci. U. S. A.* 103: 16894–16899.

Cheung, A. H. K., Wong, J. H. and Ng. T. B. (2009). *Musa acuminata* (Del Monte banana) lectin is a fructose-binding lectin with cytokine-inducing activity. *Phytomed.* 16: 594–600.

Ciraci, C., Janczy, J. R., Jain, N., Haasken, S., Silva, C. P., Benjamim, C. F., Sadler, J. J., Olivier, A. K., Iwakura, Y., Shayakhmetov, D. M., Sutterwala, F. S. and Cassel, S. L. (2016). Immune complexes indirectly suppress the generation of Th17 responses in vivo. *PLoS One* 11: e0151252.

Coelho, L. C. B. B., Silva, P. M. S., Lima, V. L. M., Pontual, E. V., Paiva, P. M. G., Napoleão, T. H. and Correia, M. T. S. (2017). Lectins, Interconnecting Proteins with Biotechnological/Pharmacological and Therapeutic Applications. *Evidence-Based Complement. Altern. Med.* 2017: 1–22.

Coltri, K. C., Oliveira, L. L., Ruas, L. P., Vendruscolo, P. E., Goldman, M. H., Panunto-Castelo, A. and Roque-Barreira, M. C. (2010). Protection against *Paracoccidioides brasiliensis* infection conferred by the prophylactic administration of native and recombinant ArtinM. *Med. Mycol.* 48: 792–799.

Comelli, E. M., Head, S. R., Gilmartin, T., Whisenant, T., Haslam, S. M., North, S. J., Wong, N. K., Kudo, T., Narimatsu, H., Esko, J. D., Drickamer, K., Dell, A. and Paulson, J. C. (2006). A focused microarray approach to functional glycomics: Transcriptional regulation of the glycome. *Glycobiol.* 16: 117–131.

Commins, S. P., Borish, L. and Steinke, J. W. (2010). Immunologic messenger molecules: Cytokines, interferons, and chemokines. *J. Allergy Clin. Immunol.* 125: S53–S72.

Cota, A. M., and Midwinter, M. J. (2015). The immune system. *Anaesth. Intensive Care Med.* 16: 353–355.

Cui, B., Li, L., Zeng, Q., Lin, F., Yin, L., Liao, L., Huang, M. and Wang, J. (2017). A novel lectin from *Artocarpus lingnanensis* induces proliferation and Th1/Th2 cytokine secretion through CD45 signaling pathway in human T lymphocytes. *J. Nat. Med.* 71: 409–421.

Dan, X., Liu, W. and Ng. T. B. (2016). Development and applications of lectins as biological tools in biomedical research. *Med. Res. Rev.* 36: 221–247.

Daniels, M. A., Hogquist, K. A. and Jameson, S. C. (2002). Sweet "n" sour: The impact of differential glycosylation on T cell responses. *Nat. Immunol.* 3:903–910.

Demetriou, M., Granovsky, M., Quaggin, S. and Dennis, J. W. (2001). Negative regulation of T-cell activation and autoimmunity by Mgat5 N-glycosylation. *Nature* 409:733–739.

Dhama, K., Saminathan, M., Jacob, S. S., Singh, M., Karthik, K., Tiwari, R., Sunkara, L. T., Malik, Y. S. and Singh, R. K. (2015). Effect of immunomodulation and immunomodulatory agents on health with some bioactive principles, modes of action and potent biomedical applications. *Int. J. Pharmacol.* 11: 253–290.

Dhuna, V. (2010). Isolation, purification and characterization of an N-acetyl-D-lactosamine binding mitogenic and anti-proliferative lectin from tubers of a cobra lily *Arisaema utile* Schott. *Adv. Biosci. Biotechnol.* 1: 79–90.

Dias-Netipanyj, M. F., Boldrini-Leite, L. M., Trindade, E. S., Moreno-Amaral, A. N. and Elifio-Esposito, S. (2016). Bjcul, a snake venom lectin, modulates monocyte-derived macrophages to a pro-inflammatory profile in vitro. *Toxicol. Vitr.* 33: 118–124.

Dinarello, C. A. (2007). Historical review of cytokines. *Eur. J. Immunol.* 37: S34–S45.

Ditamo, Y., Rupil, L. L., Sendra, V. G., Nores, G. A., Roth, G. A. and Irazoqui, F. J. (2016). In vivo immunomodulatory effect of the lectin from edible mushroom *Agaricus bisporus*. *Food Funct.* 7: 262–269.

El Enshasy, H. A. and Hatti-Kaul, R. (2013). Mushroom immunomodulators: Unique molecules with unlimited applications. *Trends Biotechnol.* 31: 668–677.

Fan, W., Liu, Q., Zhu, X., Wu, Z., Li, D., Huang, F. and He, H. (2016). Regulatory effects of anesthetics on nitric oxide. *Life Sci.* 151:76–85.

Fang, E. F., Lin, P., Wong, J. H., Tsao, S. W. and Ng, T. B. (2010). A lectin with anti-HIV-1 reverse transcriptase, antitumor, and nitric oxide inducing activities from seeds of *Phaseolus vulgaris* cv. Extralong Autumn Purple Bean. *J. Agric. Food Chem.* 58: 2221–2229.

Faria, B. A., Silva, S. M., Abreu, M. T. C. L. and Napimoga. M. H. (2008). Ação dos linfócitos T regulatórios em transplantes. *Rev. Bras. Hematol. Hemoter.* 30: 309–315.

Figueiredo, J. G., Bitencourt, F. S., Mota, M. R. L., Silvestre, P. P., Aguiar, C. N., Benevides, R. G., Nascimento, K. S., Moura, T. R., Dal-Secco, D., Assreuy, A. M. S., Cunha, F. Q., Vale, M. R., Cavada, B. S. and Alencar, N. M. N. (2009). Pharmacological analysis of the neutrophil migration induced by *D. rostrata* lectin: Involvement of cytokines and nitric oxide. *Toxicon* 54: 736–744.

Gainey, M. D., Rivenbark, J. G., Cho, H., Yang, L. and Yokoyama, W. M. (2012). Viral MHC class I inhibition evades CD8+ T-cell effector responses in vivo but not CD8+ T-cell priming. *Proc. Natl. Acad. Sci. U.S.A.* 109: E3260-7.

Galluzzi, L., Aaronson, S. A., Abrams, J. et al. (2009). Guidelines for the use and interpretation of assays for monitoring cell death in higher eukaryotes. *Cell Death Differ.* 16: 1093–1107.

Gao, W., Sun, Y., Chen, S., Zhang, J., Kang, J., Wang, Y., Wang, H., Xia, G., Liu, Q. and Kang, Y. (2013). Mushroom lectin enhanced immunogenicity of HBV DNA vaccine in C57BL/6 and HBsAg-transgenic mice. *Vaccine* 31: 2273–2280.

Gattass, C. R., Ghobrial, I. and Bunn-Moreno, M. M. (1988). Specific inhibition of OKT8 binding to peripheral blood mononuclear cells by jacalin. *Immunol. Lett.* 17:133–138.

Girão, D. K. F. B., Cavada, B. S., Pires, A. F., Martins, T. V., Franco, Á. X., Morais, C. M., Nascimento, K. S., Delatorre, P., Silva, H. C., Nagano, C. S., Assreuy, A. M. S. and Soares, P. M. G. (2015). The galactose-binding lectin isolated from *Bauhinia bauhinioides* Mart seeds inhibits neutrophil rolling and adhesion via primary cytokines. *J. Mol. Recognit.* 28: 285–292.

Gowdy, K. M., Martinu, T., Nugent, J. L., Manzo, N. D., Zhang, H. L., Kelly, F. L., Holtzman, M. J. and Palmer, S. M. (2015). Impaired CD8+ T cell immunity after allogeneic bone marrow transplantation leads to persistent and severe respiratory viral infection. *Transpl. Immunol.* 32: 51–60.

Hancock, R. E. W., Nijnik, A. and Philpott, D. J. (2012). Modulating immunity as a therapy for bacterial infections. *Nat. Rev. Microbiol.* 10: 243–254.

Hoff, P. L., Brill, L. M. and Hirsch, A. M. (2009). Plant lectins: The ties that bind in root symbiosis and plant defense. *Mol. Genet. Genomics.* 282: 1–15.

Iwasaki, A., and Medzhitov, R. (2015). Control of adaptive immunity by the innate immune system. *Nat. Immunol.* 16: 343–353.

Jackson, S. R., Yuan, J. and Teague, R. M. (2014). Targeting CD8+ T-cell tolerance for cancer immunotherapy. *Immunother.* 6: 833–852.

Jantan, I., Ahmad, W. and Bukhari, S. N. A. (2015). Plant-derived immunomodulators: an insight on their preclinical evaluation and clinical trials. *Front. Plant Sci.* 6: 655–672.

Johannssen, T., and Lepenies, B. (2017). Glycan-based cell targeting to modulate immune responses. *Trends Biotechnol.* 35: 334–346.

Kaech, S. M., and Cui, W. (2012). Transcriptional control of effector and memory CD8+ T cell differentiation. *Nat. Rev. Immunol.* 12: 749–761.

Kapral, M., Wawszczyk, J., Sosnicki, S. and Weglarz, L. (2015). Down-regulation of inducible nitric oxide synthase expression by inositol hexaphosphate in human colon cancer cells. *Acta Pol. Pharm.* 72: 705–711.

Kato, J., Okamoto, T., Motoyama, H., Uchiyama, R., Kirchhofer, D., Van Rooijen, N., Enomoto, H., Nishiguchi, S., Kawada, N., Fujimoto, J. and Tsutsui, H. (2013). Interferon-gamma-mediated tissue factor expression contributes to T-cell-mediated hepatitis through induction of hypercoagulation in mice. *Hepatol.* 57: 362–372.

Katrlík, J., Švitel, J., Gemeiner, P., Kožár, T. and Tkac, J. (2010). Glycan and lectin microarrays for glycomics and medicinal applications. *Med. Res. Rev.* 29: 394–441.

Kim, J. J., Hwang, Y. H., Kang, K. Y., Kim, I., Kim, J. B., Park, J. H., Yoo, Y. C. and Yee, S. T. (2014). Enhanced dendritic cell maturation by the B-chain of Korean mistletoe lectin (KML-B), a novel TLR4 agonist. *Int. Immunopharmacol.* 21: 309–319.

Kim, J. J., Hwang, Y. H., Kang, K. Y., Lee, S. J., Kim, J. B., Choi, J. and Yee, S. T. (2017). Antitumor effect of KML-B-treated dendritic cells via induction of lymphocyte activation. *J. Immunol. Res.* 2017: 1–7.

Kitagaki, J., Yang, Y., Saavedra, J. E., Colburn, N. H., Keefer, L. K. and Perantoni, A. O. (2009). Nitric oxide prodrug JS-K inhibits ubiquitin E1 and kills tumor cells retaining wild-type p53. *Oncogene* 28: 619–624.

Krifa, M., Bouhlel, I., Ghedira-Chekir, L. and Ghedira, K. (2013). Immunomodulatory and cellular anti-oxidant activities of an aqueous extract of *Limoniastrum guyonianum* gall. *J. Ethnopharmacol.* 146: 243–249.

Kulshreshtha, R., Ferracin, M., Wojcik, S. E., Garzon, R., Alder, H., Agosto-Perez, F. J., Davuluri, R., Liu, C. G., Croce, C. M., Negrini, M., Calin, G. A. and Ivan, M. (2007). A MicroRNA Signature of Hypoxia. *Mol. Cell. Biol.* 27: 1859–1867.

Kumar, H., and Bot, A. (2016). In this issue: cellular and molecular mechanisms orchestrating the innate immunity during infectious and non-infectious disease. *Int. Rev. Immunol.* 35: 369–371.

Lacerda, R. R., Nascimento, E. S., Lacerda, J. T. J. G., Pinto, L. S., Rizzi, C., Bezerra, M. M., Pinto, I. R., Filho, S. M. P., Pinto, V. P. T., Filho, G. C., Gadelha, C. A. A. and Gadelha, S. T. (2016). Lectin from seeds of a Brazilian lima bean variety (*Phaseolus lunatus* L. var. cascavel) presents antioxidant, antitumour and gastroprotective activities. *Int. J. Biol. Macromol.* 95: 1072–1081.

Lacerda, R. R., Moreira, I. C., Nascimento, J. S. J., Lacerda, A. C. S., Cabral, N. L., Lucetti, D. L., Viana, G. S. B., Felipe, C. F. B., Pessoa, H. L. F., Gadelha, C. A. A. and Santi-Gadelha, T. (2015). Lectin isolated from Brazilian seeds of velvet bean (*Mucuna pruriens* (L) DC. presents analgesic, anti-inflammatory and antihemolytic action. *J. Med. Plants Res.* 9: 231–242.

Lannoo, N., and Van Damme, E. J. (2014). Lectin domains at the frontiers of plant defense. *Front. Plant Sci.* 5: 397.

Laroux, F. S., Bharwani, S., Pavlick, K. P., Hoffman, J. M., Grisham, M. B., Hines, I. N., Kawachi, S. and Harada, H. (2001). Role of nitric oxide in inflammation. *Acta Physiol. Scand.* 173: 113–118.

Läubli, H., Alisson-Silva, F., Stanczak, M. A., Siddiqui, S. S., Deng, L., Verhagen, A., Varki, N. and Varki, A. (2014). Lectin galactoside-binding soluble 3 binding protein (LGALS3BP) is a tumor-associated immunomodulatory ligand for CD33-related siglecs. *J. Biol. Chem.* 289: 33481–33491.

Lavanya, V., Ahmed, N., Khan, M. K. A. and Jamal, S. (2016). Sustained mitogenic effect on K562 human chronic myelogenous leukemia cells by dietary lectin, jacalin. *Glycoconj. J.* 33: 877–886.

Lee, S., Kim, H. S., Lee, H. J., Lee, J., Jeon, B., Jun, C., Lee, S. and Kim, E. (2006). Dual effect of nitric oxide in immortalized and malignant human oral keratinocytes: induction of apoptosis and differentiation. *J. Oral Pathol. Med.* 35: 352–360.

Li, X., Liu, H. C., Yao, Q. Y., Xu, B. L., Zhang, S. C. and Tu, C. T. (2016). Quercetin ConA-induced hepatitis by inhibiting HMGB1-TLR expression and down-regulating the nuclear factor kappa B pathway. *Inflammat.* 39: 96–106.

Lima, T. A., Fernandes, K. M., Oliveira, A. P. S., Dornelles, L. P., Martins, G. F., Napoleão, T. H. and Paiva, P. M. G. (2017). Termiticidal lectins from *Myracrodruon urundeuva* (Anacardiaceae) cause midgut damage when ingested by *Nasutitermes corniger* (Isoptera: Termitidae) workers. *Pest Manag. Sci.* 73: 991–998.

Liu, B., Bian, H. J. and Bao, J. K. (2010). Plant lectins: Potential antineoplastic drugs from bench to clinic. *Cancer Lett.* 287: 1–12.

Lo Faro, M. L., Fox, B., Whatmore, J. L., Winyard, P. G. and Whiteman, M. (2014). Hydrogen sulfide and nitric oxide interactions in inflammation. *Nitric Oxide - Biol. Chem.* 41: 38–47.

Luo, Q., Zhu, L., Ding, J., Zhuang, X., Xu, L. and Chen, F. (2015). Protective effect of galangin in Concanavalin A-induced hepatitis in mice. *Drug Des. Devel. Ther.* 9: 2983–2992.

Lyons, J. J., Milner, J. D. and Rosenzweig, S. D. (2015). Glycans Instructing Immunity: The Emerging Role of Altered Glycosylation in Clinical Immunology. *Front. Pediatr.* 3: 54.

Mahima, A., Rahal, R., Deb, S. K., Latheef, H. A., Samad, R., Tiwari, A., Verma, K., Kumar, A. and Dhama, K. (2012). Immunomodulatory and therapeutic potentials of herbal, traditional/indigenous and ethnoveterinary medicines. *Pakistan J. Biol. Sci.* 15: 754–774.

Mantovani, A., Cassatella, M. A., Costantini, C. and Jaillon, S. (2011). Neutrophils in the activation and regulation of innate and adaptive immunity. *Nat. Rev. Immunol.* 11: 519–531.

Marinković, E., Lukić, I., Kosanović, D., Inić-Kanada, A., Gavrović-Jankulović, M. and Stojanović, M. (2016). Recombinantly produced banana lectin isoform promotes balanced pro-inflammatory response in the colon. *J. Funct. Foods* 20: 68–78.

Marth, J. D. and Grewal, P. K. (2008). Mammalian glycosylation in immunity. *Nat. Rev. Immunol.* 8: 874–887.

Melo, C. M. L., Paim, B. A., Zecchin, K. G., Morari, J., Chiaratti, M. R., Correia, M. T. S., Coelho, L. C. B. B. and Paiva, P. M. G. (2010a). Cramoll 1,4 lectin increases ROS production, calcium levels, and cytokine expression in treated spleen cells of rats. *Mol. Cell. Biochem.* 342: 163–169.

Melo, C. M. L., Melo, H., Correia, M. T. S., Coelho, L. C. B. B., Silva, M. B. and Pereira, V. R. A. (2011). Mitogenic response and cytokine production induced by Cramoll 1,4 lectin in splenocytes of inoculated mice. *Scand. J. Immunol.* 73: 112–121.

Melo, C. M. L., Castro, M. C. A. B., Oliveira, A. P., Gomes, F. O. S., Pereira, V. R. A., Correia, M. T. S., Coelho, L. C. B. B. and Paiva, P. M. G. (2010b). Immunomodulatory response of Cramoll 1,4 lectin on experimental lymphocytes. *Phytother. Res.* 24: 1631–1636.

Mirchandani, A. S., Besnard, A. G., Yip, E., Scott, C., Bain, C. C., Cerovic, V., Salmond, R. J. and Liew, F. Y. (2014). Type 2 innate lymphoid cells drive CD4+ Th2 cell responses. J. *Immunol.* 192: 2442–2448.

Mirchandani, A. S., Salmond, R. J. and Liew, F. Y. (2012). Interleukin-33 and the function of innate lymphoid cells. *Trends Immunol.* 33: 389–396.

Mohsenzadegan, M., Fayazi, M.R., Abdolmaleki, M., Bakhshayesh, M., Seif, F. and Mousavizadeh, K. (2015). Direct immunomodulatory influence of IFN-beta on human astrocytoma cells. *Immunopharmacol. Immunotoxicol.* 37: 214–219.

Morgan, R., Gao, G., Pawling, J., Dennis, J. W., Demetriou, M. and Li, B. (2004). N-acetylglucosaminyltransferase V (Mgat5)-mediated N-glycosylation negatively regulates Th1 cytokine production by T cells. *J. Immunol.* 173: 7200–7208.

Moura, M. C., Trentin, D. S., Napoleão, T. H., Primon-Barros, M., Xavier, A. S., Carneiro, N. P., Paiva, P. M. G., Macedo, A. J. and Coelho, L. C. B. B. (2017). Multi-effect of the water-soluble *Moringa oleifera* lectin against *Serratia marcescens* and *Bacillus* sp.: antibacterial, antibiofilm and anti-adhesive properties. *J. Appl. Microbiol.* 123: 861–874.

Mustapha, N., Mokdad-Bzéouich, I., Sassi, A., Abed, B., Ghedira, K., Hennebelle, T. and Chekir-Ghedira, L. (2016). Immunomodulatory potencies of isolated compounds from *Crataegus azarolus* through their antioxidant activities. *Tumor Biol.* 37: 7967–7980.

Napimoga, M. H., Cavada, B. S., Alencar, N. M. N., Mota, M. L., Bittencourt, F. S., Alves-Filho, J. C., Grespan, R., Gonçalves, R. B., Clemente-Napimoga, J. T., Freitas, A., Parada, C. A., Ferreira, S. H. and Cunha, F. Q. (2007). *Lonchocarpus sericeus* lectin decreases leukocyte migration and mechanical hypernociception by inhibiting cytokine and chemokines production. *Int. Immunopharmacol.* 7: 824–835.

Nóbrega, R. B., Rocha, B. A. M., Gadelha, C. A. A., Santi-Gadelha, T., Pires, A. F., Assreuy, A. M. S., Nascimento, K. S., Nagano, C. S., Sampaio, A. H., Cavada, B. S. and Delatorre, P. (2012). Structure of *Dioclea virgata* lectin: Relations between carbohydrate binding site and nitric oxide production. *Biochimie* 94: 900–906.

Oliveira, C. F. R., Moura, M. C., Napoleão, T. H., Paiva, P. M. G., Coelho, L. C. B. B. and Macedo, M. L. R. (2017). A chitin-binding lectin from *Moringa oleifera* seeds (WSMoL) impairs the digestive physiology of the Mediterranean flour larvae, *Anagasta kuehniella*. *Pestic. Biochem. Physiol.* 142: 67–76.

Oliveira, P. S. S., Rêgo, M. J. B. M., Silva, R. R., Cavalcanti, M. B., Galdino, S. L., Correia, M. T. S., Coelho, L. C. B. B. and Pitta, M. G. R. (2013). *Cratylia mollis* 1, 4 lectin: A new biotechnological tool in IL-6, IL-17A, IL-22, and IL-23 induction and generation of immunological memory. *Biomed Res. Int.* 2013: 1–6.

Paiva, P. M. G., Pontual, E. V., Napoleão, T. H. and Coelho, L. C. B. B. (2013). *Lectins and trypsin inhibitors from plants: biochemical characteristics and adverse effects on insect larvae*. 1th ed. Nova Publishers Inc., New York.

Paiva, P. M. G., Gomes, F. S., Napoleão, T. H., Sá, R. A., Correia, M. T. S. and Coelho, L. C. B. B. (2010). Antimicrobial activity of secondary metabolites and lectins from plants. *Curr. Res. Technol. Educ. Top. Appl. Microbiol. Microb. Biotechnol.* 1: 396–406.

Paiva, P. M. G., Napoleão, T. H., Santos, N. D. L., Correia, M. T. S., Navarro, D. M. A. F. and Coelho, L. C. B. B. (2011). Plant compounds with *Aedes Aegypti* larvicidal activity and other biological properties. *In:* Liong, M.T. (Ed.), *Bioprocess Sciences and Technology*, 1st ed., Nova Publishers Inc., New York, pp. 271–296.

Parham, P. (2015). Co-evolution of lymphocyte receptors with MHC class I. *Immunol. Rev.* 267: 1–5.

Park, H. J., Hong, J. H., Kwon, H. J., Kim, Y., Lee, K. H., Kim, J. B. and Song, S. K. (2010). TLR4-mediated activation of mouse macrophages by Korean mistletoe lectin-C (KML-C). *Biochem. Biophys. Res. Commun.* 396: 721–725.

Partridge, E. A., Le Roy, C., Di Guglielmo, G. M., Pawling, J., Cheung, P., Granovsky, M., Nabi, I. R. J., Wrana, L. and Dennis, J. W. (2004). Regulation of cytokine receptors by golgi N-glycan processing and endocytosis. *Science* 306: 120–124.

Patriota, L. L. S., Procópio, T. F., Brito, J. S., Sebag, V., Oliveira, A. P. S., Soares, A. K. A., Moreira, L. R., Albuquerque, L. T., Soares, T., Silva, T. D., Paiva, P. M. G., Lorena, V. M. B., Melo, C. M. L., Albuquerque, L. P. and Napoleão, T. H. (2017). *Microgramma vacciniifolia* (Polypodiaceae) fronds contain a multifunctional lectin with immunomodulatory properties on human cells. *Int. J. Biol. Macromol.* 103: 36–46.

Paul, W. E. and Zhu, J. (2010). How are Th2-type immune responses initiated and amplified?. *Nat. Rev. Immunol.* 10: 225–235.

Pervin, M., Koyama, Y. and Isemura, M. (2015). Plant lectins in therapeutic and diagnostic cancer research. *Int. J. Plant Biol. Res.* 3: 1030–1035.

Peumans, W. J., Proost, P., Swennen, R. L. and Van Damme, E. J. M. (2002). The abundant class III chitinase homolog in young developing banana fruits behaves as a transient vegetative storage protein and most probably serves as an important supply of amino acids for the synthesis of ripening-associated proteins. *Plant Physiol.* 130: 1063–1072.

Polli, C. D., Ruas, L. P., Veronez, L. C., Geraldino, T. H., Morais, F. R., Roque-Barreira, M. C. and Pereira-da-Silva, G. (2016). Jacalin-activated macrophages exhibit an antitumor phenotype. *Biomed Res. Int.* 2016: 2925657.

Ponomareva, E. G., Emelyanova, N. V. and Nikitina. V. E. (2012). Effects of *Azospirillum brasilense* Sp7 lectin and *Phaseolus vulgarus* phytohemagglutinin on cytokine activity of lymphocytes in vitro. *Bull. Exp. Biol. Med.* 152: 333–335.

Prasanna, V. K. and Venkatesh, Y. P. (2015). Characterization of onion lectin (*Allium cepa* agglutinin) as an immunomodulatory protein inducing Th1-type immune response in vitro. *Int. Immunopharmacol.* 26: 304–313.

Procópio, T. F., Patriota, L. L. S., Moura, M. C., Silva, P. M., Oliveira, A. P. S., Carvalho, L. V. N., Lima, T. A., Soares, T., Silva, T. D., Coelho, L. C. B. B., Pitta, M. G. R., Rêgo, M. J. B. M., Figueiredo, R. C. B. Q., Paiva, P. M. G. and Napoleão, T. H. (2017). CasuL: A new lectin isolated from *Calliandra surinamensis* leaf pinnulae with cytotoxicity to cancer cells, antimicrobial activity and antibiofilm effect. *Int. J. Biol. Macromol.* 98: 419–429.

Ramani, T., Auletta, C. S., Weinstock, D., Mounho-Zamora, B., Ryan, P. C., Salcedo, T. W. and Bannish, G. (2015). Cytokines: The Good, the Bad, and the Deadly. *Int. J. Toxicol.* 34: 355–365.

Rao, C.V., Gupta, P. K., Arti, R. and Vijayakumar, M. (2007). Anti-inflammatory and anti-nociceptive activities of *Fumaria indica* whole plant extract in experimental animals. *Acta Pharm.* 57: 491–498.

Reilly, E. C., Thompson, E. A., Aspeslagh, S., Wands, J. R., Elewaut, D. and Brossay, L. (2012). Activated iNKT cells promote memory CD8+ T cell differentiation during viral infection. *PLoS One* 7: e37991.

Rogerio, A. P., Cardoso, C. R., Fontanari, C., Souza, M. A., Afonso-Cardoso, S. R., Silva, É. V. G., Koyama, N. S., Basei, F. L., Soares, E. G., Calixto, J. B., Stowell, S. R., Dias-Baruffi, M. and Faccioli, L. H. (2007). Anti-asthmatic potential of a D-galactose-binding lectin from *Synadenium carinatum* latex. *Glycobiol.*17: 795–804.

Russell, J. H. and Ley, T. J. (2002). Lymphocyte-mediated cytotoxicity. *Ann. Rev. Immunol.* 20, 323–370.

Saha, C., Das, M., Stephen-Victor, E., Friboulet, A., Bayry, J. and Kaveri, V. S. (2016). Differential effects of *Viscum album* preparations on the maturation and activation of human dendritic cells and CD4+ T cell responses. *Molecules* 21: 912.

Salanga, C. L. and Handel, T. M. (2011). Chemokine oligomerization and interactions with receptors and glycosaminoglycans: The role of structural dynamics in function. *Exp. Cell Res.* 317: 590–601.

Sang, X. X., Wang, R. L., Zhang, C. E., Liu, S. J., Shen, H. H., Guo, Y. M., Zhang, Y. M., Niu, M., Wang, J. B., Bai, Z. F. and Xiao, X. H. (2017). Sophocarpine protects mice from ConA-Induced hepatitis via inhibition of the IFN-gamma/STAT1 pathway. *Front. Pharmacol.* 8: 140.

Sansone, A. C. M. B., Sansone, M., Dias, C. T. S., Nascimento, J. R. O., Claudia, A., Brito, M., Sansone, M., Tadeu, C. and Roberto, J. (2016). Oral administration of banana lectin modulates cytokine profile and abundance of T-cell populations in mice. *Int. J. Biol. Macromol.* 89, 19–24.

Sartim, M. A., Riul, T. B., Cistia-Andrade, C. D., Stowell, S. R., Arthur, C. M., Sorgi, C. A., Faccioli, L. H., Cummings, R. D., Dias-Baruffi, M. and Sampaio, S. V. (2014). Galatrox is a C-type lectin in *Bothrops atrox* snake venom that selectively binds LacNAc-terminated glycans and can induce acute inflammation. *Glycobiol.* 24: 1010–1021.

Sato, Y., Morimoto, K., Kubo, T., Yanagihara, K. and Seyama, T. (2012). High mannose-binding antiviral lectin PFL from *Pseudomonas fluorescens* Pf0-1 promotes cell death of gastric cancer cell MKN28 via interaction with α2-integrin. *PLoS One* 7: e45922.

Scheller, J., Chalaris, A., Schmidt-Arras, D. and Rose-John, S. (2011). The pro- and anti-inflammatory properties of the cytokine interleukin-6. *Biochim. Biophys. Acta - Mol. Cell Res.* 1813: 878–888.

Schumacher, U. and Pfüller, U. (2015). From Berlin and Witten to Southampton and Hamburg: 25 years of mistletoe research cooperation. *Transl. Res. Biomed.* 4: 39–47.

Shi, Z., Chen, J., Li, C., An, N., Wang, Z., Yang, S., Huang, K. and Bao, J. (2014). Antitumor effects of concanavalin A and *Sophora flavescens* lectin in vitro and in vivo. *Acta Pharmacol. Sin.* 35: 248–256.

Silva, H. C., Bari, A. U., Rocha, B. A. M., Nascimento, K. S., Ponte, E. L., Pires, A. F., Delatorre, P., Teixeira, E. H., Debray, H., Assreuy, A. M. S., Nagano, C.S. and Cavada, B. S. (2013). Purification and primary structure of a mannose/glucose-binding lectin from *Parkia biglobosa* Jacq. seeds with antinociceptive and anti-inflammatory properties. *J. Mol. Recognit.* 26: 470–478.

Silva, H.C., Bari, A. U., Pereira-Jénior, F. N., Simões, R. C., Barroso-Neto, I. L., Nobre, C. B., Pereira, M. G., Nascimento, K. S., Rocha, B. A. M., Delatorre, P., Nagano, C. S., Assreuy, A. M. S. and Cavada, B. S. (2011). Purification and partial characterization of a new pro-inflammatory lectin from *Bauhinia bauhinioides* Mart (Caesalpinoideae) seeds. *Protein Pept. Lett.* 18: 396–402.

Silva, L. C. N. and Correia, M. T. S. (2014). Plant lectins and Toll-like receptors: implications for therapy of microbial infections. *Front. Microbiol.* 5: 293–307.

Silva, L. C. N., Alves, N. M. P., Castro, M. C. A. B., Pereira, V. R. A., Paz, N. V. N., Coelho, L. C. B. B., Figueiredo, R. C. B. Q. and Correia, M. T. S. (2015). Immunomodulatory effects of pCramoll and rCramoll on peritoneal exudate cells (PECs) infected and non-infected with *Staphylococcus aureus*. *Int. J. Biol. Macromol.* 72: 848–854.

Silva, P. M., Moura, M. C., Gomes, F. S., Trentin, D. S., Oliveira, A. P. S., Mello, G. S. V., Pitta, M. G. R., Rêgo, M. J. B. M., Coelho, L. C. B. B., Macedo, A. J., Figueiredo, R. C. B. Q., Paiva, P. M. G. and Napoleão, T.H. (2018). PgTeL, the lectin found in *Punica granatum* juice, is an antifungal agent against *Candida albicans* and *Candida krusei*. *Int. J. Biol. Macromol.* 108: 391–400.

Silva, P. M., Napoleão, T. H., Silva, L. C. P. B. B., Fortes, D. T. O., Lima, T. A., Zingali, R. B., Pontual, E. V., Araújo, J. M., Medeiros, P. L., Rodrigues, C. G., Gomes, F. S. and Paiva, P. M. G. (2016 a). The juicy sarcotesta of *Punica granatum* contains a lectin that affects growth, survival as well as adherence and invasive capacities of human pathogenic bacteria. *J. Funct. Foods* 27: 695–702.

Silva, T. A., Fernandes, F. F. and Roque-Barreira, M. C. (2016 b). Data on IL-17 production induced by plant lectins. *Data Br.* 7: 1584–1587.

Silva, T. A., Oliveira-Brito, P. K. M., Gonçalves, T. E., Vendruscolo, P. E., and Roque-Barreira, M. C. (2017). ArtinM Mediates Murine T Cell Activation and Induces Cell Death in Jurkat Human Leukemic T Cells. *Int. J. Mol. Sci.* 18: 1400.

Singh, R., Nawale, L., Sarkar, D. and Suresh, C. G. (2016). Two chitotriose-specific lectins show anti-angiogenesis, induces caspase-9-mediated apoptosis and early arrest of pancreatic tumor cell cycle. *PLoS One* 11: e0146110.

Singh, R. S., Kaur, H. P. and Singh, J. (2015). Purification and characterization of a mycelial mucin specific lectin from *Aspergillus panamensis* with potent mitogenic and antibacterial activity. *Process Biochem.* 50: 2251–2258.

Singh, R. S., Kaur, P. H., Rana, V. and Kennedy, J. F. (2017). Immunomodulatory and therapeutic potential of a mucin-specific mycelial lectin from *Aspergillus panamensis*. *Int. J. Biol. Macromol.* 96: 241–248.

Souza, M. A., Carvalho, F. C., Ruas, L. P., Ricci-Azevedo, R. and Roque-Barreira, M. C. (2013). The immunomodulatory effect of plant lectins: A review with emphasis on ArtinM properties. *Glycoconj. J.* 3: 641–657.

Spits, H., and Di Santo, J. P. (2011). The expanding family of innate lymphoid cells: Regulators and effectors of immunity and tissue remodeling. *Nat. Immunol.* 12: 21–27.

Stenken, J. A. and Poschenrieder, A. J. (2015). Bioanalytical chemistry of cytokines - A review. *Anal. Chim. Acta* 853: 95–115.

Stillmark, H. (1988). About ricin, a poisonous fragment from the seeds of *Ricinus comm.* L and some other Euphorbiaceae. *Pharmacol. Inst. Univers. Dorp.* 3: 59–151.

Stojanović, M. M., Živković, I. P., Petrušić, V. Ž., Kosec, D. J., Dimitrijević, R. D., Jankov, R. M., Dimitrijević, L. A. and Gavrović-Jankulović, M. D. (2010). In vitro stimulation of Balb/c and C57 BL/6 splenocytes by a recombinantly produced banana lectin isoform results in both a proliferation of T cells and an increased secretion of interferon-gamma. *Int. Immunopharmacol.* 10: 120–129.

Sung, N., Byun, E., Song, D., Jin, Y., Kim, J., Park, J., Song, B., Jung, P., Byun, M., Lee, J., Park, S. and Kim, J. (2013). Effect of gamma irradiation on mistletoe (*Viscum album*) lectin-mediated toxicity and immunomodulatory activity. *FEBS Open Bio.* 3: 106–111.

Trefzer, U., Gutzmer, R., Wilhelm, T., Schenck, F., Kähler, K.C., Jacobi, V., Witthohn, K., Lentzen, H. and Mohr, P. (2014). Treatment of unresectable stage IV metastatic melanoma with aviscumine after anti-neoplastic treatment failure: A phase II, multi-centre study. *J. Immunother. Canc.* 2: 27.

Unitt, J., and Hornigold, D. (2011). Plant lectins are novel Toll-like receptor agonists. *Biochem. Pharmacol.* 81:1324–1328.

Van Damme, E. J. M., Smith, D. F., Cummings, R. and Peumans, W. J. (2011). Glycan arrays to decipher the specificity of plant lectins. *Adv. Exp. Med. Biol.* 705: 757–767.

Van Kooyk, Y. and Rabinovic, G. A. (2008). Protein-glycan interactions in the control of innate and adaptive immune responses. *Nat. Immunol.* 9: 593–601.

Van Vliet, S. J., Gringhuis, S. I., Geijtenbeek, T. B. H. and Van Kooyk, Y. (2006). Regulation of effector T cells by antigen-presenting cells via interaction of the C-type lectin MGL with CD45. *Nat. Immunol.* 7: 1200–1208.

Varki, A. (2006). Nothing in glycobiology makes sense, except in the light of evolution. *Cell.* 126: 841–845.

Vazquez, M. I., Catalan-Dibene, J. and Zlotnik, A. (2015). B cells responses and cytokine production are regulated by their immune microenvironment. *Cytokine* 74: 318–326.

Vidarsson, G., Dekkers, G. and Rispens, T. (2014). IgG subclasses and allotypes: From structure to effector functions. *Front. Immunol.* 5: 1–17.

Vigneshwaran, V., Thirusangu, P., Vijay Avin, B. R., Krishna, V., Pramod, S. N. and Prabhakar, B. T. (2017). Immunomodulatory glc/man-directed *Dolichos lablab* lectin (DLL) evokes anti-tumour response in vivo by counteracting angiogenic gene expressions. *Clin. Exp. Immunol.* 189: 21–35.

Walker, J. A., Barlow, J. L. and McKenzie, A. N. J. (2013). Innate lymphoid cells-how did we miss them? *Nat. Rev. Immunol.* 13: 75–87.

Weber, A. N. R., Morse, M. A. and Gay, N. J. (2004). Four N-linked glycosylation sites in human toll-like receptor 2 cooperate to direct efficient biosynthesis and secretion. *J. Biol. Chem.* 279: 34589–34594.

Wolfert, M. A. and Boons, G. J. (2013). Adaptive immune activation: glycosylation does matter. *Nat. Chem. Biol.* 9: 776–784.

Wong, J. H. and Ng, T. B. (2006). Isolation and characterization of a glucose/mannose-specific lectin with stimulatory effect on nitric oxide production by macrophages from the emperor banana. *Int. J. Biochem. Cell Biol.* 38: 234–243.

Wu, J., Wang, J., Wang, S. and Rao, P. (2016). Lunatin, a novel lectin with antifungal and antiproliferative bioactivities from *Phaseolus lunatus* billb. *Int. J. Biol. Macromol.* 89: 717–724.

Yau, T., Dan, X., Ng, C. C. W. and Ng, T. B. (2015). Lectins with potential for anti-cancer therapy. *Molecules* 20: 3791–3810.

Yu, H. L., Zhao, T. F., Wu, H., Pan, Y. Z., Zhang, Q., Wang, K. L., Zhang, C. C. and Jin, Y. P. (2015). *Pinellia ternata* lectin exerts a pro-inflammatory effect on macrophages by inducing the release of pro-inflammatory cytokines, the activation of the nuclear factor-κB signaling pathway and the overproduction of reactive oxygen species. *Int. J. Mol. Med.* 36:1127–1135.

Zernecke, A. and Weber, C. (2014). Chemokines in atherosclerosis: Proceedings resumed. *Arterioscler. Thromb. Vasc. Biol.* 34: 742–750.

Zhu, J., Yamane, H. and Paul, W.E. (2010). Differentiation of Effector CD4 T Cell Populations. *Annu. Rev. Immunol.* 28: 445–489.

Zuo, Z., Fan, H., Wang, X., Zhou, W., Zhou, W. and Li, L. (2012). Purification and characterization of a novel plant lectin from *Pinellia ternate* with antineoplastic activity. *Spring. Plus* 1: 13.

In: Hemagglutinins
Editors: T. Bun Ng, J. Wong, R. Tse et al.
ISBN: 978-1-53615-708-6
© 2019 Nova Science Publishers, Inc.

Chapter 4

ALGAL HEMAGGLUTININS WITH ANTICANCER, IMMUNOMODULATORY, ANTIVIRAL, ANTI-BACTERIAL, ANTINOCICEPTIVE, ANTI-INFLAMMATORY AND PRO-HEALING ACTIVITIES

Jack Ho Wong[1,2], Tzi Bun Ng[3,*], Charlene Cheuk Wing Ng[4], Chit Tam[3], Yau Sang Chan[5], Ryno Naude[6], Ryan Tse[1], Tak Fu Tse[1], Helen Chan[1,†], Ajay Kumar Gautam[7], Neha Gupta[7], Dakshita Tanaji Narvekar[7], Amita Bhadkaria[7], Shubhlukshmi Kumari[7], Nidhi Srivastava[8] and Sameer Suresh Bhagyawant[7]

[1,2]Vita Green Pharmaceutical (Hong Kong Limited), Genning Partners Company Limited, and Institute of Medical Research, Hong Kong, China
[2]The Third Affiliated Hospital of Shenzhen University, Shenzhen, China
[3]School of Biomedical Sciences, Faculty of Medicine, The Chinese University of Hong Kong, China
[4]GKT School of Medical Education, King's College London, United Kingdom
[5]Department of Obstetrics and Gynecology, LKS Faculty of Medicine, University of Hong Kong, China

[*] Corresponding Author's E-mail: tzibunng@cuhk.edu.hk.
[†] Corresponding Author's E-mail: helen.chan@vitagreen.com.

[6]Department of Biochemistry and Microbiology,
Nelson Mandela Metropolitan University, Port Elizabeth, South Africa

ABSTRACT

Algal lectins manifest an array of exploitable biological activities encompassing anticancer, cytokine inducing, neutrophil migration inducing, antibacterial, antinociceptive, anti-inflammatory, wound healing, anti-influenza virus, anti-porcine reproductive and respiratory syndrome virus, anti- hepatitis C virus, and anti-HIV-1 activities.

Keywords: algal lectins, anticancer, immunomodulatory, antimicrobial, antinociceptive, anti-inflammatory, wound healing

INTRODUCTION

Lectins are non-immune glycoproteins that exhibit reversible binding to glycans of glycoproteins, glycolipids and polysaccharides. Lectins own at least one non-catalytic domain, and in some cases a few carbohydrate binding domains, which enable them to effect cell agglutination or precipitation of glycoconjugates. Lectins are found in a constellation of organisms. Algae, especially red algae, produce lectins with unique properties and they are of value in pharmaceutics for the design of protein expression systems and nutraceutics (Singh et al., 2015). The various biological activities of algal lectins are reviewed in this chapter.

ANTI-CANCER ACTIVITY

The lectin from freshwater cyanobacterium/blue-green alga *Microcystis viridis* NIES-102 strain, with a unique sequence, consists of two tandemly repeated homologous domains each with 54 amino acids residues (Yamaguchi et al., 1999). The recombinant form of *Microcystis viridis* lectin demonstrated IC50 values of 40.20, 42.67, 49.87, and 53.40 μg/ml for HT-29 colon cancer cells, HepG2 liver cancer cells, SGC-7901 gastric cancer cells and SK-OV-3 ovarian cancer cells, respectively (Li and Zhang, 2010) and 24.12 μg/ml for A549 lung cancer cells (Li et al., 2011). The recombinant lectin upregulated aldehyde dehydrogenase 1 and β-actin, and downregulated plastin 3, beta-tubulin, tropomyosin 3, heat shock 60, and heat shock protein 90 (Li et al., 2011).

Cytokine Inducing Activity

Caulerpa cupressoides (CcL), *Pterocladiella capillacea* (PcL) and *Solieria filiformis* (SfL) lectins stimulated interleukin-10 (IL-10) formation but were devoid of cytotoxicity toward murine spleen cells at 1-100 μg/mL, and did not elevate interlekin-2 and interferon-γ levels. CcL, PcL and SfL induced IL-6 production after treatment for 24 and 72 h. CcL and PcL but not SfL augmented nitrite release (Monteiro Abreu et al., 2012).

Neutrophil Migration Inducing Activity

Amansia multifida, Bryothamnion seaforthii, Bryothamnion triquetrum and *Gracilaria caudata* lectins induced neutrophil migration *in vivo* and *in vitro*. These activities were inhibited by D-mannose but not by glycoproteins (Neves et al., 2001).

Anti-Influenza Virus Activity

There are two disulfide-linked tetrameric 46.56-kDa isforms (HRL40-1 and HRL40-2) of the type I high-mannose specific lectin HRL40 from the green alga *Halimeda renschii*, containing at least 13 half-cystines. The lectin potently inhibited infection of NCI-H292 cells by influenza virus (A/H3N2/Udorn/72) with a half-maximal effective dose (ED50) of 2.45 nM via binding to a viral envelope hemagglutinin. HRL40 consisted of (Mu et al., 2017).

Anti-Porcine Reproductive and Respiratory Syndrome Virus (Anti-PRRSV) Activity

Griffithsin, a lectin from red algae of the *Griffithsia* genus with 121amino acids, exerted strong saccharide-dependent antiviral activity against PRRSV, by interacting with PRRSV surface glycans which interfere with viral entry. The mechanism entails blocking viral adsorption, without affecting viral penetration. Griffithsin may interfere with cell-to-cell spread to halt transmission of PRRSV (Li et al., 2018).

ANTI- HEPATITIS C VIRUS ACTIVITY

Griffithsin undermines viral infectivity, binding to the hepatitis C virus envelope glycoproteins E1 and E2, and thwarts entry of the virus into human liver cells. The lectin suppresses the titer of the virus in a murine model of engrafted human liver cells (Takebe et al., 2013).

ANTI-HIV-1 ACTIVITY

Carbohydrate-binding proteins isolated from algae, cyanobacteria, sea corals, plants, invertebrates, vertebrates, especially those which bind N-acetylglucosamine and mannose, demonstrate potent anti-HIV activity *in vitro*. These carbohydrate binding proteins exist in the form of monomers, dimers, trimers or tetramers. Their molecular weights range from 8.5 kDa to over 50 kDa, and the number of carbohydrate binding sites per molecule ranges from 2 to 12. They prevent infection of cells by cell-free virus and virus transmission from virus-infected cells to uninfected T-lymphocytes. Their antiviral mechanism of action involves inhibition of viral entry (i.e., fusion) into target cells by direct binding to the HIV-1 gp120 envelope glycans. They may cross-link several glycans during virus/cell interaction and/or freeze the conformation of gp120 consequently preventing further interaction with the coreceptor. Several carbohydrate binding proteins manifest a high genetic barrier since five or more glycan deletions in the HIV envelope are required to bring about a moderate level of drug resistance (Balzarini et al., 2006).

Algal lectins such as cyanovirin-N, griffithsin, microvirin, *Microcystis viridis* lectin, *Oscillatoria agardhii* agglutinin and scytovirin represent potential topical microbicides for inhibiting sexual transmission of HIV. They halt infection of cells by cell-free virus and also interfere with viral transmission from virus-infected cells to uninfected CD4(+) target T-cells and DC-SIGN-directed capture of HIV-1 and transmission to CD4(+) T cells. The lectins also express a suppressive action on several other enveloped viruses (Huskens and Schols, 2012).

Griffithsin exhibits high affinity binding to high mannose N-linked glycans in virus surface glycoproteins, and thus inhibits the entry of primary isolates and laboratory strains of T- and M-tropic HIV-1. Bacterial, cyanobacterial, and marine algal high mannose-binding lectins exert an antitumor action as well as anti-HIV and anti-influenza activities through binding to high-mannose glycans on virus spike (Sato et al., 2015).

Grifonin-1, an antiviral peptide derived from griffithsin, had an EC(50) of 546.6 ± 66.1 nM in p24gag antigen release assays and an EC50 of 190.8 ± 11.0 nM in *in vitro* TZM-bl assays. Grifonin-1 is capable of structural plasticity and presents different conformations in solvents with different polarities and hydrophobicities. Oligomer formation due to intermolecular β-sheet interactions is observed at elevated concentrations of grifonin-1. Grifonin-1 as well as griffithsin bind viral glycoproteins gp41 and g120 through the surface N-linked glycans. Hence grifonin-1 and griffithsin could be developed into topical and/or systemic anti-HIV-1therapeutic agents (Micewicz et al., 2010).Griffithsin is a tight dimer and in order to study the role of the dimer in anti-HIV activity, an obligate dimer with almost structural identity to wild-type griffithsin was designed by expressing griffithsin (Grft) with a peptide linker between the two subunits. An "one-armed" obligate dimer Grft-linker-Grft OneArm was also designed, with three mutated carbohydrate binding sites of one subunit and the other subunit remaining unaltered. It was found that although both dimers were capable of binding gp120 and the viral surface, Grft-linker-Grft One Arm was 84- to 1,010-fold lower HIV inhibitory activity than wild-type Grft which exhibited similar activity to Grft-linker-Grft. Wild-type protein, but not Grft-linker-Grft One Arm, modified the gp120 structure by exposing the CD4 binding site. Hence the dimer form of griffithsin is paramount to its HIV inhibitory activity (Xue et al., 2013).

Electrospun fibers serve as a platform for the delivery of anti- HIV agents. Griffithsin-modified electrospun fibers having different surface modification densities of griffithsin are safe and effective against HIV-1 infection *in vitro* (Grooms et al., 2016). Poly(lactic-co-glycolic) acid (PLGA), an FDA-approved polymer with good chemical and biocompatible characteristics and thus extensively deployed for delivery of drugs, is electrospun to generate a fiber scaffold. Fibers are surface-modified with the anti-HIV-1 lectin griffithsin by utilizing 1-ethyl-3-(3-dimethylaminopropyl) carbodiimide) and N-hydroxy succinimide. The resulting griffithsin-modified PLGA-electrospun fibers display high HIV-1_inactivating activity *in vitro* (Vuong et al., 2017).

The recombinant lectin scytovirin from the cyanobacterium *Scytonema varium* exhibits binding to HIV-1 gp160 and curtails the HIV- cytopathic effect in R5 and X4 HIV-1 infected TZM-bl cells induced by HIV. The recombinant strain of human vaginal L. *plantarum* for extracellular expression of recombinant scytovirin may be employed in inhibiting HIV entry to the vaginas of women at high risk (Janahi et al., 2018).

Anti-HIV algal lectins comprise *Boodlea coacta* lectin, griffithsin, and *Oscillatoria agardhii* agglutinin. Anti-HIV cyanobacterial lectins encompass cyanovirin-N, scytovirin, *Microcystis viridis* lectin, and microvirin. Actinohivin is an anti-HIV actinomycete lectin. In addition, there is a host of lectins with anti-HIV activity from other organisms e.g., plants such as *Artocarpus heterophyllus* (jacalin) lectin, Con A, *Galanthus nivalis* (snowdrop) agglutinin-related lectins, *Musa acuminata* (banana) lectin, *Myrianthus holstii* lectin, *Narcissus pseudonarcissus* lectin, and *Urtica diocia* lectin.

Anti-HIV worm lectins include lectins from polychaete marine worm *Chaetopterus variopedatus*, sea worm *Serpula vermicularis*, and nematode (*Laxus oneistus*). The anti-HIV nonpeptidic lectin mimics comprise pradimicins and benanomicins (Akkouh et al., 2015).

ANTIBACTERIAL ACTIVITY

Saline extracts and aqueous ethanol extracts of marine algae exhibited activity against *Vibrio vulnificus* and *V. pelagius*, but were devoid of activity toward *V. neresis*. Algal lectins from *Eucheuma serra* and *Galaxaura marginata* lectins from algae strongly inhibited *V. vulnificus* but were devoid of activity toward *V. neresis* and *V. pelagius*. The expression of antibacterial activity was prevented by glycoproteins and sugars (Liao et al., 2003).

ANTINOCICEPTIVE AND ANTI-INFLAMMATORY ACTIVITIES

Lectin from the green marine alga *Caulerpa cupressoides* reduced the number of acetic acid-induced writhing. It also suppressed paw edema. However, it was devoid of antinociceptive activity in the hot plate test, implying that it acts through a peripheral mechanism to produce its antinociceptive effect which can be nullified after pre-incubation of the lectin with mucin. There was a fall in neutrophil count when the lectin was given intravenously prior to administration of carrageenan. The lectin did not bring about hepatic or renal damage (Vanderlei et al., 2010).

Modulation of the Pharmacological Effects of Enzymatically-Active Rattlesnake *Phospholipase A2*

A small isolectin BTL-2 from *Bryothamnion triquetrum* formed a stable heterodimer with *phospholipase A2* (PLA2) isolated from rattlesnake (*Crotalus durissus cascavella*) venom. BTL-2 augmented the activity of *C. durissus phospholipase A2* by 23%. BTL-2, PLA2 and PLA2-BTL-2 suppressed the growth of Gram-negative *Xanthomonas axonopodis pv passiflorae* by 9.8%, 27.3% and 98.5%, and the growth of Gram-positive *Clavibacter michiganensis michiganensis* by 98%, 98.5% and 98.5%, respectively. PLA2 but not BTL-2 triggered aggregation in washed platelets. BTL-2 significantly inhibited

PLA2-induced platelet aggregation. BTL-2 alone did not induce edema but reduced PLA2-induced edema without affecting 48/80 compound-induced edema (Oliveira et al., 2008).

Wound-Healing Activity

Lectin from the marine red alga *Bryothamnion seaforthii* expedited wound healing. For healing tests using mice, the lectin was applied topically on skin wounds induced in the dorsal thoracic region for 12 days. The wound area, wound closure and histological appearance were studied. Treatment with the lectin facilitated wound closure and produced total restructuring of the luminal epithelium. The existence of dermal collagen and young skin annexes disclosed the potential use of the lectin for acute wounds (do Nascimento-Neto et al., 2012).

Discussion

Algal lectins, like lectins of other origins, demonstrate HIV inhibitory activity, and have been extensively studied. In addition, algal lectins display antifungal, antibacterial, antiviral, anticancer, immunomodulatory and activities also found in other lectins. The antinociceptive, anti-inflammatory and wound healing activities reported from some algal lectins are unique and not common among lectins of other origins. Algal lectins have many potential applications and await development into drugs for medical applications.

Acknowledgments

We gratefully acknowledge the award of Health and Medical Research Fund (no. 12131221) from Food and Health Bureau, The Government of Hong Kong Special Administrative Region, and a grant from National Natural Science Foundation of China (no. 81471927).

References

Akkouh O, Ng TB, Singh SS, Yin C, Dan X, Chan YS, Pan W, Cheung RC. Lectins with anti-HIV activity: a review. *Molecules*. 2015 Jan 6;20(1):648-68.

Balzarini J. Inhibition of HIV entry by carbohydrate-binding proteins. *Antiviral Res.* 2006;71(2-3):237-47.

do Nascimento-Neto LG, Carneiro RF, da Silva SR, da Silva BR, Vassiliepe Sousa Arruda F, Carneiro VA, do Nascimento KS, Saker-Sampaio S, da Silva VA Jr, Porto AL, Cavada BS, Sampaio AH, Teixeira EH, Nagano CS. Characterization of isoforms of the lectin isolated from the red algae *Bryothamnion seaforthii* and its pro-healing effect. *Mar Drugs.* 2012;10(9):1936-54.

Grooms TN, Vuong HR, Tyo KM, Malik DA, Sims LB, Whittington CP, Palmer KE,Matoba N, Steinbach-Rankins JM. Griffithsin-modified electrospun fibers as a delivery scaffold to prevent HIV Infection. *Antimicrob Agents Chemother.* 2016 Oct 21;60(11):6518-6531.

Huskens D, Schols D. Algal lectins as potential HIV microbicide candidates. *Mar Drugs.* 2012 Jul;10(7):1476-97.

Janahi EMA, Haque S, Akhter N, Wahid M, Jawed A, Mandal RK, Lohani M, Areeshi MY, Almalki S, Das S, Dar SA. Bioengineered intravaginal isolate of *Lactobacillus plantarum* expresses algal lectin scytovirin demonstrating anti-HIV-1 activity. *Microb Pathog.* 2018 Sep; 122:1-6.

Li Y, Zhang B, Wang X, Yan H, Chen G, Zhang X. Proteomic analysis of apoptosis induction in human lung cancer cells by recombinant MVL. *Amino Acids.* 2011 Oct;41(4):923-32.

Li Y, Zhang X. Recombinant *Microcystis viridis* lectin as a potential anticancer agent. *Pharmazie.* 2010 Dec; 65(12) :922-3.

Li L, Tian X, Chen J, Li P, Zheng Q, Hou J. Griffithsin inhibits porcine reproductive and respiratory syndrome virus infection *in vitro. Arch Virol.* 2018;163(12):3317-25.

Liao WR, Lin JY, Shieh WY, Jeng WL, Huang R. Antibiotic activity of lectins from marine algae against marine vibrios. *J Ind Microbiol Biotechnol.* 2003;30(7):433-9.

Micewicz ED, Cole AL, Jung CL, Luong H, Phillips ML, Pratikhya P, Sharma S, Waring AJ, Cole AM, Ruchala P. Grifonin-1: a small HIV-1 entry inhibitor derived from the algal lectin, Griffithsin. *PLoS One.* 2010 Dec 16;5(12):e14360.

Monteiro Abreu T, Castelo Melo Silva LM, Vanderlei ES, de Melo CM, Pereira VR, Barros Benevides NM. Cytokine production induced by marine algae lectins in BALB/c mice splenocytes. *Protein Pept Lett.* 2012;19(9):975-81.

Mu J, Hirayama M, Sato Y, Morimoto K, Hori K. A novel high-mannose specific lectin from the green alga *Halimeda renschii* exhibits a potent anti-influenza virus activity through high-affinity binding to the viral hemagglutinin. *Mar Drugs.* 2017 Aug 16;15(8). pii: E255.

Neves SA, Dias-Baruff M, Freitas AL, Roque-Barreira MC. Neutrophil migration induced *in vivo* and *in vitro* by marine algal lectins. *Inflamm Res.* 2001;50(10):486-90.

Oliveira SC, Fonseca FV, Antunes E, Camargo EA, Morganti RP, Aparício R, Toyama DO, Beriam LO, Nunes EV, Cavada BS, Nagano CS, Sampaio AH, Nascimento KS, Toyama MH. Modulation of the pharmacological effects of enzymatically-active

PLA2 by BTL-2, an isolectin isolated from the *Bryothamnion triquetrum* red alga. *BMC Biochem.*2008;9:16.

Sato Y. Structure and function of a novel class of high mannose-binding proteins with anti-viral or anti-tumor activity. *Yakugaku Zasshi.*2015;135(11):1281-9.

Silva LM, Lima V, Holanda ML, Pinheiro PG, Rodrigues JA, Lima ME, Benevides NM. Antinociceptive and anti-inflammatory activities of lectin from marine red alga *Pterocladiella capillacea. Biol Pharm Bull.* 2010; 33(5):830-5.

Singh RS, Thakur SR, Bansal P. Algal lectins as promising biomolecules for biomedical research. *Crit Rev Microbiol.* 2015;41(1):77-88.

Suttisrisung S, Senapin S, Withyachumnarnkul B, Wongprasert K. Identification and characterization of a novel legume-like lectin cDNA sequence from the red marine algae *Gracilaria fisheri. J Biosci.* 2011;36(5):833-43.

Takebe Y, Saucedo CJ, Lund G, Uenishi R, Hase S, Tsuchiura T, Kneteman N, Ramessar K, Tyrrell DL, Shirakura M, Wakita T, McMahon JB, O'Keefe BR. Antiviral lectins from red and blue-green algae show potent *in vitro* and *in vivo* activity against hepatitis C virus. *PLoS One.* 2013 May 21;8(5):e64449.

Vanderlei ES, Patoilo KK, Lima NA, Lima AP, Rodrigues JA, Silva LM, Lima ME, Lima V, Benevides NM. Antinociceptive and anti-inflammatory activities of lectin from the marine green alga__*Caulerpa cupressoides. Int Immunopharmacol.* 2010 Sep;10(9):1113-8.

Vuong HR, Tyo KM, Steinbach-Rankins JM. Fabrication and characterization of griffithsin-modified fiber scaffolds for prevention of sexually transmitted infections. *J Vis Exp.* 2017 Oct 31;(128).

Xue J, Hoorelbeke B, Kagiampakis I, Demeler B, Balzarini J, Liwang PJ. The griffithsin dimer is required for high-potency inhibition of HIV-1: evidence for manipulation of the structure of gp120 as part of the griffithsin dimer mechanism. *Antimicrob Agents Chemother.* 2013;57(8):3976-89.

Yamaguchi M, Ogawa T, Muramoto K, Kamio Y, Jimbo M, Kamiya H. Isolation and characterization of a mannan-binding lectin from the freshwater cyanobacterium (blue-green algae) *Microcystis viridis. Biochem Biophys Res Commun.* 1999 Nov 30;265(3):703-8.

In: Hemagglutinins
Editors: T. Bun Ng, J. Wong, R. Tse et al.
ISBN: 978-1-53615-708-6
© 2019 Nova Science Publishers, Inc.

Chapter 5

ANTICANCER AND OTHER ACTIVITIES OF LECTINS FROM JACALIN, WHEAT, RICE AND POTATOES

Jack Ho Wong[1,2], Tzi Bun Ng[3,*], Charlene Cheuk Wing Ng[4], Chit Tam[3], Yau Sang Chan[5], Ryno Naude[6], Ryan Tse[1], Tak Fu Tse[1] and Helen Chan[1,†]

[1]Vita Green Pharmaceutical (Hong Kong Limited), Genning Partners Company Limited, and Institute of Medical Research, Hong Kong, China
[2]The Third Affiliated Hospital of Shenzhen University, Shenzhen, China
[3]School of Biomedical Sciences, Faculty of Medicine, The Chinese University of Hong Kong, China
[4]GKT School of Medical Education, King's College London, UK
[5]Department of Obstetrics and Gynecology, LKS Faculty of Medicine, University of Hong Kong, China
[6]Department of Biochemistry and Microbiology, Nelson Mandela Metropolitan University, Port Elizabeth, South Africa

ABSTRACT

Jacalin, jacalin-related lectin from *Artocarpus nitidus* subsp. Lingnanensis, wheat germ lectin and rice bran lectins, banana lectin and potato lectin has anticancer activity.

* Corresponding Author's E-mail: tzibunng@ cuhk.edu.hk.
† Corresponding Author's E-mail: helen.chan@vitagreen.com.

Jacalin, jacalin-related lectin from *Artocarpus nitidus* subsp. Lingnanensis, and banana lectin show immunomodulatory activity. Banana lectin has anti-HIV-1 reverse transcriptase activity and jacalin has antiviral and antiparasite activities. The recombinant wheat jacalin-like lectin His(6)-HFR1 protein and rice jacalin-like lectin orysata have anti-insect activity. Potato lectin has antifungal and antibacterial activities.

Keywords: jacalin, jacalin-related lectins, anticancer, immunomodulatory, anti-insect, antiviral, antimicrobial

INTRODUCTION

Mannose-specific lectins are structurally related to jacalin such as lectins from the Jerusalem artichoke, banana or rice. They manifest a beta-prism organization. Jacalin is redefined as a polyspecific lectin which, in addition to galactose, also interacts with mannose and mannose-containing glycans (Barre et al., 2001). In this review, jacalin and jacalin-related proteins from banana, rice and wheat will be discussed. In addition, potato lectin will also be mentioned.

JACALIN

Jacalin, a 65-kDa major protein from jackfruit (*Artocarpus heterophyllus*) seeds, is a lectin combining a heavy alpha chain of 133 amino acid residues with a light beta chain of 20-21 amino acid residues. It is highly specific for the alpha-O-glycoside of the disaccharide Thomsen-Friedenreich antigen (Gal beta1-3GalNAc), even in its sialylated form. This characteristic renders jacalin useful for investigating O-linked glycoproteins, especially human IgA1. Its strong mitogenicity for human CD4+ T cells also renders it valuable for assessment of the immune status of HIV-1 infected individuals. Jacalin is used for the purification of human plasma glycoproteins, investigation of IgA-nephropathy, analysis of O-linked glycoproteins and tumor detection (Aucouturier et al., 1989, Kabir et al., 1998).

In Vitro Goat Primordial Follicle Activation, Survival and Gene Expression

Jacalin enhanced both the percentages and viability of developing follicles in cultured caprine ovarian tissue (Ribeiro et al., 2015).

1,2-Dimethylhydrazine-Induced Colon Carcinogenesis

Phytic acid -loaded jacalin-pectin-gold nanoparticles normalized the abnormal colonic architecture, oxidative, and hemodynamic parameters in 1,2-dimethylhydrazine -induced colon carcinogenesis in Wistar rats (Arya et al., 2018).

Chemopreventive Effects on Development of Colon Cancer

Jacalin specifically recognized the tumor- associated Thomsen-Friedenreich antigen and exerted antiproliferative effects on human colon cancer cells. Oral administration of jacalin increased intestinal production of TNF-α, decreased proliferation of colonic epithelial cells, downregulated stromal COX-2 expression and curtailed the number of preneoplastic lesions in carcinogen (methyl-N'-nitro -N-nitroso-guanidine) -exposed Balb/c mice (Geraldino et al., 2017).

Induction of Apoptosis in Human Colon Cancer HT-29 cElls and Breast Cancer MCF-7 cells

Jacalin bound to the surface of macrophages and upregulated the expression and/or release of proinflammatory cytokines through NF-κB signaling, and also expression of iNOS mRNA. Supernatants from jacalin- stimulated macrophages induced apoptosis of human breast cancer MCF-7 cells and colon cancer HT-29 cells (Danella Polli et al., 2016).

Activity toward T-Cell Lymphoma Cell Lines Eb and ESb

A lectin purified from the seeds of the Vietnamese *Artocarpus integrifolia* distinguishes between the mouse T-cell lymphoma cell lines Eb and ESb, with low and high metastatic potential, respectively. It agglutinates Eb cells as well as human erythrocytes, but not ESb cells or the human colon carcinomas cells HT29. The haemagglutinin is specific for alpha-galactosyl residues and has a molecular mass of 62 kDa (Fischer et al., 1997).

Antiproliferative Activity toward Human Carcinoma Cell Line A431

Although both jacalin and *Arachis hypogea* (peanut) agglutinin elicit antiproliferative activity toward human epidermoid carcinoma A431 cells, jacalin but not peanut agglutinin brings about cell rounding with preservation of membrane integrity. Cell rounding is attributed to stress since jacalin enhances phosphorylation of caveolin-1 (at tyr14), p38 but not c-Jun N-terminal kinase. Jacalin inhibits phosphorylation of the epidermal growth factor receptor and extracellular signal regulated. No internalization of jacalin is observed. The110- kDa oxygen-regulated protein 150 (ORP150) (endoplasmic reticulum chaperone) homologous to Hsp70 is one of the six to eight proteins that binds to jacalin.A431 cells overexpressing ORP150 are protected from jacalin-induced toxicity. Jacalin attenuates the functions of ORP150 and increases the susceptibility of A431 cells to stress-induced death (Sahasrabuddhe et al., 2006).

Phosphorylation of p38 MAPK and IL-4 Synthesis-I

The intracellular signaling events, following stimulation of CD4+ T cells with jacalin plus CD28 cross-linking (CD28XL) anti-CD28 antibody, enhanced phosphorylation of Tec and linker for activation of T cells when compared with stimulation with jacalin alone or CD28XL alone. CD4+ T cell activation with jacalin only or with CD28XL only cannot induce interleukin 4(IL-4) secretion. Nevertheless, activation with CD28XL and then with jacalin increased phosphorylation of p38 mitogen-activated protein kinase (MAPK) and IL-4 secretion. Addition of SB203580, a p38 inhibitor, suppressed p38 phosphorylation and IL-4 secretion. Thus jacalin acts in conjunction with CD28XL to induce Th2 differentiation through p38 MAPK activation (Tamma et al., 2006).

Proliferative Response in CD4+ T Cells

Jacalin induces mitogenic responses in CD4+ T cells and thwarts human immunodeficiency virus type 1 (HIV-1) infection in a T lymphoid cell line. Pre-exposure of normal T cells to native HIV-1 envelope glycoprotein gp 120, but not to deglycosylated gp 120 which does not bind to CD4, attenuates the proliferative and interleukin-2 secretory responses to jacalin. Jacalin competes with gp 120 for binding to CD4 molecules. In HIV-infected patients, the proliferative response to peripheral blood mononuclear cells to jacalin shows a correlation with the percentage of CD4+ T cells but is defective compared with the response of healthy volunteers. Jacalin is a useful surrogate marker for deficiency /defect of CD4+ T cells in HIV-1 infected individuals (Tamma et al., 1996).

Immunity against Multiple *Aeromonas* Infection in Zebrafish

Jacalin-capped platinum nanoparticles exerted a protective action on zebrafish against multiple infections with *Aeromonas hydrophila*.This is noteworthy because platinum salt is devoid of antibacterial activity and jacalin does not have immunogenic activity (Ayaz Ahmed et al., 2018).

IgA Production by Peripheral Blood Mononuclear Cells

The ratio of CD4 (+) CD25(+) CD152(+) Tregs:effector T cells in peripheral blood mononuclear cell cultures rose 60-fold in response to jacalin. Th2-stimulated IgA production by B cells may have been downregulated by this rise. Immune complexes containing IgA1 in blood are deposited on the glomerular mesangium in IgA nephropathy, bringing about proliferation of mesangial cells and more rapid extracellular matrix production.

Immune tolerance induced by jacalin suppresses IgA production (Miyamoto et al., 2012).

Immunoenhancing and Adjuvant Activities in Immunization against *Neospora caninum*

ArtinM and Jacalin are jackfruit (*Artocarpus integrifolia*) lectins. ArtinM manifests higher immunoenhancing and adjuvant potencies than jacalin in immunization of mice against neosporosis caused by *Neospora caninum*, a parasite which is an etiologic agent of canine neuromuscular disease and bovine reproductive disorders. This is brought about by elicitng a protective Th1-biased pro-inflammatory immune response and greater protection following exposure to the parasite (Cardoso et al., 2011).

Jacalin binds core glycoprotein desmoglein-1 (Dsg1) and attenuates Dsg1/PF IgG interaction. Subcutaneously administered jacalin prevents PF IgG deposition at the surface of epidermal cells and inhibits formation of PF IgG-induced skin blisters in neonatal mice. Jacalin binds to O-linked TF carbohydrate motifs on Dsg1, impairs the Dsg1/PF autoantibody interactions and abrogates its pathogenicity *in vivo*. TF-specific binding ligands may have a potential therapeutic value for PF (Li et al., 2010).

Mitogenicity and the Inhibition of Herpes Virus Infection

Jacalin exerts a cytopathic action towards herpes simplex virus type 2, varicella-zoster virus, and cytomegalovirus, as shown in the plaque reduction assay.The lectin stimulates proliferation of CD16(+)/CD56(+) cells natural killer cells (Wetprasit et al., 2000).

Lack of Anthelmintic Activity

Jacalin was tested, together with a number of other plant lectins, for the ability to inhibit feeding of the first stage larvae of the sheep gastrointestinal nematodes *Haemonchus contortus, Teladorsagia circumcincta,* and *Trichostrongylus colubriformis.* Some of the plant lectins, but not jacalin, demonstrated inhibitory activity (Ríos-de Álvarez et al., 2012).

Lectin from *Artocarpus nitidus* subsp. Lingnanensis, a Jacalin-Related Lectin Induction of Apoptosis in Human B-Lymphoma cells

Lectin from *Artocarpus nitidus* subsp. Lingnanensis, a jacalin-related lectin, exhibited mitogenic activity and N-acetyl-D-galactosamine inhibitable binding to human B-lymphoma Raji cells. The apoptotic activity of the lectin was witnessed by externalization of phosphatidylserine, cleavage of poly ADP-ribose polymerase, downregulation of Bcl-2, and activation of caspase-3 and p38 mitogen-activated protein kinase signaling pathways (Luo et al., 2018).

Banana Lectins

Immunomodulatory Activity

Oral administration of banana (*Musa acuminata*) lectin (BanLec) to mice led to a rise in levels of interleukin-10, interleukin-17 and tumor necrosis factor, and a decline in levels of interleukin-6 and interferon-gamma in peripheral blood. An upregulation of CD4+ T-cells and a downregulation of CD8+ T cells were detected in the thymus (Sansone et al., 2016).

Modulation of the Proliferation Response in CD3+, CD4+, and CD8+ Populations of Human PBMCs

Recombinant banana lectin stimulated proliferation of CD3+, CD4+, and CD8+ populations of human PBMCs (Gavrovic-Jankulovic et al., 2008).

Macrophage-Stimulating, Antiproliferative and HIV-1 Reverse Transcriptase Inhibiting Activities

The glucose/mannose-specific emperor banana (*Musa basjoo* cv. 'Emperor Banana') lectin enhanced nitric oxide production by murine macrophages and [3H-methyl]-thymidine uptake by murine spleen cells. It exerted antiproliferative activity toward leukemia L1210 cells and inhibited HIV-1 reverse transcriptase activity (Wong and Ng, 2006).

Cytokine Induction

The homodimeric, 30-kDa des-Gly thermostable and pH-stable fructose-binding lectin from Del Monte bananas exhibited mitogenic activity toward mouse spleen cells and upregulated the expression of the cytokines interferon-gamma, interleukin-2 and tumor necrosis factor-alpha in the spleen cells. The lectin also displayed antiproliferative activity toward leukemia L1210 cells and hepatoma HepG2 cells and inhibited HIV-1 reverse transcriptase activity (Cheung et al., 2009).

Enhancement of Hematopoietic Stem and Progenitor Cell Pool

Banana and garlic lectins produced an increase in the hematopoietic stem and progenitor cell pool and may thus serve as therapeutic micronutrients for diseases of hematological deficiencies (Hinge et al., 2010).

Rice Lectins

Protection of Plants against Biting-Chewing and Piercing-Sucking Insects

A mannose-specific jacalin-related lectin from rice, orysata, demonstrated entomotoxic activity against three important agricultural: insect pests including the piercing-sucking pea aphid *Acyrthosiphon pisum* and green peach aphid *Myzus persicae* the biting-chewing beet armyworm *Spodoptera exigua*. Pea aphids feeding on an artificial diet supplemented with different amounts of recombinant Orysata, and green peach aphids and beet armyworm larvae feeding on detached transgenic tobacco leaves overexpressing Orysata, displayed retarded larval development, diminished weight gain in larvae and higher mortality (Al Atalah et al., 2014).

Rice bran lectin RBA1 was a covalently linked heterodimer of 20- and 12-kDa subunits, and rice bran lectin RBA2 was a noncovalently linked dimer of 12-kDa subunits.The rice bran lectins were thermostable and relatively pH -stable.Both RBA1 and RBA2 bound to desialylated complex glycoproteins including α1-acid glycoprotein, fetuin, and transferrin, their galactosylated forms, and chitooligosaccharides. RBA1was transported to the basolateral side of the human intestinal Caco-2 cell monolayers by endocytosis. monolayers by endocytosis. RBA1 enhanced the transport of the fluorescent marker, rhodamine 123, which is transported via the P-glycoprotein-mediated efflux pathway across the monolayers (Nakata et al., 2017).

Inhibition of Cholesterol Micellar Solubility *In Vitro* and Reduction of Serum Cholesterol Level in Rats

Rice bran protein which suppressed cholesterol micellar solubility *in vitro* and reduced blood cholesterol concentrations in rats was a lectin and non-specific lipid-transfer protein (Wang et al., 2017).

Induction of Apoptosis in Human Monoblastic Leukemia U937 Cells

Rice bran lectin and wheat germ agglutinin trigger apoptosis in human monoblastic leukemia U937 cells as evidenced by chromatin condensation, DNA ladder formation and membrane phosphatidylserine externalization. Both of them brought about G2/M phase cell cycle arrest with increased expression of waf1/p21 (Miyoshi et al., 2001).

Defense against Fungal Pathogens

The rice protein OsJAC1 is composed of a jacalin-related lectin domain predicted to bind mannose-containing oligosaccharides, and a dirigent domain which probably acts to stereoselectively couple monolignols. Transgenic overexpression of OsJAC1 in rice enables the plants to acquire resistance against pathogenic bacteria, fungi and oomycetes. Overexpression of OsJAC1 gene or the related wheat TAJA1 gene in barley augments resistance against the powdery mildew fungus. Both jacalin-related lectin domain and dirigent domain of OsJAC1 are necessary for the resistance (Weidenbach et al., 2016).

Wheat Germ Agglutinin (WGA)

WGA considerably influenced the cell growth of all tested human breast cancer cells *in vitro* (MCF-7, T 47D, HBL 100, BT 20) (Valentiner et al., 2003).

The activities of lectins characterized by different carbohydrate- binding specificities on mouse melanoma (B16), human choriocarcinoma (JAr), human hepatoma (H3B), and rat osteosarcoma (ROS) cell lines were compared. Wheat germ agglutinin demonstrated the most potent inhibitory action on the viability of H3B, JAr and ROS cells. The cytotoxicity of *Maackia amurensis* lectin and *Solanum tuberosum* lectin with similar sugar-binding specificity to wheat germ agglutinin, was less potent than that of wheat germ agglutinin (Wang et al., 2000).

Wheat Jacalin-Like Lectin and Hessian fly Infestation

Hessian fly (*Mayetiola destructor*) larvae are very destructive to wheat plants. Elevated levels of Hfr-1(Hessian fly-responsive gene 1), a wheat jacalin-like lectin gene, were detected in infested pest-resistant wheat plants compared with susceptible or uninfested plants. Recombinant His (6)-HFR1 protein has affinity for Manalpha1-6(Manalpha1-3) Man trisaccharides. Fruitfly larvae receiving dietary administration of His(6)-HFR1 displayed retarded growth and pupation and elevated death rate (Subramanyam et al., 2008).Resistant wheat plants accumulated high levels of HFR1 at the larval feeding sites, as revealed by imunodetection, but the avirulent larvae were deterred from feeding and consumed only small amounts of the lectin. Behavioral studies revealed that avirulent Hessian fly larvae on resistant plants exhibited prolonged searching and writhing behaviors as they unsuccessfully attempted to establish feeding sites. During His(6)-HFR1 feeding bioassays, *Drosophila melanogaster* larvae experienced significant delays in growth and pupation, while percentage mortality increased with progressively higher concentrations of

His(6)-HFR1 in the diet. Thus, HFR1 is an antinutrient to dipteran larvae and may play a significant role in deterring Hessian fly larvae from feeding on resistant wheat plants (Subramanyam et al., 2008).

TaJRLL1and Plant Defence against Fungal Pathogens

TaJRLL1 is a mannose-specific jacalin-related lectin-like gene expressed in wheat spikes and stems. It codes for a protein possessing two jacalin-like lectin domains. The expression of TaJRLL1 is upregulated after treatment with phytohormones and pathogen infection. *Arabidopsis thaliana* transformed with TaJRLL1 displayed increased resistance to the fungi *Fusarium graminearum* and *Botrytis cinerea*. Virus-induced gene silencing of TaJRLL1 renders the plants more susceptible to the facultative fungal pathogen *Fusarium graminearum* and the biotrophic fungal pathogen *Blumeria graminis* (Xiang et al., 2011).

Potato Lectin

The potato lectin possesses two extremely similar chitin-binding modules, composed of two in-tandem arrayed hevein domains linked by an extensin-like domain with about 60 amino acids (Van Damme et al., 2004).

Anticancer Activity

Potato (*Solanum tuberosum*) tuber lectin is a 65.5-kDa chitin-binding glycoprotein with an abundance of hydroxyproline (Allen et al., 1966). Recently, a Bangladesh potato lectin at 2.5-160 μg/ml concentration produced 8-36% inhibition of growth of Ehrlich ascites carcinoma (EAC) cells after incubation for 24 h. The inhibition was reduced to 7.6 and 22.3% respectively by inhibitors of caspase-3 and caspase-8.Upregulation of Bax and p53 and cell cycle arrest at the G2/M phase occurred in the presence of the lectin (Kabir et al., 2016).

Antibacterial and Antifungal Activities

A new chitin-binding potato lectin from the Bangladeshi cultivar 'Deshi' exhibited antifungal activity against *Aspergillus niger, Penicillium spp.* and *Rhizopus spp.* and antibacterial activity against *Escherichia coli, Listeria monocytogenes Salmonella enteritidis* and *Shigella boydii*. It also showed antibiofilm activity against *Pseudomonas aeruginosa* (Hsan et al., 2014).

Discussion

Proteins containing a jacalin-related lectin domain and a dirigent domain are found in monocots. Rice, wheat, and barley plants overexpressing OsJAC1, a member of the family of jacalin-related lectins, exhibit resistance against phytopathogensic fungi (Esch andSchaffrath, 2017).

Jacalin has a diversity of activities including anticancer (Fischer et al., 1997; Sahasrabuddhe et al., 2006;Danella Polli et al., 2016;Geraldino et al., 2017; Arya et al., 2018),, immunomodulatory (Tamma et al., 1996, 2006; Li et al., 2010;Cardoso et al., 2011; Miyamoto et al., 2012; Ayaz Ahmed et al., 2018)., antiviral (Wetprasit et al., 2000) and antiparasitic (Ríos-de Álvarez et al., 2012) activities.

The jacalin-related lectin from *Artocarpus nitidus* subsp. Lingnanensis, exhibited mitogenic and anticancer activities (Luo et al., 2018).

The recombinant wheat jacalin-like lectin His(6)-HFR1 protein has anti-insect activity (Subramanyam et al., 2008). The rice jacalin-like lectin orysata also has anti-armyworm and anti-aphid activities (Al Atalah et al., 2014). The rice (Weidenbach et al., 2016) and wheat (Xiang et al., 2011) jacalin-related lectins manifest antifungal activity. Wheat germ lectin (Wang et al., 2000; Miyoshi et al., 2001;Valentiner et al., 2003) and rice bran lectins (Miyoshi et al., 2001) demonstrate anticancer activity. Banana lectin has immunomodulatory, anticancer and anti-HIV-1 reverse transcriptase activities (Wong and Ng, 2006; Gavrovic-Jankulovic et al., 2008; Cheung et al., 2009; Sansone et al., 2016).Potato lectin has anticancer, antifungal and antibacterial activities (Hsan et al., 2014; Kabir et al., 2016).

References

Allen AK, Bolwell GP, Brown DS, Sidebottom C, Slabas AR. Potato lectin: a three-domain glycoprotein with novel hydroxyproline-containing sequences and sequence similarities to wheat-germ agglutinin. *Int J Biochem Cell Biol.* 1996 Nov;28(11):1285-91.

Arya M, Singh P, Tripathi CB, Parashar P, Singh M, Kanoujia J, Guleria A, Kaithwas G, Gupta KP, Saraf SA. Pectin-encrusted gold nanocomposites containing phytic acid and jacalin: 1,2-dimethylhydrazine-induced colon carcinogenesis in Wistar rats, PI3K/Akt,

COX-2, and serum metabolomics as potential targets. *Drug Deliv Transl Res.* 2018 Nov 27. doi: 10.1007/s13346-018-00605-y. [Epub ahead of print]

Al Atalah B, Smagghe G, Van Damme EJ. Orysata, a jacalin-related lectin from rice, could protect plants against biting-chewing and piercing-sucking insects. *Plant Sci.* 2014 May;221-222:21-8.

Ayaz Ahmed KB, Raman T, Veerappan A. Jacalin capped platinum nanoparticles confer persistent immunity against multiple Aeromonas infection in zebrafish. *Sci Rep.* 2018 Feb 2;8(1):2200.

Bartley DJ, Grant G, Huntley JF. In vitro screening of plant lectins and tropical plant extracts for anthelmintic properties. *Vet Parasitol.* 2012 May 25; 186(3-4):390-8.

Cardoso MR, Mota CM, Ribeiro DP, Santiago FM, Carvalho JV, Araujo EC, Silva NM, Mineo TW, Roque-Barreira MC, Mineo JR, Silva DA. ArtinM, a D-mannose-binding lectin from *Artocarpus integrifolia,* plays a potent adjuvant and immunostimulatory role in immunization against Neospora caninum. *Vaccine.* 2011Nov 15;29(49):9183-93.

Cheung AH, Wong JH, Ng TB. *Musa acuminata* (Del Monte banana) lectin is a fructose-binding lectin with cytokine-inducing activity. *Phytomedicine.* 2009 Jun;16(6-7):594-600.

Danella Polli C, Pereira Ruas L, Chain Veronez L, Herrero Geraldino T, Rossetto de Morais F, Roque-Barreira MC, Pereira-da-Silva G. Jacalin-activated macrophages exhibit an antitumor phenotype. *Biomed Res Int.* 2016;2016:2925657.

Esch L, Schaffrath U. An update on jacalin-like lectins and their role in plant defense. *Int J Mol Sci.* 2017 Jul 22;18(7). pii: E1592.

Fischer E, Khang NQ, Brossmer R. The alpha-galactosyl specific lectin from *Artocarpus integrifolia* distinguishes between two lymphoma lines with different metastatic potential. *Biochem Cell Biol.* 1997;75(2):171-5.

Gavrovic-Jankulovic M, Poulsen K, Brckalo T, Bobic S, Lindner B, Petersen A. A novel recombinantly produced banana lectin isoform is a valuable tool for glycoproteomics and a potent modulator of the proliferation response in CD3+, CD4+, and CD8+ populations of human PBMCs. *Int J Biochem Cell Biol.* 2008;40(5):929-41.

Geraldino TH, Modiano P, Veronez LC, Flória-Santos M, Garcia SB, Pereira-da-Silva G. Jacalin has chemopreventive effects on colon cancer development. *Biomed Res Int.* 2017;2017: 4614357.

Hasan I, Ozeki Y, Kabir SR. Purification of a novel chitin-binding lectin with antimicrobial and antibiofilm activities from a bangladeshi cultivar of potato (*Solanum tuberosum*). *Indian J Biochem Biophys.* 2014 Apr;51(2):142.

Hinge A, Bajaj M, Limaye L, Surolia A, Kale V. Oral administration of insulin receptor-interacting lectins leads to an enhancement in the hematopoietic stem and progenitor cell pool of mice. *Stem Cells Dev.* 2010 Feb;19(2):163-74.

Jagtap UB, Bapat VA. Artocarpus: a review of its traditional uses, phytochemistry and pharmacology. *J Ethnopharmacol.* 2010 May 27;129(2):142-66.

Kabir SR, Rahman MM, Amin R, Karim MR, Mahmud ZH, Hossain MT. *Solanum tuberosum* lectin inhibits Ehrlich ascites carcinoma cells growth by inducing apoptosis and G2/M cell cycle arrest. *Tumour Biol.* 2016 Jun;37(6):8437-44.

Li N, Park M, Zhao M, Hilario-Vargas J, McInnes DM, Prisayanh PS, Liu Z, Diaz LA. The Thomsen-Friedenreich antigen-binding lectin jacalin interacts with desmoglein-1 and abrogates the pathogenicity of pemphigus foliaceus autoantibodies *in vivo. J Invest Dermatol.* 2010 Dec;130 (12):2773-80.

Luo Y, Liu X, Lin F, Liao L, Deng Y, Zeng L, Zeng Q. Cloning of a novel lectin from *Artocarpus lingnanensis* that induces apoptosis in human B-lymphoma cells. *Biosci Biotechnol Biochem.* 2018 Feb;82(2):258-267.

Miyamoto K, Chiba T, Shinohara N, Nagata Y, Asakawa N, Kato S, Mizutani N, Murata T, Nomura S, Horiuchi T. Jacalin regulates IgA production by peripheral blood mononuclear cells. *Immunotherapy.* 2012;4(12):1823-34.

Miyoshi N, Koyama Y, Katsuno Y, Hayakawa S, Mita T, Ohta T, Kaji K, Isemura M. Apoptosis induction associated with cell cycle dysregulation by rice bran agglutinin. *J Biochem.* 2001 Dec; 130(6):799-805.

Nakata H, Lin CY, Abolhassani M, Ogawa T, Tateno H, Hirabayashi J, Muramoto K. Isolation of rice bran lectins and characterization of their unique behavior in Caco-2 Cells. *Int J Mol Sci.* 2017 May 13;18(5). pii: E1052.

Ríos-de Álvarez L, Jackson F, Greer A, Bartley Y, Ribeiro RP, Portela AM, Silva AW, Costa JJ, Passos JR, Cunha EV, Souza GB, Saraiva MV, Donato MA, Peixoto CA, van den Hurk R, Silva JR. Effects of jacalin and follicle-stimulating hormone on in vitro goat primordial follicle activation, survival and gene expression. *Zygote.* 2015 Aug;23(4):537-49.

Sahasrabuddhe AA, Ahmed N, Krishnasastry MV. Stress- induced phosphorylation of caveolin-1 and p38, and down-regulation of EGFr and ERK by the dietary lectin jacalin in two human carcinoma cell lines. *Cell Stress Chaperones.* 2006;11(2):135-47.

Sansone AC, Sansone M, Dos Santos Dias CT, Oliveira do Nascimento JR.Oral administration of banana lectin modulates cytokine profile and abundance of T-cell populations in mice. *Int J Biol Macromol.* 2016 Aug;89:19-24.

Subramanyam S, Smith DF, Clemens JC, Webb MA, Sardesai N, Williams CE.Functional characterization of HFR1, a high-mannose N-glycan-specific wheat lectin induced by Hessian fly larvae. *Plant Physiol.* 2008 Jul;147(3):1412-26.

Tamma SM, Balan SP, Chung KW, Pahwa S. The lectin jacalin plus costimulation with anti-CD28 antibody induces phosphorylation of p38 MAPK and IL-4 synthesis-I. *J Leukoc Biol.* 2006 Apr;79(4):876-80.

Tamma SM, Oyaizu N, McCloskey TW, Kalyanaraman VS, Pahwa S. HIV-1 gp 120 blocks jacalin-induced proliferative response in CD4+ T cells: jacalin as a useful surrogate marker for qualitative and quantitative deficiency of CD4+ T cells in HIV-1 infection. *Clin Immunol Immunopathol.* 1996 Sep;80(3 Pt 1):290-7.

Valentiner U, Fabian S, Schumacher U, Leathem AJ. The influence of dietary lectins on the cell proliferation of human breast cancer cell lines *in vitro*. *Anticancer Res.* 2003 Mar-Apr;23(2B):1197-206.

Van Damme EJ, Barre A, Rougé P, Peumans WJ. Potato lectin: an updated model of a unique chimeric plant protein. *Plant J.* 2004 Jan;37(1):34-45.

Wang H, Ng TB, Ooi VE, Liu WK. Effects of lectins with different carbohydrate-binding specificities on hepatoma, choriocarcinoma, melanoma and osteosarcoma cell lines. *Int J Biochem Cell Biol.* 2000 Mar;32(3):365-72.

Wang J, Shimada M, Nagaoka S. Identification of the active protein in rice bran protein having an inhibitory activity of cholesterol micellar solubility. *Biosci Biotechnol Biochem.* 2017 Jun;81(6):1216-1219.

Weidenbach D, Esch L, Möller C, Hensel G, Kumlehn J, Höfle C, Hückelhoven R, Schaffrath U. Polarized defense against fungal pathogens is mediated by the jacalin-related lectin domain of modular Poaceae-specific proteins. *Mol Plant.* 2016 Apr 4;9 (4) :514-27

Wetprasit N, Threesangsri W, Klamklai N, Chulavatnatol M. Jackfruit lectin: properties of mitogenicity and the inhibition of herpesvirus infection. *Jpn J Infect Dis.* 2000 Aug;53(4):156-61.

Wong JH, Ng TB. Isolation and characterization of a glucose/mannose-specific lectin with stimulatory effect on nitric oxide production by macrophages from the emperor banana. *Int J Biochem Cell Biol.* 2006 Feb;38(2):234-43.

Xiang Y, Song M, Wei Z, Tong J, Zhang L, Xiao L, Ma Z, Wang Y. A jacalin-related lectin-like gene in wheat is a component of the plant defence system. *J Exp Bot.* 2011 Nov;62(15):5471-83.

In: Hemagglutinins
Editors: T. Bun Ng, J. Wong, R. Tse et al.

ISBN: 978-1-53615-708-6
© 2019 Nova Science Publishers, Inc.

Chapter 6

CHICKPEA LECTIN: PURIFICATION STRATEGIES, CHARACTERISTICS AND POTENTIAL APPLICATIONS

Ajay Kumar Gautam[1], Neha Gupta[1], Dakshita Tanaji Narvekar[1], Amita Bhadkaria[1], Shubhlukshmi Kumari[1], Nidhi Srivastava[2], Sameer Suresh Bhagyawant[1,*], Jack Ho Wong[3,4], Tzi Bun Ng[5], Yau Sang Cha[6], Chit Tam[5], Charlene Cheuk Wing Ng[7], Ryno Naude[8], Ryan Tse[3], Tak Fu Tse[3] and Helen Chan[3]

[1]School of Studies in Biotechnology, Jiwaji University, Gwalior, India
[2]Department of Bioscience & Biotechnology, Banasthali Vidhyapeeth, Banasthali, India
[3]Vita Green Pharmaceutical (Hong Kong Limited), Genning Partners Company Limited, and Institute of Medical Research, Hong Kong, China
[4]The Third Affiliated Hospital of Shenzhen University, Shenzhen, China
[5]School of Biomedical Sciences, Faculty of Medicine, The Chinese University of Hong Kong, China
[6]Department of Obstetrics and Gynecology, LKS Faculty of Medicine, University of Hong Kong, China
[7]GKT School of Medical Education, King's College London, United Kingdom
[8]Department of Biochemistry and Microbiology, Nelson Mandela Metropolitan University, South Africa

---

* Corresponding Author's E-mail: sameerbhagyawant@gmail.com.

Abstract

Legume seeds contain carbohydrate binding proteins of non-immune origin that are termed as lectins. Chickpea belongs to the family *Leguminosae* and is a food legume. The seeds of chickpea provide an exceptional source of dietary proteins and is one of the important legumes in both developed and developing countries around the world. Desi and kabuli chickpea seed types are the most ubiquitous. Significant amounts of lectins have been reported to be present in them. Some of the biological roles of chickpea lectin have yet to be fully elucidated. Antifungal, antibacterial, anticancer, and anti-HIV reverse transcriptase activities have recently been reported. Owing to the aforesaid characteristics of chickpea lectin, there is an urgent need to develop the lectin as a therapeutic candidate in the pharmaceutical industry. Up-to-date research findings on different aspects of chickpea lectins *vis-à-vis* purification strategies and health impacts are presented systematically herein.

Keywords: chickpea lectin, purification, characteristics, potential applications

Introduction

Nowadays, legumes constitute an important source of dietary protein in many developed as well as developing countries including India. In addition to nutritional aspects, most of the legumes possess biologically active compounds. Despite being regarded as antinutritional, legumes are consumed by humans or animals (Liener 1979) and many such compounds in legumes have been reported (Chrispeels 1984). Antinutritional compounds may be either heat-labile e.g., lectins, protease inhibitors and cyanogens or heat-stable such as antigenic proteins, condensed tannins and quinolizidine alkaloids, etc. Commonly observed antinutritional factors in grain legumes are phyotlectins, polyphenols (tannins) and certain other proteins with inhibiting activity (Singleton 1981; Ryan 1990).

Extensive studies have been conducted on protease inhibitors (PIs) in legumes such as winged bean, pigeon pea, cowpea and chickpea. It was observed that these inhibitors are localized in storage tissues such as seeds and tubers. Both chickpea and pigeon pea contain considerably higher levels of PIs than other commonly consumed Indian edible legumes. However, the levels are still much lower than those of soybeans. Protease inhibitors in chickpea seeds are considered as antinutritional factors and reported in many seed types (Liener 1976; Price et al., 1980). In addition to PIs, lectins or phytohemagglutinins are also found in legumes. These proteins agglutinate red blood cells which have specific glycoprotein receptor sites on the cell surface. Their antinutritional activity interferes with the action of digestive enzymes and reducing the availability of nutrients transported into the cells. These antinutritional substances are toxic for human consumption but can be completely degraded by moist heat treatment (Liener 1979). However lectins are synthesized by plants for their own defense and signal recognition of symbiotic nitrogen fixing bacteria.

The seeds of chickpea (*Cicer arietinum* L.) provide an exceptional source of dietary proteins and are the most extensively exploited food legume in the world. Chickpea is used for both food and medicinal purposes. The progenitor wild relative of domestic chickpea is *Cicer reticulatum,* the only relative in the primary gene pool. Seeds being the main edible part of the plant are a rich source of proteins, carbohydrates and minerals, especially for the vegetarian population. Two major seed types of chickpea are recognized, the white-seed 'kabuli' and the brown coloured 'desi' types (Figure 1). Kabuli chickpea with a thinner seed coat are relatively bigger in size while the desi type seeds are relatively smaller in size and have a thicker seed coat. The desi type chickpea contributes to around 80% and the kabuli type accounts for approximately 20% of the total production of chickpea. Diversity in the chickpea seed phenotype can be observed as mentioned below.

Some of the lectins are glycoproteins. They bind reversibly with specific sugars on the surface of cells in the gut wall, hence causing interference in nutrient breakdown and absorption. Lectins agglutinate blood cells and this reaction has been demonstrated *in vitro,* by agglutination of red blood cells from different animal species. Lectins are unusually specific in action. Traditionally, lectins have been quantified by their hemagglutinating activity (Grant 1991). Nowadays, to achieve high specificity, enzyme linked immunosorbent assay (ELISA)-based immunological methods are used (Hajos et al., 1996; Muzquiz et al., 2001). Kolberg et al., (1983) isolated a lectin from extracts of the seeds of the chickpea *Cicer arietinum* and N-terminal amino acid sequence analysis showed that it consisted of only one kind of polypeptide chain. The lectin is probably dimeric in nature. Papain-treated human erythrocytes of different ABO blood groups could be agglutinated very well by the lectin whereas untreated erythrocytes reacted weakly only in the presence of bovine serum albumin.

Figure 1. Different accessions of desi and wild chickpea varieties.

Caudrado et al., (Madrid 2005, unpublished) analyzed the desi and kabuli varieties of chickpea. Native red blood cells and blood cells treated with trypsin were used for a more sensitive assay of hemagglutinating activity. They used lectin from *Phaseolus vulgaris* cvs Processor and Pinto as standard in the assay. One HU (hemagglutination unit) is defined as the amount of material in the last dilution which agglutinates 50% of the red blood cells. These results showed that chickpea had low reactivity toward all red blood cells with HU values similar to *Lupius sp.* and it was found that among the 15 legumes studied, chickpea seeds had the lowest hemagglutining activity and were non-toxic (Grant et al., 1983). Dietary lectins vary in the nature and extent of their antinutritional effects. Lectins can be toxic: they can cause interference in hormonal balance and deplete nutrient reserves leading to severe growth abnormality and hence high mortality. However, lectins may be beneficial by stimulating gut function, limiting tumor growth and helping to lower the level of obesity.

Haemagglutinating Activity of Chickpea Lectin

Lectins are proteins, displaying a diversity in structure and carbohydrate specificity, and are responsible for bringing about agglutination of red blood cells. Chickpea (*C. reticulatum*) seeds are a potent source of lectin. Hemagglutinating activity was reported in mature chickpea seeds, which can be used as a model for studying lectins and their attributes. The specific activity achieved after purification was found to be about 2600 titer/mg. Initially, hemagglutinating activity was performed using normal human blood (A, B, AB and O) and rabbit erythrocytes followed by trypsin-treated erythrocytes. However, *C. reticulatum* lectin had a subunit molecular mass of 35 kDa and agglutinated trypsin-treated rabbit erythrocytes (Bhagyawant et al., 2015). The hemagglutinating activity of chickpea lectin C-25 was inhibited by N-acetyl-D-galactosamine and C-25 was shown to be a N-acetyl-D-galactosamine-specific lectin (Kumar et al., 2014). Chickpea lectin did not cause hemagglutination of red blood cells from humans (A, B, O types), rats or monkeys (Ayyagari et al., 1989). However, chickpea lectin agglutinated red blood cells from rabbits and cows although not at toxic levels (Contreras and Tagle 1974). The lectin concentration in raw lentil and pea samples was 0.1-1 g/kg and 0.2-2 g/kg, respectively (Peumans and Van Damme 1996). However, the lectin concentration in any chickpea variety remains unclear due to the diversity of chickpeas. Therefore, the concentration of lectin in chickpea is lower in comparison to pea or lentil and this may be responsible for its lower hemagglutinating activity.

Purification Strategies for Chickpea Lectin

The presence of lectins in numerous organisms has been widely recognized. Before the 1970s, very few lectins had been purified. One of the major breakthroughs in lectin purification came with the use of affinity chromatography for the capture of lectins with cross-linked dextran gels by Goldstein and his coworkers (Agrawal and Goldstein 1965). Since then, most purification strategies include an affinity chromatography step on columns containing polysaccharides to take advantage of the specificity of lectins for sugars and the ready availability of sugar-based gel matrix for chromatography, especially the glucose-based Sephadex-TM (cross-linked dextran) and the galactose-based Sepharose-TM. Initially, the purification of crude extracts was performed using several techniques followed by affinity chromatography, ion-exchange chromatography, hydrophobic interaction chromatography, and gel filtration chromatography (Tsuruta et al., 2008). Accordingly, the most commonly used unit operation is chromatography, followed by precipitation and the purification procedure includes two to four steps. A small decrease in the recovery yield can be observed as the number of purification steps increases from one to three, even though the purification fold decreases sharply for processes involving more than three steps. Before the loading of crude extract on to the chromatographic column, it is necessary to remove the extraneous proteins by partial purification employing ammonium sulfate precipitation and dialysis.

Partial Purification

Ammonium Sulfate Fractionation

The removal of unwanted protein is the first step of partial purification. This is worked out by adding ammonium sulfate. Ammonium sulfate is commonly used to precipitate the proteins (salting out) and maintain the native structures of the proteins (Voet and Voet 1995). Different saturation levels of ammonium sulfate i.e., 0-20, 20-40, 40-60, 60-80 and 80-100% are applied for the partial stages of purification. The suspension is generally centrifuged at 10,000 rpm for 20 minutes at 4°C, depending on the nature of the lectin and the supernatant is saved for further purification steps. Wong et al., (2012) purified the lectin using *Phaseolus vulgaris* that was precipitated at 70% saturation level of ammonium sulfate. In other studies, fractionation with 80% (Gautam et al., 2018a), 30-60%, (Qureshi et al., 2006) and 60% (Kolberg et al., 1983) saturation levels of ammonium sulfate was used for partial purification of lectin from *Cicer arietinum*.

Dialysis

Dialysis is the process of removal or exchange of small molecules from macromolecules in solution based on differential diffusion using an artificial semi-permeable membrane bag/tubing. Different dialysis tubings have different molecular-weight cutoffs ranging from 1-1,000,000 Da depending on the number and average size of the pores. These membranes are composed of modified and cross-linked cellulose fibers derived from wood pulp or cotton fibers of different pore sizes. For the removal of lectin, the partially purified lectin is placed in the dialysis bag (with a definite pore size) and dialyzed against the respective buffer with stirring on a magnetic stir plate overnight at 4°C. Molecules that are larger than the pore size are retained on the inner side of the membrane while small molecules pass through the membrane. A dialysis membrane with a 10-kDa molecular-weight cutoff retains lectins having a molecular mass of at least 10 kDa. A dialysis membrane with a molecular-weight cut-off of 3500 Da was used for purification of lectins from *Cochlospermum tinctorium* (Nergard et al., 2005) and *Ornithogalum caudatum* (Chen et al., 2010).

Chromatography

The diversity of functions of lectin and their mechanisms have been observed. The information on purification and characterization of lectin observed in leguminous plants are specific. Currently, many types of lectins are well characterized with many applications in immunology, biochemistry, glycobiology and clinical research. Common types of techniques used for the isolation and purification of lectin such as affinity chromatography, ion-exchange chromatography and gel filtration are discussed briefly below:

Lectin Affinity Chromatography

The use of affinity chromatography for the purification of lectins was a turning point in the field of lectinology and enormously increased the rapidity of purification of lectins in a short period of time (Agrawal and Goldstein 1965; Goldstein and Hays 1978; Goldstein et al., 1980). Affinity chromatography depends on the specific interaction between the lectin and a carbohydrate structure attached to an inert matrix. Carbohydrate affinity chromatography is a universally used method for purifying lectins. It is a simple method that takes advantage of their specific properties. Selection of the most appropriate matrix, ligand and elution technique is necessary for purification. Presently, lectins are widely used as sorbents and lectin affinity chromatography is a common method for the isolation and purification of lectins. Affinity chromatography was first developed by Donnelly and Goldstein for the specific isolation of lectins (Donnelly et al., 1970). The interaction of lectins with carbohydrates depends on non-covalent binding and the specificity of some lectins may involve binding to two or more glycans during affinity chromatography (Durham et al., 2006). The better resolution of lectins can be provided by involving a multiple elution approach.

Any organism or tissue of interest from humans, animals, plants, fungi to microorganisms including viruses can be selected for isolation of lectins. The lectin has to be released first from the organism into the solution used for extraction, generally by liberating it from the cells. The isolation of lectin can be achieved by using various methods such as homogenization by mechanical disruption or by lysis (Voet and Voet 1995). All steps of lectin extraction must be carried out at 4°C to prevent denaturation and degradation by proteolytic enzymes. In addition, proteolytic activity is inhibited by addition of a protease inhibitor to the extraction buffer (Kvennefors et al., 2008; Matsumoto et al., 2011; Watanabe et al., 2007) and glycosidase inhibitors (glucosidase inhibitor 1-deoxynojirimycin and mannosidase inhibitor 1-deoxymannojirimycin) may also be added to the extract to prevent cleavage of immobilized carbohydrate ligands in the affinity column (Watanabe et al., 2007). Impurities can be removed from the extract before application to the carbohydrate affinity column by using fractional ammonium sulphate precipi tation. The hemagglutinating activity can be checked throughout the purification procedure by using erythrocytes that are specific for the lectin. Once the extract exhibiting hemagglutinating activity is prepared, the carbohydrate affinity chromatography method appropriate for purifying the lectin has to be selected. This involves the use of an adsorbent that can bind the lectin and separate it from the extract using an elution technique.

Ion Exchange Chromatography

Ion-exchange chromatography is an important analytical technique for the separation and determination of ionic compounds (Haddad and Jackson 1990). The separation of molecules is based on ionic interactions between ionic and polar analytes, ions present in the eluent and ionic functional groups fixed to the chromatographic support. This technique is generally used for the separation of peptides, nucleic acids, proteins and biopolymers which are charged molecules (Okada 1998; Levison 2003; Ishihara et al., 2007; Bruch et al., 2009). The formation of ionic bonds between the charged groups of biomolecules and an ion-exchange resin which carries the opposite charge is the basis of separation of molecules (Stanton 2004).

Selection of Ion-Exchange Resins

The choice of a suitable ion-exchange matrix is probably the single most important aspect of any ion-exchange protocol and is based on various factors, which include (i) desired ion exchanger charge/strength; (ii) linear flow rate/sample volume; and (iii) sample properties. As mentioned above, ion-exchange functional groups fall into two charge categories. Positively charged diethylaminoethyl (DEAE) and quarternary ammonium (Q) functional groups for example, are routinely employed in anion-exchange chromatography, whilst negatively charged carboxymethyl (CM), sulphomethyl (S) and sulphopropyl (SP) groups are typical cation-exchangers (Figure 2). Both exchanger types can be further categorized as either "strong" or "weak". Strong ion-exchangers are fully ionized over a broad working pH range (i.e., show no loss or gain of charge with varying pH), whilst weak ion-exchangers are only partially ionized over a narrow pH range (i.e., charge can vary significantly with pH). The ion exchanger transports a surplus electric charge distributed over the surface. There are two types of ion exchangers. Cation exchangers carry negatively charged functional groups and possess exchangeable cations, while anion exchangers carry anions because of the positive charge (Zagorodni 2007). The protein binding capacity of a resin depends on its specificity and strength of its charged group along with polarity and density (Kastner 2000).

Column Preparation

DEAE-cellulose and SP-Sephadex are applied for pre-swollen in distilled water in cold conditions. The slurry may be further washed with respective buffer for 2-4 hours to remove any particulate matter. The column is then packed with binding buffer (Figure 3).

Sample Loading

The isolated lectins from the above steps will be applied to ion exchange resin columns one by one. The samples will be loaded on a reservoir with a controlled flow rate of 0.1 mL/minute.

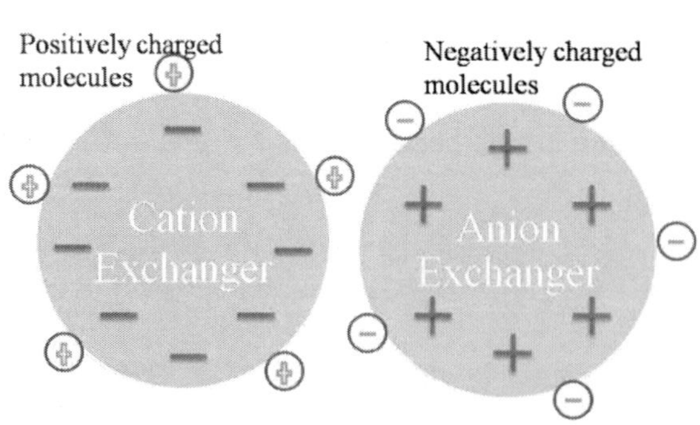

Figure 2. Positively and negatively charged ions on the surface of a chromatographic resin.

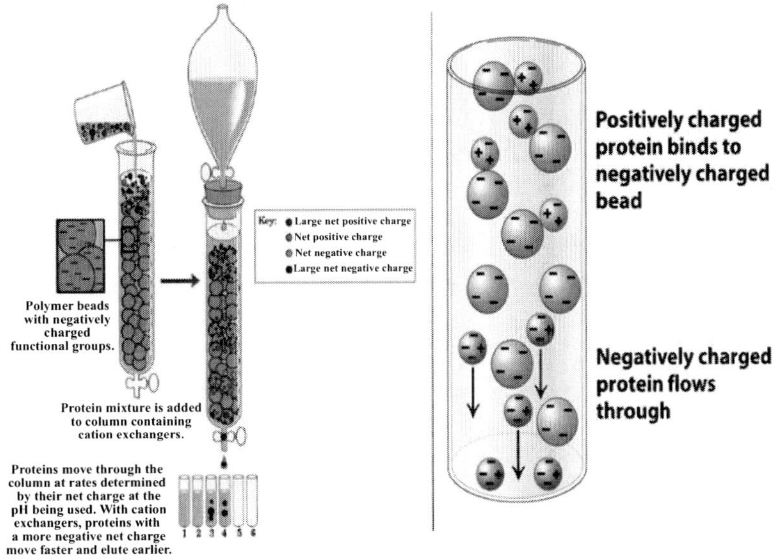

Figure 3. Purification step showing the separation of charged molecules with an anion exchanger resin.

Elution of Lectin

Elution of bound molecules can be achieved by using an increasing salt gradient. The 0.1M to 1M NaCl gradient is generally used as an elution buffer and fractions collected with a flow rate of 0.1 mL/minute were monitored with absorbance at 280 nm and the desired fractions are pooled. Subsequently these fractions are tested in an assay of hemagglutinating activity to detect the presence of lectin. The lectin sample in liquid form can then be subjected to freeze drying (lyophilization).

Lyophilization is the term used for freeze-drying, a process used for preserving biological materials by removing water from the sample, which involves first freezing the sample and then drying, under a vacuum, at very low temperatures. The basis for this sublimation process involves the absorption of heat by the frozen sample in order to vaporize the ice. The sample (lectin) is freeze dried at -20°C or in liquid nitrogen. Lyophilization is carried at 130 Pa vacuum pressure and a condenser temperature maintained at -48°C. The lyophilized samples in powder form are stable at room temperature and used for further analysis.

The capacity of an ion exchanger depends upon the ionic magnitude of the resin that has the number of sites available for the exchanger. The exchanger capacity of a resin (polyfunctional cation exchangers-carboxylic and hydroxyl group) depends on the pH. As the pH increases, the exchange capacity of a cation exchanger increases while the capacity of an anion exchanger decreases. The exchange capacity is expressed in gram equivalent of the ions retained by one cubic meter of the swollen ion exchange resin. In general, resins are more stable towards strong acids, strong bases, while the capacity of weakly acidic and basic ion exchangers depends on pH 4.

PURIFICATION OF CHICKPEA LECTIN

The study of Kolberg et al., (1983) demonstrated the isolation and purification of lectin from the seeds of *C. arietinum* L. using ion exchange chromatography (DEAE Sepharose C1-6B) and gel filtration (CM-Sepharose C1-6B) columns. They also performed affinity chromatography on desialated human IgM coupled to AH-Sepharose. However, the yield was very low. The lectin exhibited a molecular mass of 44 kDa as determined by ultracentrifugation and gel filtration, and a molecular mass of 26 kDa as determined by SDS-PAGE. The lectin is dimeric due to the presence of only one N-terminal amino acid sequence.

Qureshi et al., (2006) focused on the purification of a novel lectin (CAA-II) from chickpea (cultivar P256) by using N-acetyl-D-galactosamine-linked agarose chromatography. The purified lectin is composed of four identical 30-kDa subunits. The native molecular mass of the lectin was likely to be 120 kDa by gel filtration chromatography and confirmed by mass spectrometry. They started the isolation of lectin from mature chickpea seeds (40 g) which had been soaked overnight in 10 mM Tris–hydrochloride buffer (pH 6.8) at 4°C followed by acid-precipitation. The clear supernatant (crude extract) was used for determining protein concentration and for testing hemagglutinating activity. The supernatant was subjected to fractionation with 0-30, 30-50 and 50-100% ammonium sulfate at 4°C followed by dialysis against the same buffer. This dialysed 30-50% fraction was loaded onto an N-acetyl-D-galactosamine-linked agarose column that had been equilibrated with Tris–hydrochloride buffer. The lectin was eluted with 0.2 M galactose in Tris–hydrochloride buffer. In another study presented by Katre et al., (2005), a lectin was purified from *C. arietinum* (CAL) cv BDN 9-3. The purified lectin was a 43-kDa dimer composed of two identical subunits (MW 21 500) as confirmed by SDS-PAGE. Its hemagglutinating activity was inhibited by fetuin and desialated fetuin but not by simple sugars. They started with 100 g dry seeds which were finely powdered and then soaked in Tris-HCl buffer containing 150 mM NaCl (pH 7.2). Thereafter, the supernatant was subjected to fractionation with ammonium sulfate precipitation using 0-80% saturated ammonium sulfate followed by dialysis against the same buffer. This clear dialyzed supernatant was further applied onto a DEAE-cellulose column pre-equilibrated with 20 mM Tris-HCl buffer (pH 7.2). The eluted fractions were then loaded onto an SP-Sephadex column (4 cm x20 cm size). Finally, the bound protein was eluted with acetate buffer (pH 5.0) containing NaCl in a stepwise gradient. The study carried out by Khan et al., (2011) resulted in the isolation of phytohemagglutinins from two chickpea cultivars (KK-1 and Hassan-2K) using shoots, leaves, roots, and reproductive organs, which presented potent lectin activity. For the isolation of lectin, two grams of roots, shoots, leaves, inflorescence, and seeds were finely ground and then stirred in 20 ml of 0.15 M NaCl. The degree of hemagglutina tion was monitored visually at room temperature for 30 min. The results were recorded as hemagglutination titer (the reciprocal of the highest dilution given visible hemagglutination).

Hemagglutination of erythrocytes was carried out with human erythrocytes of different blood groups to test the hemagglutinating activity. Furthermore, another study was reported by Wakankar et al., (2013a). They purified the recombinant lectin from *C. arietinum* (rCAL) seeds which showed sugar binding specificity with asialo- triantennary glycan from fetuin. They prepared the recombinant lectin from the gene of *C. arietinum* seed lectin cloned in pET 28a$^+$ vector and expressed in the BL21 CodonPlus (DE3)-RIL cells of *E. coli.* They purified the recombinant protein by using ion-exchange (unadsorbed on DEAE-cellulose) and gel filtration chromatography (Superose 12 prep grade, FPLC) (Wakankar et al., 2013b). In another study, Kumar et al., (2014) described the isolation and purification of a protein(C-25) from *C. arietinum* which acted as a lectin and had a molecular weight of 25 kDa confirmed by 12% SDS-PAGE. They used seeds, homogenized in 10mM Tris-Cl buffer (pH7.2) in cold conditions followed by the ammonium sulfate with 30% saturation and a precipitant. The salt was removed by a dialysis membrane (10 kDa) against the same buffer. The final dialyzed sample was loaded onto a Sephadex G-100 gel filtration column pre-equilibrated with 10 mM Tris-HCl buffer (pH 7.2) buffer containing 150 mM NaCl and proteins were eluted using the same buffer.

Wang et al. (2017) reported the isolation, identification and characterization of a new type of lectin from *C. arietinum* L. seeds that possess α-amylase inhibitory activity. The isolation and purification was done using ammonium sulfate precipitation, ion exchange chromatography (IEC) and reversed phase liquid chromatography (RPLC). They also identified *C. arietinum* lectin by using ultra-performance liquid chromatography-tandem mass spectrometry (UPLC/MS-MS) which indicated that the purified proteinaceous α-amylase inhibitor was a *C. arietinum* lectin in GenBank (accession No. AGL46982.1). *C. arietinum* lectin had 227 amino acids containing a hemopexin- like repeat domain. *C. arietinum* lectin is a new type of lectin with inhibitory activity against α-amylases, which may serve to increase pest resistance through genetic engineering for breeding. Recently, Gautam et al. (2018a) studied the isolation and purification of a 35-kDa lectin from chickpea (*Cicer arietinum* L.) by employing ion exchange chromatography on DEAE-cellulose and SP-Sephadex.

APPLICATIONS OF CHICKPEA LECTIN

In plants, lectins play a significant role in defense against predation. The isolated lectins from legume seeds also exhibit some additional biological activities. Previous studies have recognized the advantages of legumes as potent therapeutic agents for improving human health. A remarkable biomedical potential of plant lectins is anti-cancer activity. The proliferation of lymphocytes is regulated by lectins through modulating the synthesis of antibody and cytokine regulation (Nasi et al., 2009). The effects of lectins include mitogenic, anti-carcinogenic, and anti-microbial activities (Sharon and Lis 2002), alpha amylase inhibiting, DNA protective, HIV-1 reverse transcriptase inhibitory and insecticidal activities. The benefits of chickpea lectin in a pharmacological perspective are discussed as follows.

Antibacterial Action

Antimicrobial resistance is the key concern in medical care at present. The chickpea lectin was therefore tested for antimicrobial activity against four species of bacteria *B. subtilis, E. coli, S. marcescens* and *P. aeruginosa*. The antimicrobial activity of chickpea lectin was observed with MICs in the range of 80-180 µg/ml (Figure 4). The diameters of the zones of inhibition produced were16, 15, 4 and 5 mm, respectively (Gautam et al., 2018a).

The antimicrobial action of lectins is consistent with the peptidoglycan layer or lipopolysaccharides of bacterial cell wall (Dziarski et al., 2000). *Lathyrus ochrus* lectin forms hydrogen bonds between the glycogen binding sites and muramic acid (Bourne et al., 1994). Growth of the four bacterial species *B. subtilis, S. aureus, E. coli* and *P. aeruginosa* was retarded by rhizome lectin at MICs values of 0.011, 0.005, 0.092 and 0.002 mg/mL, respectively (Petnual et al., 2010).

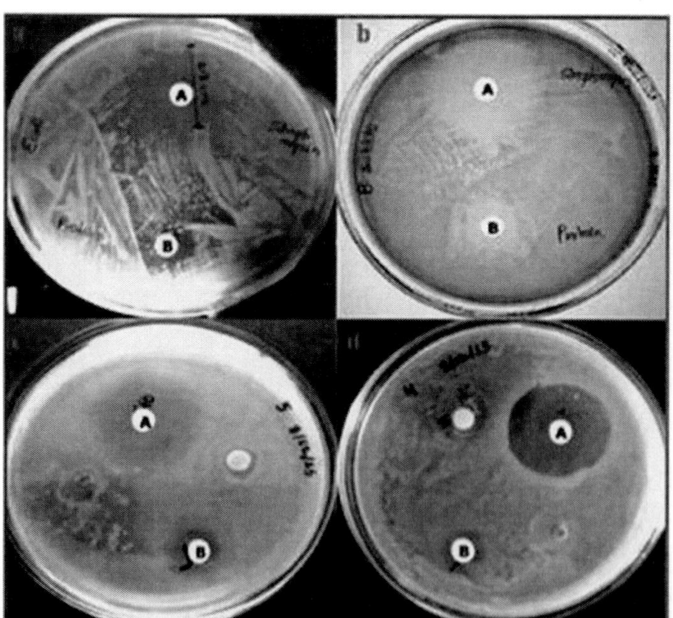

Figure 4. Chickpea lectin exhibiting antibacterial activity towards (a) *E. coli* (b) *B. subtilis* (c) *P. aeruginosa* and (d) *S. marcescens.*

Antifungal Activity

Lectins isolated from plants possessed antifungal properties towards plant pathogens. They are commonly used as antimicrobials and pesticides in agriculture (Jebali et al., 2014; Ye et al., 2001; Xia et al., 2005; Zhang et al., 2010; Charungchitrak et al., 2011). A 25-kDa lectin was isolated from chickpea by Kumar et al., (2014). It exhibited antifungal property against a variety of human pathogenic fungi; *Candida krusei, C. tropicalis* and *C. parapsilosis,* with MIC values of 1.56-12.5 μg/l. Similarly, it dose- dependently inhibited the growth of fluconazole-resistant and -susceptible fungal strains. Antifungal activity was also assessed by Gautam et al., (2018a) against various fungal species; *Candida krusei, Fusarium oxysporum, C. albicans* and *Saccharomyces cerevisiae* employing the disc diffusion method (Figure 5). CAL suppressed the growth of *C. krusei* and *F. oxysporum* by 85% and 81%, respectively. The positive control amphotericin B produced 88.8% reduction of growth. A 14-mm zone of inhibition of *S. cerevisiae* and an 8 mm zone of inhibition of *C. albicans* were observed.

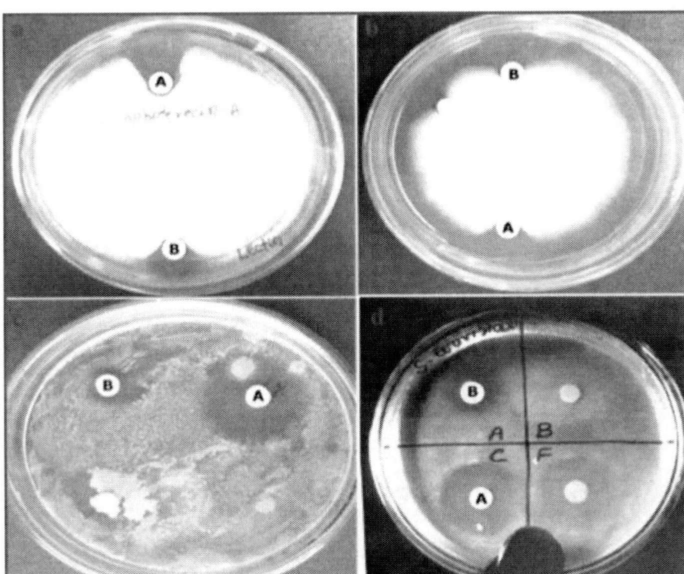

Figure 5. Chickpea lectin exhibiting antifungal activity towards (a) *F. oxysporium* (b) *C. krusei* (c) *C. albicans* and (d) *S. cerevisiae*.

The interaction between lectins and fungi occurs through chitin and glucans, the major constituents of the fungal cell wall (Adams 2004). Lectins act predominantly on the cell wall of *Candida* species by disrupting the cell wall and distorting the cellular morphology. Earlier, it was also observed by scanning electron microscopy that lectins disrupted the cell wall and caused cytoplasmic leakage (Wang et al., 2004). Thus, chickpea lectins displayed exploitable antifungal activity.

A small number of lectins interfere with spore formation by inhibiting chitin synthesis in the cell wall (Lis and Sharon 1981; Selitrennikoff 2001; Trindade et al., 2006). For example hevein (4.7 kDa) and poutein (14 kDa) reached the plasma membrane of fungal cell wall by penetration and blocked the catalytic domain of enzymes (Boleti et al., 2007; Gomes et al., 2012). Current studies on the potential of lectins from different origins and with different carbohydrate moieties demonstrate that they are antifungal agents. Plant lectins showing such antifungal effects bind to fungal hyphae and cause inhibition of growth and prevent spore germination. Lectins from *Phaseolus vulgaris* seeds inhibited the growth of *Coprinus comatus* and *Rhizoctonia solani* (Ye et al., 2001). Lectins from red kidney bean and wheat germ lectin bound specifically with the sugar N-acetylglucosamine and exhibited antifungal activity (Sitohy et al., 2007). Thus chickpea lectin as a biopeptide can be used as a fungicide and bactericide for treating infections.

MITOGENIC ACTIVITY

In human, lectins are intensely mitogenic for $CD4^+$ T lymphocytes and therefore lectins can be implemented as a tool for evaluation of the immune status of patients infected with HIV-1 (Lafont et al., 1996). Lectin from *C. reticulatum* augmented the mitogenic response of mouse spleen cells with a maximum at 10 μg/ml concentration. A dose-dependent escalation was reported when the lectin concentration was increased from 2.5 μg/ml to 15 μg/ml. However, an additional increase in lectin concentration elicited a dose- dependent decrease in the response (Gautam et al., 2018b). The mitogenic activity of lectin from *C. reticulatum* is of immense importance in the light that mitogenic lectins, in comparison to commercially available plant mitogenic lectins like PHA and Con A, can be useful in a study of lymphocyte transformation as a model for antigen activation, initiation of cell division and growth. *Arisaema flavum* lectin exhibited effective mitogenic activity towards BALB/C splenocytes and human lymphocytes in comparison with Con A, a well-known plant mitogen (Singh et al., 2004).

DNA DAMAGE PROTECTION

In humans, the release of free radicals may cause DNA damage and initiate several other diseases. Partial damage to DNA can cause cells to become cancerous. Plants are natural sources of antioxidants that provide protection against free radical-mediated DNA damage (Kalita et al., 2012; Guha et al., 2011). Lectins provide protection against oxidative DNA strand breakage. The capability of lectins to protect pUC18 plasmid against Fenton's reagent was estimated through the DNA nicking assay. *C. reticulatum* lectin exhibited an interesting feature of DNA damage inhibitory activity. The DNA protecting activity on pUC18 plasmid was examined at various dosages viz. 20, 30 and 40 μg. However, the 40 μg dose displayed significant protective activity. This result thus, therefore highlights impending of chickpea lectin in cancer therapy (Gautam et al., 2018b).

Anticancer Potential

Lectins possess anticancer activity and intensive research is under way in this regard to explore the cytotoxic effects of different plant lectins. Perusal of the literature reveals different modes of cytotoxic effects by different plant lectins (Lin et al., 2008; Li et al., 2010). Previous studies revealed that the ethanol/acetone extract of *C. arietinum* inhibited CaCo-2 cells (Giron-Calle et al., 2004), while C-25 was an antifungal lectin that effected reduction in human oral carcinoma cell proliferation (Kumar et al., 2014). The other examples include concanavalin A (Schwarz et al., 1999), *Canavalia ensiformis* lectin (Faheina-Martins et al., 2012), *Typhonium divaricatum* lectin (Luo et al., 2007), phytohaemagglutinin-L (Schwarz et al., 1999), *Griffonia simplicifolia* agglutinin and wheat germ agglutinin (Kiss et al., 1997).

The anti-proliferative activity of chickpea lectin and its effect on cell cycle arrest by initiating the apoptotic cycle within the cancerous cells were studied by Gupta et al., (2018) in MCF-7 cells. A significant decline in the number of breast cancer cells in a dose dependent manner of chickpea lectin for 24 hr. Chickpea lectin induced cell cycle arrest at the S and G2 phase as analyzed by flow cytometry. Thus chickpea lectin induces apoptosis in human breast cancer cells (MCF-7) resulting in cell cycle arrest, lactate dehydrogenase leakage, and ROS generation. Therefore, chickpea lectin possesses anti-cancer property and can serve as a supportive agent in the treatment of breast cancer.

The inhibitory effects of *C. arietinum* lectin on the growth of different human cancer cell lines (Ishikawa, HepG2, MCF-7 and MDA-MB-231) were tested by Gautam et al., (2018a). The sulforhodamine B assay, which is sensitive for large-scale application in evaluating cytotoxicity, was employed. In this assay, fixed cells having basic amino acids bind electrostatically to the sulforhodamine-B dye. The dose-dependent morphological changes were observed at different concentrations of the lectin revealing a remarkable inhibition of cell viability. To test the lectin efficacy on these cell lines, a series of concentrations (20, 40, 60, 80, and 100 μg/mL) was employed. The IC_{50} values on Ishikawa, HepG2, MCF-7 and MDA-MB-231 cell lines were disclosed to be 46.67, 44.20, 53.58 and 37.46 μg/mL respectively, after 24 h of lectin treatment. Similar findings were reported in earlier studies on other plant hemagglutinins (Ye and Ng 2011). Therefore lectins induce apoptosis and are also cooperative in regulating the expression of apoptotic genes in cancerous cells (Lam and Ng 2010; Ramnath et al., 2009).

Lectin C-25 isolated from *C. arietinum* possessed antiproliferative activity in oral cancer cell line (KB cell line) (Kumar et al., 2014). A significant dose-dependent decline in the growth rate of oral cancer cell line was observed after treatment with lectin C-25. This lectin also inhibited p38αMAP kinase by binding with p38α. The p38α induces the production of cytokines interleukin-1β, tumor necrosis factor-α and interleukin -6; and overproduction of cytokines leads to tumor development (Gill et al., 2012). Therefore, a significant decline in overexpression of p38α was reported after treatment with the lectin. Thus it can be concluded that the lectin can modulate many biological processes in cells such as growth, adhesion, apoptosis, and anticancer activity which will further assist in formulating new drugs in modern medicine (Singh et al., 2004).

Anti-HIV Reverse Transcriptase Activity

As every lectin has its own characteristic sugar specificity, lectins from different sources may possess anti-HIV-1 reverse transcriptase activity. This possibility needs to be assessed. Anti-HIV-1 reverse transcriptase activity has been observed in lectins isolated from different plant resources (Ye et al., 2002). The anti-HIV-1 reverse transcriptase activity of *C. reticulatum* lectin is relatively potent compared to other anti-HIV-1 plant isolates. The ability to inhibit human immunodeficiency virus (HIV-1) reverse transcriptase activity was assayed with a non-radioactive ELISA kit by Gautam et al., (2018b). This mechanism of inhibition is likely dependent upon protein-protein interaction, similar to what has been shown for the inhibition of HIV-1 reverse transcriptase by HIV-1 protease (Zheng et al., 2007). Therefore, this seems to be an interesting feature that can prove to be one of the useful tools in AIDS research in the coming days.

Alpha-Amylase Inhibitory Activity

C. arietinum lectin is a known cytoplasmic protein made up of 227 amino acids and having a hemopexin-like repeat domain. The purified lectin demonstrated an inhibitory action against different α-amylases. The inhibitory activity of *C. arietinum* lectin against various α-amylases was affected by temperature, pH, duration of incubation, concentration of substrate and concentration of *C. arietinum* lectin. A proteinaceous α-amylase inhibitor was isolated and purified from mature chickpea seeds by ammonium sulfate precipitation, ion-exchange chromatography and reversed phase liquid chromatography (RPLC) by Wang et al., (2017). The purified α-amylase inhibitor was identified by ultra-performance liquid chromatography-tandem mass spectrometry (UPLC/MS-MS) to be *C. arietinum* lectin in the GenBank (accession No. AGL46982.1).

Chickpea lectin is also involved in lowering the blood glucose level by binding to and disrupting the intestinal mucosal cells resulting in a hindrance to nutrient absorption. The lectins from lentil, pea and chickpea were found to be non-toxic (Gonzalez and Prisecaru, 2005). Previous studies on the lectin activity of various leguminous crops demonstrated that chickpea and pigeon pea have lectin activity, though the activity was reported to be below toxic levels (Singh 1988).

Antioxidant Capacity

The antioxidant activity of chickpea lectin was studied by Gautam et al. (2018a). *In vitro* antioxidant activity of the lectin was assessed by employing reduction of DPPH that contributes in free radical scavenging activity (Llorach et al., 2003). DPPH scavenging activity of the lectin was manifested in a concentration-dependent manner (20–100 µg/mL). Chickpea lectin exhibited an IC_{50} of 0.88 µg/mL, compared to the antioxidant activity of the positive control ascorbic acid which showed an IC_{50} of 0.52 µg/mL. Thus chickpea lectin can serve as a potent antioxidant.

Demonstration of Insecticidal and Pesticidal Activities

A high level of resistance has developed in insects and pests against conventional insecticides due to indiscriminate use of insecticides for control of insects and pests in the crop field (Kranthi et al., 2002). The ability of the cultivated species to protect themselves against predatory insects is relatively low (Sharma 2001). It is crucial to establish the potent noble genes through genetically modifying the plants in order to increase the resistance to control the pests. The body weights of potato beetle larvae were reduced by 27.21% after they had been fed chickpea lectin. Therefore chickpea lectin can be used as a candidate in genetic engineering for breeding for pest resistance (Wang et al., 2017).

Out of the seven plant lectins tested by Shukla et al., (2005) against the cotton bollworm/legume pod borer *Helicoverpa armigera*, snowdrop and chickpea lectins presented a remarkable anti-insect effect concerning insect survival and development. The larvael survival percentage was significantly reduced in larvae reared on artificial diet impregnated with snowdrop (64%) and chickpea (65%) lectins in comparison to untreated control diet (90%). The pupal weight was also adversely affected in larvae reared on a diet containing chickpea lectin (272.6 mg) as compared to that of larvae reared on an untreated control diet (335.4 mg). Among the tested lectins, the pupation rate was lower in larvae reared on diets containing chickpea (50%), snowdrop (60%) and soyabean lectin (70%). The total adult emergence as a percentage of released larvae was observed to be the lowest in larvae reared on diet comprising snowdrop lectin (20%), followed by other diets containing chickpea lectin (30%), peanut lectin (38%), jacalin (48%), concanavalin A (58%) and soya bean (65%) lectins, compared to 90% adult emergence in untreated larvae. In the diet impregnation assay, the larval survival rate was significantly lower in larvae reared on diets impregnated with snowdrop lectin, chickpea lectin, and jacalin. There was no such adverse effect of the plant lectins on pupal weight, except in larvae reared on a diet with chickpea lectin. Snowdrop lectin, chickpea lectin and jacalin produced adverse effects on the survival and development of *H. armigera*. However, plant lectins can be used in conjunction with *Bt* genes to elevate the levels, and diversify the basis of resistance to *H. armigera* in transgenic plants.

In Silico Characterstics

Lectins from legumes are useful as a model to understand the folding pattern of oligomeric proteins (Dev 2006). The interesting unfolding behavior of many of the legume lectins has been characterized. The N-terminal sequence of the chickpea lectin C-25 bears some similarity to the formerly isolated lectins from other plants sources with different molecular weights. BLAST search of database revealed 100% sequence homology of lectins from *Pisum sativum, Lathyrus sativus* and *C. arietinum* (Kumar et al., 2014). The nucleotide sequence of the lectin was translated into the protein sequence to study the secondary structure of recombinant lectin (Wakankar et al., 2013b). The prediction of secondary structure of recombinant *C. reticulatum* lectin (CAL) by PDBSum and VADAR servers significantly showed a relatively higher percentage of beta sheets (about 40%). However, the BLAST search of database of the lectin sequence showed 85% of homology with the sequence of LS-24 albumin from *Lathyrus sativus* (PDB ID:3LP9). This homology model of recombinant *C. reticulatum* lectin was made up using the MODELER 9v10 suite representing the two monomers with only Trp in each monomer. Trp seems to appear towards the outer surface in both monomers. The homology model of recombinant *C. reticulatum* lectin containing the Trp residue was imaged by PyMOL and the residue within 4̊A of the Trp (W220) residue were identified. The Trp residue was surrounded by hydrophobic amino acids (L217, L223, T216, Y32, Y188, AND Y219) forming hydrogen bonds with Y32 and T216. Two Asp residues (D39 and D218) were found around W220 and positively charged residues were absent. Thus, due to the presence of hydrophobic amino acids in its microenvironment, Trp appears to be on the outer surface with an emission maximum at 342 nm. Ramachandran plot prediction of the above model from PROCHECK presented 86.4% residues in the favored regions and 0% in the disallowed region. Quality factor was checked by ERRAT plot to the modeled structure and found to be 89.391. Pro-SAII analyzed Z-score (-5.17) for the model was comparable to that of the template 3LP9 (-5.86). Thus the predicted results proposed that the expected model is a reliable 3D structure of recombinant *C. reticulatum* lectin (Figure 6).

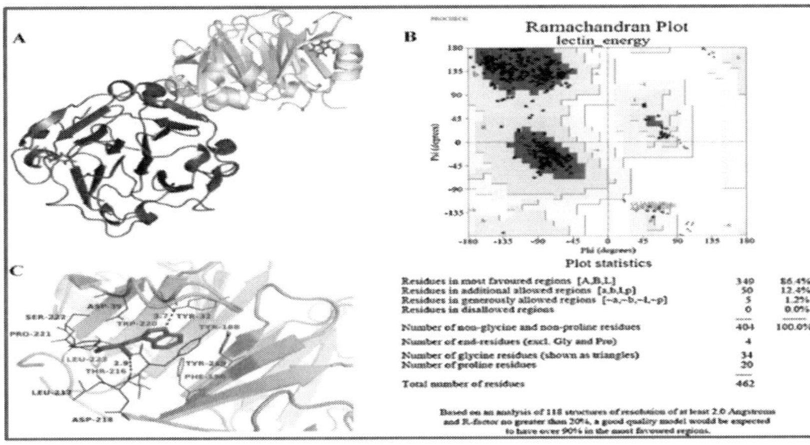

Figure 6. Homology model of recombinant *C. arietinum* lectin (rCAL). (A) Cartoon representation of rCAL homology model constructed by the MODELER 9v 10 software using 3LP9 (PA2 albumin from *Pisum sativum*) as the template. The single Trp in each monomer is shown in red as stick representation. (B) Ramachandran plot of rCAL model. (C) Cartoon representation of the Trp microenvironment in the monomer of rCAL homology model.

Conclusion

Chickpea lectin is a protein with distinctive properties *vis-à-vis* mitogenic, antifungal/bacterial and HIV-1 reverse transcriptase inhibiting activities. Further research is warranted to elucidate the mechanism of action of such inhibiting effects of chickpea lectin. It is possible for chickpea lectin to be developed into a pharmaceutical for AIDS therapy.

References

Adams, David J. 2004. "Fungal Cell Wall Chitinases and Glucanases." *Microbiology* 150:2029-35. doi: 10.1099/mic.0.26980-0.

Agrawal, B. B. L., and Irwin J. Goldstein. 1965. "Specific Binding of Concanvalin A to Cross-linked Dextran Gels." *Biochem J* 96:23C-35C. doi: 10.1042/bj0960023C.

Ayyagari, Radha, Bagepalli S. Narasinga Rao, and Dwijendra Nath Roy. 1989. "Lectins, Trypsin Inhibitors, BOAA and Tannins in Legumes and Cereals and the Effect of Processing." *Food Chemistry* 34:229-38. doi: 10.1016/0308-8146(89)90143-X.

Boleti, Ana Paula de A, Maria das Graças M. Freire, Mirela B. Coelho, Walciane da Silva, P. A. Baldasso, Valdirene M Gomes, Sérgio Marangoni, José Camillo Novello, and

Ligia Macedo 2007. "Insecticidal and Antifungal Activity of a Protein from *Pouteria torta* Seeds with Lectin-like Properties". *J Agric Food Chem* 55:2653-58. doi: 10.1021/jf0636317.

Bourne, Yves, Ahidjo Ayouba, Pierre Rouge, and Christian Cambillau. 1994. "Interaction of a Legume Lectin with Two Components of the Bacterial Cell Wall." *J Biol Chem* 269:9429-35.

Bruch, Thomas, Heiner Graalfs, Lothar R. Jacob, and Christian Frech. 2009. "Influence of Surface Modification on Protein Retention in Ion-Exchange Chromatography Evaluation using Different Retention Models." *Journal of Chromatography A* 1216(6):919-26. doi: 10.1016/j.chroma.2008.12.008.

Charungchitrak, Sarinya, Amorn Petsom, Polkit Sangvanich, and Aphichart Karnchanatat. 2011. "Antifungal and Antibacterial Activities of Lectin from the Seeds of *Archidendron jiringa* Nielsen." *Food Chemistry* 126: 1025-32. doi: 10.1016/j.foodchem.2010.11.114.

Chen, Ruizhan, Fanlei Meng, Zhiqiang Liu, Ruiping Chen, and Min Zhang. 2010. "Antitumor Activities of Different Fractions of Polysaccharide Purified from *Ornithogalum caudatum* Ait." *Carbohydrate Polymers* 80:845-51. doi: 10.1016/j.carbpol.2009.12.042.

Chrispeels, Maarten J. 1984. Biosynthesis, processing and transport of storage proteins and lectins in cotyledons of developing legume seeds". *Phil Trans R Soc Lond B* 304:309-322. doi: 10.1098/rstb.1984.0026.

Contreras S., and Marfa A. Tagle. 1974. "Toxic factors in Chilean legumes." *Arch Latinoa mer Nutrition* 24:191-99.

Gonzalez, De Mejía Elvira, and Valentin Ion Prisecaru. 2005. "Lectins as Bioactive Plant Proteins: A Potential in Cancer Treatment". *Crit Rev Food Sci Nutr* 45(6):425-45. doi: 10.1080/10408390591034445.

Dev, Sagarika, Nirmala Devi K., Sharmistha Sinha, and Avadesha Surolia. 2006. "Thermodynamic Analysis of Three State Denaturation of Peanut Agglutinin." *IUBMB Life* 58:549-55. doi: 10.1080/ 15216540600902228.

Donnelly, Edwin H., and Irvin J. Goldstein. 1970. "Glutaraldehyde-insolubilized Concanavalin A: An Adsorbent for the Specific Isolation of Polysaccharides and Glycoproteins." *Biochem J* 118(4):679-80. doi:10.1042/bj1180679.

Durham, Malaika, and Fred E. Regnier. 2006. "Targeted Glycoproteomics: Serial Lectin Affinity Chromatography in the Selection of O-glycosylation Sites on Proteins from the Human Blood Proteome." *J Chromatogr A* 1132(1-2): 165-73. doi:10.1016/j.chroma.2006.07.070.

Dziarski, Roman, Mark Rasenick, and Dipika Gupta. 2000. "Bacterial Peptidoglycan Binds to Tubulin." *Biochim Biophys Acta* 1524:17-26. doi: 10.1016/S0304-4165(00)00137-9.

Faheina-Martins, Glaucia, Aletheia da Silveira, Bruno C. Cavalcanti, Marcio V. Ramos, Manoel O Moraes, Claudia Pessoa, and, Demetrius A M Araújo. 2012. "Antiproliferative Effects of Lectins from *Canavalia ensiformis* and *Canavalia brasiliensis* in Human Leukemia Cell Lines." *Toxicology In Vitro* 1161-69. doi: 10.1016/j.tiv.2012.06.017.

Gautam, Ajay K., Neha Gupta, Dakshita T. Narvekar, Rajni Bhadkariya, and, Sameer S Bhagyawant. 2018a. "Characterization of Chickpea (*Cicer arietinum* L.) Lectin for Biological Activity." *Physiol Mol Biol Plants.* 24(3):389-97. doi: 10.1007/s12298-018-0508-5.

Gautam, Ajay K., Nidhi Srivastava, D. P. Nagar, and Sameer S. Bhagyawant. 2018b. "Biochemical and Functional Properties of a Lectin Purified from the Seeds of *Cicer arietinum* L." *3 Biotech* 8:272. doi: 10.1007/s13205-018-1272-5.

Gill, Kamal, Bidhu K. Mohanti, Saquib. Ashraf, Abhay K. Singh, and, Sharmistha Dey. 2012. "Quantification of p38αMAP kinase: A Prognostic Marker in HNSCC with Respect to Radiation Therapy," *Clinica Chimica Acta* 413(12):219-25. doi: 10.1016/j.cca.2011.09.031.

Giron-Calle, Julio, Javier Vioque, Maria del Mar Yust, Justo Pedroche, and, Sharmistha Dey. 2004. "Effect of Chickpea Aqueous Extracts, Organic Extracts, and Protein Concentrates on Cell Proliferation." *J Med Food* 7(2):122-29. doi: 10.1089/1096620041224175.

Goldstein, Irvin J., and Colleen E. Hayes. 1978. "The lectins: carbohydrate binding proteins of plants and animals." *Adv Carbohydr Chem Biochem* 35:127-340. doi: 10.1016/S0065-2318(08)60220-6.

Goldstein, Irvin J., R.C. Colin Hughes, Michel Monsigny, Toshiaki Osawa, and, Nathan Sharon. 1980. "What Should be Called A Lectin?" *Nature* 285:66. doi: 10.1038/285066b0.

Gomes, Fabiana Oliveira Dos S., Thamara F. Proco′pio, Thiago H. Napolea˜o, Luana Cassandra Breitenbach Barosso Coelho, and, P.M.G. Paiva 2012. "Antimicrobial Lectin from *Schinus terebinthifolius* Leaf." *J Appl Microbiol* 114:672-679. doi: 10.1111/jam.12086.

Grant George. 1991. "In: Toxic Substances in Crop Plants." D'Mello, J. P. F., Duffus, C. M. and Duffus, J. H. (eds). Royal Society of Chemistry, Cambridge 49–67.

Grant George, Linda J. More, Norma H. McKenzie, James C. Stewart, and, Arpad Pusztai. 1983. "A Survey of the Nutritional and Haemagglutination Properties of Legume Seeds Generally Available in the UK." *British Journal of Nutrition* 50:207–14. doi: 10.1079/BJN19830090.

Guha, Gunjun, Venkatadri Rajkumar, R. Ashok Kumar, and Lazar Mathew. 2011. "Antioxidant Activity of *Lawsonia inermis* Extracts Inhibits Chromium(VI)-Induced Cellular and Dna toxicity." *Evidence-Based Complementary and Alternative Medicine* Volume 2011, Article ID 576456, 9 pages. doi: 10.1093/ecam/nep205.

Gupta, Neha, Prakash S. Bisen, and Sameer S. Bhagyawant. 2018. "Chickpea Lectin Inhibits Human Breast Cancer Cell Proliferation and Induces Apoptosis through Cell Cycle Arrest." *Protein & Peptide Letters* 25:1-8. doi: 10.2174/0929866525666180406142900.

Haddad, Paul R., and Peter E. Jackson. 1990. *Ion chromatography principles and applications*. Amsterdam: Elsevier Science.

Hajós, Gyongyi, Eva Gelencsér, Arpad Pusztai, and Susan Bardocz. 1996. "*Measurements of the antinutrient contents of legume samples*." Office for Official Publications of the European Communities, Luxembourg, 1:130–134.

Ishihara Takashi, Toshihiko Kadoya, and Shuichi Yamamoto. 2007. "Application of a Chromatography Model with Linear Gradient Elution Experimental Data to the Rapid Scale-up in Ion-exchange Process Chromatography of Proteins." *Journal of Chromatography A* 1162:34-40. doi: 10.1016/j.chroma.2007.03.016.

Jebali, Jed, Emna Fakhfekh, Maram Morgen, Najet Srairi-Abid, Hafedh Majdoub, Ali Faouzi Gargouri, Elayeb Mohamed, Joes Luis, Naziha Marrakchi, and Sameh Sarray. 2014. "Lebecin, A New C-type Lectin like Protein from *Macrovipera lebetina* Venom with Antitumor Activity Against the Breast Cancer Cell Line MDA-MB231." *Toxicon* 86:16-27. doi: 10.1016/j.toxicon.2014.04.010.

Kalita, Sanjeeb, Gaurav Kumar, Karthik Loganathan, and Bhaskar Rao. 2012. "*In vitro* Antioxidant and DNA Damage Inhibition Activity of Aqueous Extract of *Lantana camara* L. (*Verbenaceae*) leaves." *Asian Pacific Journal of Tropical Biomedicine* 5(3):S1675-79. doi: 10.1016/S2221-1691(12)60476-6.

Kastner, Michael. 2000. *Protein Liquid Chromatography. Journal of Chromatography Library*. Amsterdam, Elsevier Science, 933.

Katre, Uma V., Sushama M. Gaikwad, Sameer S. Bhagyawant, Uday D. Deshpande, M. I. Khan, and C. G. Suresh. 2005. "Crystallization and preliminary X-ray characterization of a lectin from *Cicer arietinum* (chickpea)." *Acta Crystallographica Section F* 61:141-143. doi: 10.1107/S1744309104032166.

Khan, Saleem, Farhad Ahmad, Farhad Ali, Hakim Khan, Peerayub Khan, and Zahoor Ahmad Swati. 2011. "Phyto-agglutinin, Total Proteins and Amino Assimilating Enzymatic Activity of Indigeneous Chickpea (*Cicer arietinum* L.) Cultivars." African Journal of Biotechnology 10(57):12276-80. doi: 10.5897/AJB11.1308.

Kiss, Robert, Isabelle Camby, C. Duckworth, R. De Decker, I. Salmon, J. L. Pasteel, Andre Danguy, and Paul Yeaton. 1997. "*In vitro* Influence of *Phaseolus vulgaris*, *Griffonia simpliciflia*, Concanavalin A, Wheat Germ, and Peanut Agglutinin on HCT-15, LoVo, and SW837 Human Colorectal Cancer Cell Growth." *Gut* 40:253-261. doi: 10.1136/gut.40.2.253.

Kolberg, Jan, Terje E. Michaelsen, and Knut Sletten. 1983. "Properties of a Lectin Purified from the Seeds of *Cicer arietinum*." *Hoppe Seylers Z Physiol Chem* 364:655-64. doi: 10.1515/bchm2.1983.364.1.655.

Kranthi Keshav, Deepak R. Jadhav, S. Kranthi, R. R. Wanjari, Seemo Ali, and D. A. Russell. 2002. "Insecticide resistance in five major insect pests of cotton in India." *Crop Protection* 21(6):449-60. doi: 10.1016/S0261-2194(01)00131-4.

Kumar Suresh, Vaishali Kapoor, Kamal Gill, Kusum Singh, Immaculata Xess, Satya N. Das, and Shamistha Dey. 2014. "Antifungal and Antiproliferative Protein from *Cicer arietinum*: A Bioactive Compound Against Emerging Pathogens." *BioMed Research International* Volume 2014, Article ID 387203, 9 pages. doi: 10.1155/2014/387203.

Kvennefors, Eva Charlotte E., William Leggat, Ovi Hoegh-Guldberg, Bernard M. Degnan, and Andrew C. Barnes. 2008. "An Ancient and Variable Mannose-Binding Lectin from the coral *Acropora millepora* Binds Both Pathogens and Symbionts." *Developmental and Comparative Immunology* 32(12):1582-1592. doi: 10.1016/j.dci.2008.05.010.

Lafont, Virginie, Jacques Dornand, Arnaud d'Angeac, Serge Monier, Andres Alcover, and Jean Favero. 1996. "Jacalin, a Lectin that Inhibits in vitro HIV-1 Infection, Induces Intracellular Calcium Increase via CD4 in Cells Lacking the CD3/TcR Complex." *Journal of Leukocyte Biology* 56(4):521-24. doi: 10.1002/jlb.56.4.521.

Lam Sze K., and Tzi B. Ng. 2010. "Isolation and Characterization of a French Bean Hemagglutinin with Antitumor, Antifungal and Anti-HIV-1 Reverse Transcriptase Activities and an Exceptionally High Yield." *Phytomed* 17:457-462. doi: 10.1016/j.phymed.2009.07.017.

Levison, Peter R. 2003. "Large-scale ion exchange column chromatography of proteins comparison of different formats". *Journal of Chromatography B* 790:17-33. doi: 10.1016/S1570-0232(03)000
87-4.

Li Yanrui, Guo-Qing Zhang, Tzi B. Ng, and Hexiang Wang. 2010. "A Novel Lectin with Antiproliferative and HIV-1 Reverse Transcriptase Inhibitory Activities from Dried Fruiting Bodies of the Monkey Head Mushroom *Hericium erinaceum*." *Journal of Biomedicine and Biotechnology* Volume 2010, Article ID 716515, 9 pages. doi: 10.1155/2010/716515.

Liener, Irvin E. 1979. "Significance for humans of biologically active factors in soybeans and other food legumes." *J Am Oil Chemists Society* 56(3):121-129. doi: 10.1007/BF02671433.

Liener, Irvin E. 1976. "Legume toxins in relation to protein digestibility-a review." *J Food Sci* 41:1076-1081. doi: 10.1111/j.1365-2621.1976.tb14391.x.

Li, Yanrui, Guoqing Zhang, Tzi B. Ng, and Hexiang Wang. 2010. "A Novel Lectin with Antiproliferative and HIV-1 Reverse Transcriptase Inhibitory Activities from Dried Fruiting Bodies of the Monkey Head Mushroom *Hericium Erinaceum*." *Journal of Biomedicine and Biotechnology* Volume 2010, Article ID 716515, 9 pages. doi: 10.1155/2010/716515.

Liener, Irvin E. 1976. "Legume Toxins in Relation to Protein Digestibility-A Review." *J Food Sci* 41:1076-1081. doi: 10.1111/j.1365-2621.1976.tb14391.x.

Liener, Irvin. 1979. "Significance for Humans of Biologically Active Factors in Soybeans and Other Food Legumes." *J Am Oil Chemists Society* 56(3):121-129. doi: 10.1007/BF02671433.

Lin, Peng, Xiujuan Ye, and Tzi B. Ng. 2008. "Purification of Melibiose-Binding Lectins from Two Cultivars of Chinese Black Soybeans." *Acta Biochimica et Biophysica Sinica* 40(12):1029-1038. doi: 10.1111/j.1745-7270.2008.00488.x.

Lis, Halina, and Nathon Sharon. 1981. *Lectins in Higher Plants, 371-447.* New York: Academic Press.

Llorach, Rafael, Juan C. Espian, Fransisco A. Tomaa N. S-Barberaa, and Fedrico Ferreres. 2003. "Valorization Of Cauliflower (*Brassica oleracea* L. var. Botrytis) By-Products as a Source of Antioxidant Phenolics." *J Agric Food Chem* 51:2181-2187. doi: 10.1021/jf021056a.

Luo, Yongting, Xiaochao Xu, Jiwei Liu, Jian Li, Yisheng Sun, Zhen Liu, Jinzhi Liu, Els Van Damme, Jan Balzarini, and Jinku Bao. 2007. "A Novel Mannose-binding Tuber Lectin from *Typhonium divaricatum* (l.) Decne (Family Araceae) with Antiviral Activity Against HSV-II and Anti-proliferative Effect on Human Cancer Cell Lines." *J Biochem Mol Biol* 40(3):358-367. doi: 10.5483/BMBRep.2007.40.3.358.

Matsumoto, Ryo, Tomoko F. Shibata, Hisanori Kohtsuka, Mamoru Sekifuji, Natsuko Sugii, Hiroaki Nakajima, Noriaki Kojima, Yuki Fujii, Sarkar M. A. Kawsar, Hidetaro Yasumitsu, Jiharu Hamako, Taei Matsui, and Yasuhiro Ozeki. 2011. "Glycomics of A Novel Type-2-Acetyllactosamine-Specific Lectin Purified From The Feather Star, *Oxycomanthus japonicus* (Pelmatozoa: Crinoidea)." *Comparative Biochemistry and Physiology B* 158(4):266-273. doi: 10.1016/j.cbpb.2010.12.004.

Muzquiz, Mercedes, Carmen Burbano, Carmen Cuadrado, and M. Martin. 2001. "Analytical Methods for Determination of Compounds with No Nutritive Value." In *Handbook Of Common Bean Related Laboratory Methods,* edited by H. J. Jacobson, M. Muzquiz, and I Iassa, 11-26. Spain: Galicia.

Nakamura-Tsuruta, Sachiko, Noboru Uchiyama, Willy J. Peumans, Els J. M. Van Damme, Kiichiro Totani, Yukishige Ito and Jun Hirabayashi. 2008. "Analysis of the Sugar-Binding Specificity of Mannose-Binding-Type Jacalin-Related Lectins by Frontal Affinity Chromatography - An Approach To Functional Classification." *The FEBS Journal* 275(6):1227-1239. doi: 10.1111/j.1742-4658.2008.06282.x.

Nasi Antonella, Gianluca Picariello, and Pasquale Ferranti. 2009. "Proteomic Approaches to Study Structure, Functions and Toxicity of Legume Seeds Lectins. Perspectives for the Assessment of Food Quality and Safety." *J of Proteomics* 72:527-538. doi: 10.1016/j.jprot. 2009.02.001.

Nergard, Cecilie S., Drissa Diallo, Kari Inngjerdingen, Terje E. Michaelsen, Tsukasa Matsumoto, Hiroaki Kiyohara, Haruki Yamada, and Berit S. Paulsen. 2005. "Medicinal Use of *Cochlospermum tinctorium* In Mali: Anti-Ulcer, Radical Scavenging and Immunomodulating Activities of Polymers in the Aqueous

Extract of the Roots." *Journal of Ethnopharmacology* 69:255-269. doi: 10.1016/j.jep.2004.09.018.

Okada, Tetsuo. 1998. "Nonaqueous Ion-Exchange Chromatography and Electrophoresis Approaches to Nonaqueous Solution Chemistry and Design of Novel Separation." *Journal of Chromatography A* 804:17-28. doi: 10.1016/S0021-9673(97)01234-X.

Petnual, Pariyaphon, Polkit Sangvanich, and Aphichart Karnchanatat. 2010. "A Lectin from the Rhizomes of Turmeric (*Curcuma longa* L.) and its Antifungal, Antibacterial, and Alpha-Glucosidase Inhibitory Activities." *Food Sci Biotechnol* 19:907-916. doi: 10.1007/s10068-010-0128-5.

Peumans, Willy J., and Els J. Van Damme. 1996. "Prevalence, Biological Activity and Genetic Manipulation of Lectins in Foods." *Trends in Food Science & Technology* 7:132-138. doi: 10.1016/0924-2244(96)10015-7.

Price, Martin L., Ann E. Hagerman, and Larry G. Butler. 1980. "Tannin Content of Cowpeas, Chickpeas, Pigeon Peas, and Mung Beans." *J Agric Food Chem* 28(2):459-461. doi: 10.1021/jf60228a047.

Qureshi, Insaf A., Prasanta Dash, Prem S. Srivastava, and Kirpar R. Koundal. 2006. "Purification and Characterization of an N-Acetyl-D-Galactosamine-Specific Lectin from Seeds of Chickpea (*Cicer arietinum* L.)." *Phytochem Anal* 17(5):350-356. doi: 10.1080/10425170601060608.

Ramnath, Viswanathan, Pobby S. Rekha, Girija Kuttan, and Ramadasan Kuttan. 2009. "Regulation of Caspase-3 and Bcl-2 Expression in Dalton's Lymphoma Ascites Cells by Abrin." *Evidence-Based Complementary and Alternative Medicine* 6(2):233-238. doi: 10.1093/ecam/nem099.

Ryan, Clarens A. 1990. "Protease Inhibitors in Plants: Genes for Improving Defenses Against Insects and Pathogens." *Annu Rev Phytopathol* 28:425-449. doi: 10.1146/annurev.py.28.090190.002233.

Schwarz, Roderich E., Donald C. Wojciechowicz, Antonio I. Picon, Margaret A. Schwarz, and Philip B. Paty. 1999. "Wheat Germ Agglutinin-Mediated Toxicity in Pancreatic Cancer Cells." *J Cancer* 80:1754-1762. doi: 10.1038/sj.bjc.6690593.

Selitrennikoff, Claude P. 2001. "Antifungal Proteins." *Appl Environ Microbiol* 67(7):2883-2894. doi: 10.1128/AEM.67.7.2883-2894.2001.

Sharma, Hari C. 2001. *Cotton Bollworm/Legume Pod Borer, Helicoverpa armigera (Hubner) (Noctuidae: Lepidoptera): Biology and Management*. Crop protection compendium: Commonwealth Agricultural Bureau International Walingford, UK. pp 72.

Sharon, Nathan., and Halina Lis. 2002. "How Proteins Bind Carbohydrates: Lessons from Legume Lectins." *Agric Food Chem* 50:6586-6591. doi: 10.1021/jf020190s.

Shukla, Sonali, Richa Arora, and Hari C. Sharma. 2005. "Biological Activity of Soybean Trypsin Inhibitor and Plant Lectins against Cotton Bollworm/Legume Pod Borer,

Helicoverpa armigera." *Plant Biotechnology* 22(1):1-6. doi: 10.5511/plantbiotechnology.22.1.

Singh, Jagmohan, Jatinder Singh and S. S. Kamboj. 2004. "A Novel Mitogenic and Antiproliferative Lectin from a Wild Cobra Lily, *Arisaema flavum*." *Biochemical and Biophysical Research Communications* 318:1057-1065. doi: 10.1016/j.bbrc.2004.04.135.

Singh, Udayan. 1988. "Antinutritional Factors of Chickpea and Pigeonpea and Their Removal by Processing." *Plant Foods Hum Nutr* 38(3):251-261. doi: 10.1007/BF01092864.

Singleton, Vernon L. 1981. "Naturally Occurring Food Toxicants: Phenolic Substances of Plant Origin Common in Foods." *Adv Food Res* 27:149-242. doi: 10.1016/S0065-2628(08)60299-2.

Sitohy, Mahmoud, Mahmoud Doheim, and Haitham Badr. 2007. "Isolation and Characterization of a Lectin with Antifungal Activity from Egyptian *Pisum Sativum* Seeds." *Food Chem* 104(3):971-979. doi: 10.1016/j.foodchem.2007.01.026.

Stanton, Peter. 2004. *HPLC of Peptides and Proteins*: *Methods in Molecular Biology*. New Jersey: Humana Press.

Trindade, Melissa B., Jose L. Lopes, Andrea Soares-Costa, Ana C. Monteiro-Moreira, Renato A. Moreira, Moreira L. Oliva, and Leila M. Beltramini. 2006. "Structural Characterization of Novel Chitin-Binding Lectins from the Genus Artocarpus and their Antifungal Activity." *Biochim Biophys Acta* 1764:146-152. doi: 10.1016/j.bbapap.2005.09.011.

Voet, Donald, and Judith G. Voet. 1995. *Biochemistry* (Second Edition). New York: John Wiley & Sons.

Wakankar, Madhurima S., Krunal A. Patel, Musti V. Krishnasastry, and Sushama M. Gaikwad. 2013a. "Solution and *In Silico* Ligand Binding Studies of *Cicer arietinum* Lectin." *Biochem Physiol S2* doi:10.4172/2168-9652.S2-002.

Wakankar, Madhurima S., Musti V. Krishnasastry, Tulika M. Jaokar, Krunal A. Patel, Sushama M. Gaikwad. 2013b. "Solution and *In Silico* Studies on the Recombinant Lectin from *Cicer arietinum* Seeds." *International Journal of Biological Macromolecules* 56:149-155. doi: 10.1016/j.ijbiomac.2013.02.015.

Wang, Shao Y., Jin H. Wu, Tzi B. Ng, Xiu Y. Ye, and Ping F. Rao. 2004. "A Nonspecifc Lipid Transfer Protein with Antifungal and Antibacterial Activities from the Mung Bean." *Peptides* 25(8):1235-1242. doi: 10.1016/j.peptides.2004.06.004.

Wang, Zhankui, Ming Chen, Yajing Zhu, Peipei Qian, Yali Zhou, Jiaping Wei, Yingzi Shen, Abudoukeyumu Mijiti, Aixing Gu, Ze Wang, Hua Zhang, and Hao Ma. 2017. "Isolation, Identification and Characterization of a New Type of Lectin with A-Amylase Inhibitory Activity in Chickpea (*Cicer arietinum* L.)." *Protein & Peptide Letters* 24(11):1008-1020. doi: 10.2174/0929866524666170711120501.

Watanabe, Taisuke, Ichiro Matsuo, Jun-ichi Maruyama, Katsuhiko Kitamoto, and Yukishige Ito. 2007. "Identification and Characterization of an Intracellular Lectin, Calnexin, from *Aspergillus oryzae* Using N-Glycan-Conjugated Beads." *Bioscience, Biotechnology, and Biochemistry*. 71(11):2688-2696. doi: 10.1271/bbb.70289.

Wong, Jack H., David C. W. Ip, Tzi B. Ng, Yat S. Chan, Francia Fang, and Wen L. Pan. 2012. "A Defensin-Like Peptide from *Phaseolus Vulgaris* cv. King Pole Bean." *Food Chemistry* 135:408-414. doi: 10.1016/j.foodchem.2012.04.119.

Xia, Lixin, and Tzi B. Ng. 2005. "An Antifungal Protein from Flageolet Beans." *Peptides* 26(12):2397-2403. doi: 10.1016/j.peptides.2005.06.003.

Ye, Xiu J., and Tzi B. Ng. 2011. "Antitumor And HIV-1 Reverse Transcriptase Inhibitory Activities of a Hemagglutinin and a Protease Inhibitor from Mini-Black Soybean." *Evidence-Based Complementary and Alternative Medicine* Article ID 851396, 12 pages. doi: 10.1155/2011/851396.

Ye, Xiu Y., and Tzi B. Ng. 2002. "Isolation of a New Cyclophilin-Like Protein from Chickpeas with Mitogenic, Antifungal and Anti-HIV-1 Reverse Transcriptase Activities." *Journal of Life Sciences* 70(10):1129-1138.

Ye, Xiu Y., Tzi B. Ng, Paul W. K. Tsang, and Jun Wang. 2001. "Isolation of a Homodimeric Lectin with Antifungal and Antiviral Activities from Red Kidney Bean (*Phaseolus vulgaris*) Seeds." *Journal of Protein Chemistry* 20(5):367-375. doi: 10.1023/A:1012276619686.

Zagorodni, Andrei A. 2007. *Ion Exchange Materials, Properties and Applications*. Amsterdam: Elsevier Science.

Zhang, Guoqing, Jia Sun, Hexiang Wang, and Tzi B. Ng. 2010. "First Isolation and Characterization of a Novel Lectin with Potent Antitumor Activity from a Russula Mushroom." *Phytomedicine* 17(10):775-781. doi: 10.1016/j.phymed.2010.02.001.

Zheng, Suyue, Cuixin Li, Tzi B. Ng, and He X. Wang. 2007. "A Lectin with Mitogenic Activity from the Edible Wild Mushroom *Boletus edulis*." *Process Biochemistry* 42:1620-1624. doi: 10.1016/ j.procbio.2007.09.004.

In: Hemagglutinins
Editors: T. Bun Ng, J. Wong, R. Tse et al.
ISBN: 978-1-53615-708-6
© 2019 Nova Science Publishers, Inc.

Chapter 7

HEMAGGLUTININS WITH ANTICANCER, IMMUNOMODULATORY AND ANTIMICROBIAL ACTIVITIES FROM MEDICINAL HERBS

Jack Ho Wong[1,2], Tzi Bun Ng[3,*], Charlene Cheuk Wing Ng[4], Chit Tam[3], Yau Sang Chan[5], Ryno Naude[6], Ryan Tse[1], Tak Fu Tse[1] and Helen Chan[1]

[1]Vita Green Pharmaceutical (Hong Kong Limited), Genning Partners Company Limited, and Institute of Medical Research, Hong Kong, China
[2]The Third Affiliated Hospital of Shenzhen University, Shenzhen, China
[3]School of Biomedical Sciences, Faculty of Medicine, The Chinese University of Hong Kong, China
[4]GKT School of Medical Education, King's College London, UK
[5]Department of Obstetrics and Gynecology, LKS Faculty of Medicine, University of Hong Kong, China
[6]Department of Biochemistry and Microbiology, Nelson Mandela Metropolitan University, Port Elizabeth, South Africa

[*] Corresponding Author's E-mail: tzibunng.cuhk.edu.hk.

Abstract

Hemagglutinins/lectins are produced by the following medicinal herbs: *Astragalus membranaceus* roots, *Astragalus mongholicus* roots, *Dendrobium findleyanum* pseudobulbs, different tissues of *Dendrobium officinale*, *Dendrobium pendulum* pseudobulbs, *Dioscorea batatas* tubers, *Euphorbia characias* latex, *Euphorbia heterophylla* seeds, *Euphorbia marginata* latex, *Euphorbia milii* latex, *Euphorbia neriifolia* latex, *Euphorbia tirucalli* latex, *Hura crepitans* latex, *Momordica charantia* seeds, *Trichoanthes anguina* seeds, *Trichosanthes cordata* seed, *Trichosanthes dioica* seeds, *Trichosanthes cucumerina* seeds, *Trichosanthes japonica* seeds, *Trichosanthes kirilowii* roots, *Trichosanthes kirilowii* seeds and *Withania somnifera* leaves. They display various biological activities including anticancer, immunomodulatory, antimicrobial, anti-insect and insulin-like activities.

Keywords: anticancer, antimicrobial, anti-insect, hemagglutinins, medicinal plants, immunomodulatory, insulin-like activities

Introduction

Hemagglutinins/lectins have been isolated from a variety of organisms. They display antipathogenic activities including anticancer, antiparasite, antibacterial, antifungal, antiviral, and anti-insect activities. In addition they have immunomodulatory activity. They are produced by edible plants as well as medicinal plants. The purpose of this chapter is to review lectins from medicinal herbs. The herbs selected comprise *Astragalus*, *Dendrobium*, *Euphorbia*, *Hura*, *Momordica*, *Sophora* and *Trichosanthes* species.

Astragalus Membranaceus Lectin

The galactose-binding lectin induced apoptosis in chronic myeloid leukemia K562 cells. Galactose and lactose abrogated the activity (Huang et al., 2012). The 67- kDa acid/alkali stable *Astragalus membranaceus* seed lectin possessed 16.4% sugar, about 30% hydrophobic amino acids, and about70% polar amino acids. It was denatured at temperatures higher than 64°C. It demonstrated anticancer activity toward H22, HepG2 and SGC-7901cancer cells, and displayed marked antibacterial effects, with half maximal inhibitory concentration for *E. coli*, *S. aureus* and *B. dysenteriae* being 0.97, 1.2 and 1.27 μM, respectively (Bai et al., 2018).

Astragalus Mongholicus Root Lectin

The 66.4- kDa homodimeric *Astragalus mongholicus* root lectin, with the N-terminal amino acid sequence ESGINLQGDATLANN, was a glycoprotein with a 19.6%. neutral carbohydrate content. It agglutinated both rabbit and human erythrocytes. The hemagglutinating activity was inhibited by D-galactose and lactose. A pH of 4.5 to 7.5 was required for optimal activity of the lectin, and the lectin was stable at temperatures till 65 degrees C. The lectin inhibited the fungi *Botrytis cincerea*, *Colletorichum* sp., *Drechslera turcia* and *Fusarium Oxysporum*, but was devoid of activity against *Mycosphaerella arachidicola* and *Rhizoctonia solani* (Yan et al., 2009a). *Astragalus mongholicus* lectin inhibited the growth of human osteoblast-like MG63 cells, human leukemia K562 cells and human cervical carcinoma HeLa cells by 48%, 84% and 92%, respectively. The treated cells showed apoptotic features including appearance of membrane-enclosed apoptotic bodies, nuclear fragmentation. Apoptosis in treated HeLa cells was demonstrated by Annexin V/FITC and propidium iodide staining, and cell cycle arrest occurred at S phase of the cell cycle (Yan et al., 2009b).

Dendrobium Findleyanum Pseudobulb Agglutinin

The 14.5-kDa mannose-binding *D. findleyanum* lectin expressed inhibitory activity against the fungi *Alternaria alternata* and *Collectotrichum* sp. The lectin precursor revealed 94% amino acid sequence homology with a precursor from *D. officinale* lectin. The site of processing occurs between residues 24 and 25 of the lectin precursor (Sattayasai et al., 2009).

A 56.2-kDa mannose-binding lectin, comprising four 14.5-kDa polypeptide subunits and three peptide fragments identical to parts of a *D. officinale* lectin precursor, was purified from *D. findleyanum* pseudobulbs. The lectin exhibited three conformational forms with an isoelectric point of 5.18, 4.87 and 4.72, respectively. The hemagglutinating and antifungal activities of the lectin were stable at elevated temperatures (Sudmoon et al., 2008).

Dendrobium Officinale Agglutinin 2

Its precursor has 170 amino acids. The lectin has three mannose-binding sites typical of mannose-binding lectin. The gene encoding the lectin, doa2, was a constitutively expressed gene: its mRNA expression occurred in leaves, stems and roots although higher in stems than in leaves. The lectin displayed inhibitory activity toward the fungus *Gibberella zeae* (Chen et al., 2005).

Dendrobium Pendulum Pseudobulb Agglutinin

The lectin demonstrates structural resemblance to snowdrop lectin. There are two native trimeric forms of the lectin, which are 44.2 kDa and 45.3 kDa in molecular mass and composed of three 14.0- kDa subunits. The lectin displayed hemagglutinating activity toward chicken red cells after the cells had been treated with trypsin. The lectin was induced under stressful conditions such as presence of abscisic acid, acetylene gas, hydrogen peroxide, and ultraviolet light (Siripipatthana et al., 2015).

Dioscorea Batatas Lectin

There are two highly homologous isolectins, DB1 (Cys86) and DB1(Leu86), each containing 108 amino acids. DB1 exhibited striking sequence resemblance to snowdrop (*Galanthus nivalis*) bulb lectin. The sugar-binding sites of snowdrop lectin were highly conserved in DB1. DB1 manifested high binding affinity to oligosaccharides with two mannose-alpha-1,3-D-mannose units.The rate of adult *Helicoverpa armigera* emergence from pupae was reduced by 67% following dietary inclusion of DB1 at 0.01% (w/w). The strong binding of DB1 to larval brush border and peritrophic membrane was revealed by immunostaining (Ohizumi et al., 2009).

Euphorbia Characias (Mediterranean Spurge) Latex Lectin

The 80-kDa *Euphorbia characias* lectin is a homodimeric galactose-binding glycoprotein with 11% carbohydrate. The lectin was devoid of mitogenic activity toward human B cells and T cells as well as translation inhibitory activity toward rabbit reticulocyte lysate (Barbieri et al., 1983).

Euphorbia Heterophylla Seed Lectin

An N-acetylgalactosamine-specific homodimeric 32-kDa lectin, which exhibits structural dissimilarity compared with other Euphorbiaceae lectins, was isolated from *Euphorbia heterophylla* seeds (Nsimba-Lubaki et al., 1983).

Euphorbia Marginata Latex Lectin

The galactose-specific lectin is a homodimeric 60-kDa glycoprotein with potent activity toward human T cells. It enhanced tumor necrosis factor-alpha and interleukin-1 beta secretion from cultured mononuclear cells (Stirpe et al., 1993).

Euphorbia Milii Latex Lectin (Euphorbin)

Neutrophil migration upregulated by euphorbin was reduced by its inhibitory sugar D-galactose. The transitory neutrophil aggregation elicited by euphorbin (0.1 mg/ml) typical of homotypic neutrophil adhesion was quickly reversible. On the other hand, longer lasting and more widespread neutrophil aggregation brought about by higher concentrations of euphorbin was characteristic of cell agglutination induced by a multivalent lectin (Dias - Baruffi et al., 2000). N- acetylgalactosamine was the strongest inhibitory sugar for *Euphorbia milii* lectin with the hydroxyl axial position of C4 and acetamido on C2 of N-aetylgalactosamine as the recognition points (Irazoqui et al., 2005).

Euphorbia Neriifolia Latex lectin

It is a 60-kDa galactose-binding dimer with about 12% carbohydrate. It manifested hemagglutinating activity toward trypsinized human and rabbit red cells and sialidase-treated ovine red cells. It stimulated the mitogenic response of mouse spleen cells. It was devoid of translation inhibitory activity in rabbit reticulocyte lysate (Seshagirirao and Prasad, 1995).

Euphorbia Tirucalli Latex Lectin (Eutirucallin)

The α-lactose-binding lectin is a type 2 ribosome inactivating protein with 32-kDa subunits. Its hemagglutinating activity was calcium-independent and remained stable till 55°C. It elicited neutrophil recruitment and macrophage cytokine production.

Its partial sequence exhibited 83% resemblance to the lectin domains of abrin and ricin both of which are type 2 ribosome inactivating proteins (Santana et al., 2014).

Hura Crepitans (Sand-Box Tree) Latex Lectin

The lectin is a mixture of three galactose-binding 140-kDa heterotetrameric isolectins with 37.5-kDa, 35.5-kDa, 31-kDa, and 29-kDa subunits. The lectin expressed mitogenic activity toward human T cells but not B cells. It exhibited translation inhibitory activity in rabbit reticulocyte lysate (Barbieri et al., 1983).

Momordica Balsamina Seed Isolectins

Three galactose-binding isolectins including 108-kDa MBaL-30, 56-kDa MBaL-60, and 160-kDa MBaL-80 were isolated from *Momordica balsamina* seeds. MBaL-30 and MBaL-60 agglutinated all human ABO blood groups. On the other hand, MBaL-80 was devoid of agglutinating activity toward AB and B blood groups. The N-terminal sequence of MBaL displayed 50% identity and 88% identity to momordin II and *Momordica charantia* lectin, a type-1 *Momordica balsamina* and a type-2 *Momordica charantia* ribosomal inactivating protein, respectively. The activity of MBaL remained following exposure to temperatures up to 50°C for 30 minutes. The optimal pH range lied between 4 and 8 for MBaL-30 and MBaL-60, and between 3 and 5 for MBaL-80. The hemagglutinating activity of MBaL-30 and MBaL-60 was abolished by EDTA. Zinc and ferric ions reversed the abolition of activity and brought about a 2 to 3 fold increase in activity. MBaL-30 agglutinated only gram negative bacteria *Escherichia coli* while MBaL-30 and MBaL-60 agglutinated gram positive bacteria *Staphylococcus aureus* and *Listeria monocytogenes*. The isolectins were devoid of antibacterial and antiproliferative activities on bladder, glioblastoma and gastric cancer cells (Awadallah et al., 2017).

Momordica Charantia Lectin

The galactose-specific lectin from *Momordica charantia* seeds, at various concentrations (6 - 24 μg/ml), brought about 8 - 23% inhibition of the growth of Ehrlich ascites carcinoma cells and induced cell cycle arrest at the G0/G1 phase. When given intraperitoneally to mice at 1.2, 2.0, and 2.8 mg/kg/day, the lectin produced 28% to 75% growth inhibition of Ehrlich ascites carcinoma cells without eliciting morphological changes characteristic of apoptosis. The red blood cell count and hematocrit increased. The lectin exhibited toxicity in brine shrimp nauplii with an LC50 value of 49.7 μg/ml, and agglutinated 4 out of the 7 pathogenic bacteria examined in the absence of the inhibitory sugar D-galactose/D-lactose (Kabir et al., 2015).

M. charantia lectin induced, in liver cancer cells, G2/M phase arrest, DNA fragmentation, mitochondrial damage, membrane potential depolarization, caspase and MAPK pathway activation, up-regulation of truncated Bid, apoptosis and autophagy. The employment of BI-6C9 (Bid inhibitor), Bid siRNA and Z-IETD-FMK (caspase 8 inhibitor) to inhibit Bid led to a decline in the effects of the lectin. The lectin in conjunction with sorafenib exerted a more potent activity against liver cancer cells *in vitro* and *in vivo* (Zhang et al., 2015).

M. charantia lectin, which is a type II ribosome inactivating protein, inhibited nasopharyngeal cancer CNE-1 and CNE-2 cells without adversely affecting normal nasopharyngeal NP 69 cells. The lectin brought about mitochondrial damage, downregulated phosphoretinoblastoma protein and cyclin D1 expression, cell cycle arrest in G1-phase, augmented cytochrome c release into the cytosol, activated caspases-3, - 8, and - 9, and upregulated formation of cleaved PARP, culminating in DNA fragmentation and apoptosis. The mitogen-activated protein kinases (MAPK; including p38 MAPK, JNK, and ERK) phosphorylation and downstream nitric oxide formation were upregulated. The p38 MAPK inhibitor SB-203580 suppressed nitric oxide formation and impaired the action of the lectin on nasopharyngeal cancer cells. The lectin administered intraperitoneally reduced growth of xenografts formed by subcutaneous inoculation of the nasopharyngeal cancer (Fang et al., 2012).

Maximal hemagglutinating activity of the lectin was observed from pH 7.4 to pH 11.0, but there was a precipitous decline at a pH below 7.0. Although the hemagglutinating activity was stable from 4 to 50 degrees, it plummeted between 50 and 60 degrees C, which could be correlated with alterations in protein structure demonstrable by circular dichroism and fluorescence spectroscopy (Sultan and Swamy, 2005). Unlike other galactose-binding and N-acetylgalactosamine-binding lectins, *M. charantia* lectin possessed antilipolytic and lipogenic activitie (Ng et al., 1989). *M. charantia* lectin inhibited IV-1 reverse transcriptase and weakly inhibited N-glycohydrolases including alpha-glucosidase, beta-glucosidase and beta-glucuronidase (Wang et al., 2001).

M. charantia lecin is a T cell-independent B cell activator and a polyclonal immunoglobulin inducer. The lectin stimulated B cell activation through membrane immunoglobulin on the B cell surface. It upregulated the cell activation marker CD86 in a B cell subpopulation different from that stimulated by lipooplysaccharide (Huang et al., 2008).

M. charantia lectin suppressed synthesis of protein and DNA in normal (mitogen-stimulated) and leukemic human peripheral blood lymphocytes at concentrations well under the dosages which undermined protein synthesis in Yoshida ascites cells (Licastro et al., 1980).

M. charantia lectin isolated by Horejsi et al. (1980) has a molecular weight of 129 kDa. and subunits with molecular weights 29 kDa, 32 kDa and 36 kDa, linked by by cystine bridges. The lectin is a glycoprotein with 4.1% neutral sugar and 0.5% glucosamine. It has a relatively high methionine and cystine content, and asparagine and valine as the two N-terminal amino acids. Two dissimilar dimeric lectins composed of 26-kDa subunits with distinct amino-terminal sequences and amino acid compositions have been reported by Li (1980).

M. charantia lectin isolated by Mazumder et al. (1981) was a tetrameric 120-kDa galactose binding glycoprotein containing 2% glucose, 8% mannose and 2.5% glucosamine. It was composed of 28-kDa and 30-kDa subunits. Its tyrosine and tryptophan residues were crucial for its sugar binding activity.

Protein synthesis by a rabbit reticulocyte lysate was inhibited by *M. charantia* lectin (Barbieri et al., 1979). A 115-kDa lectin from *M. charantia* seeds, which was composed of 30.5-kDa, 29-kDa, 28.5 -kDa and 27.2-kDa subunits, inhibited protein synthesis translation in a rabbit reticulocyte lysate with an ID50 of about 5 ug/ml and partially suppressed protein synthesis in Yoshida ascites cells at 100 ug/ml (Barbieri et al., 1980).

Ng et al. (1986) isolated a 124- kDa lectin with 5% carbohydrate and 37-kDa, 35-kDa and 33-kDa subunits from *M. charantia* seeds. *M. charantia* lectin exhibited some sequence resemblance to the ribosome inactivating protein beta-momorcharin from *M. charantia* and other ribosome inactivating proteins, and some degree of homology in sequence to lectins from *Cucurbita argyrosperma*, *C. maxima*, *Ricinus communis* and *Sambucus nigra* (Wang and Ng, 1998). *Momordica charantia* lectin inhibited HIV-1 reverse transcriptase (Wang and Ng, 2001).

The lectin manifested potent hemagglutinating activity and antilipolytic and lipogenic activities in isolated rat adipocytes. The antilipolytic activity was destructible by galactose, glutathione, heat, chymotrypsin, and trypsin, signifying that the inhibiting sugar galactose, the disulfide bonds, and integrity of the protein moiety, all play a central role in interaction with fat cells to bring about the insulinomimetic activity (Ng et al., 1986). The galactose-binding lectins from *M. charantia* seeds and *Trichosanthes kirilowii* tubers did not affect corticosterone production in the absence of adrenocorticotropin or corticosterone production in the presence of a maximally active dose of adrenocorticotropin. The lectins enhanced corticosterone production stimulated by a submaximal dose of adrenocorticotropin (Ng et al., 1987). *M. charantia* lectin, unlike most N-acetylgalactosamine-binding and galactose-binding lectins, manifested high antilipolytic and lipogenic activities (Ng et al., 1989). *Momordica charantia* lectin, as well as *Trichosanthes kirilowii* lectin, had no effect on corticosterone production in the absence of corticotropin or a maximally active concentration of corticotropin, but enhanced corticosterone production stimulated by a submaximal dose of corticotropin (Ng et al., 1987). The lectin remained as a tetramer even at low pH. Addition of lactose or adenine to lectin-8- anilinonaphthalene-1-sulfonate mixture did not alter the change in 8-anilinonaphthalene-1-sulfonate fluorescence suggesting that lactose, adenine and 8-anilinonaphthalene 1-sulfonate bind to the lectin at independent and non-interacting sites (Kavitha et al., 2009).

Crude *M. charantia* seed lectin injected intramuscularly into BALB/c for up to four immunizations at biweekly intervals elevated the titer of antibodies in blood (Huang et al., 2008).

Research on lectin-porphyrin binding is important in view of the lectin-cancer cell interaction and utilization of porphyrins in photodynamic cancer therapy. The tetrameric *M. charantia* lectin binds four molecules of porphyrin. Cationic and anionic porphyrins bind with similar affinities. The secondary and tertiary structures of the lectin are not perturbed by porphyrin binding The sugar inhibiting the hemagglutinating activity, lactose, had no effect on the stoichiometry of binding, and the porphyrins were devoid of any effect on the hemagglutinating activity. It appears that the lectin has distinct porphyrin and sugar binding sites. at different sites. The lectin has approximately 13% alpha-helix, 36% beta-sheet, 21% beta-turn, and 40% unordered structures (Sultan et al., 2004).

SOPHORA JAPONICA LECTINS

A new D-mannose/D-glucose-specific lectin (B-SJA-II) from the bark of the Japanese pagoda tree, *Sophora japonica* possesses a 19.4-kDa subunit a-1, a 18.2-kDa subunit a-2, a 15-kDa subunit b-1, and a 13.2-kDa subunit b-2. The subunits a-1, a-2, b-1, and b-2 have N-linked oligosaccharides characteristic of plant glycoproteins. All of the subunits bind carbohydrates containing asialofetuin and fetuin. In the absence of a specific sugar the lectin aggregated to form precipitates and upon addition of the specific sugar it became soluble. Each subunit of the lectin has a sugar-binding site for the mannosyl core of N-linked oligosaccharide chains and recognizes each other sugar specifically to form aggregates. Subunits a-1 and a-2 exhibit an N-terminal sequence with 50% identity to that of other *S. japonica* lectins. Subunits b-1 and b-2 demonstrate N-terminal sequence homology to concanavalin A (Ueda et al., 1991). Sophoragrin, a mannose/glucose-specific bark lectin exhibited the phenomenon of self-aggregation dependent on the concentrations of calcium ions and specific sugars: mannose, glucose or sucrose (Ueda et al., 2004).

Five N-acetyl-galactosamine-specific tetrameric lectins, which are glycoproteins with close structural and immunological relationships to the *Sophora* seed and leaf lectins, are present only in the bark. They represent different combinations of 30-kDa, 30.1-kDa and 33-kDa subunits with 3% to 5% carbohydrate (Hankins et al., 1988).

The 32-kDa leaf lectin I and 34-kDa leaf lectin II from leaves and 29.5 -kDa seed lectin are tetrameric glycoproteins with identical amino acid composition, sugar binding specificity and specific hemagglutinating activity and N-terminal amino acid sequence homology. However, the leaf and seed lectins are distinct immunologically and in subunit molecular weight, carbohydrate content, and pH sensitivity of hemagglutinating activities (Hankins et al., 1987). Molecular cloning studies disclosed that the seeds produce only a GalNAc-specific lectin with pronounced homology to the GalNAc-specific bark lectin and other legume lectins (Van Damme. et al., 1997). The leaf and bark lectins are found in protein-storage vacuoles (Herman et al., 1988).

Trichosanthes Anguina (Snake Gourd) Seed Lectin

A 56 kDa lectin, composed of a 22- kDa subunit and a 34- kDa subunit, was isolated from snake guard seeds (Kabir et al., 2012). The lectin, a nontoxic homolog of type II ribosome-inactivating proteins, possesses a lectin chain composed of two β-trefoil domains. The catalytic chain, linked via a disulfide bond to the lectin chain in type II ribosome-inactivating proteins, is cleaved into two. The lack of toxicity of the lectin is a consequence of alterations in the catalytic and carbohydrate binding sites (Sharma et al., 2013). The lectin manifested hemagglutinating activity which was stable in the presence of denaturants, from 30°C to 70°C, and from pH 3.0 to 12.0. The lectin displayed agglutinating activity toward 6 bacterial pathogens but was devoid of antifungal activity. The lectin administered intraperitonally at 1 mg/kg/day for 5 days produced 28% inhibition, and at double the dosage, 72% inhibition of the growth of Ehrlich ascites carcinoma cells in tumor bearing mice. At the double dosage, the life span of the treated mice was prolonged by approximately 40%. The lectin boosted the population of macrophages in normal mice (Kabir et al., 2012). The crystallization of the lectin in two hexagonal forms employing the hanging-drop method has been reported. Molecular replacement studies discloses homology of the lectin to type II ribosome inactivating proteins (Manoj et al., 2001).

Trichosanthes Anguina (Snake Gourd) Phloem Lectin

Heterodimeric snake gourd phloem lectin, with a 48-kDa subunit and a 53-kDa subunit connected by disulfide bonds, specifically bound chitooligo- saccharides. The lectin revealed, in circular dichroism studies, 9% α-helix, 39% β-sheet, 20% β-turns and 32% unordered structures. The secondary and tertiary structures were not altered by saccharide binding to any major extent. The lectin bound at 25°C with an association constant of 1.7 × 10(5) M(-1) M(-1) to chitotriose (N-actylglucosamine trimer) and 3.6 × 10(5)M(-1) to chitohexaose (N-acetylglucosamine hexamer). The stability of interactions between the lectin and saccharide was enhanced by hydrogen bonding and van der Waals interactions (Narahari et al., 2011).

Trichosanthes Cordata Seed Lectin

Two galactose-specific glycoproteins isolectins, including heterodimeric 59-kDa TCA-I with 32-kDa and 27- kDa subunits, and 52-kDa TCA-II, were isolated. They closely resembled type 2 ribosome inactivating proteins in carbohydrate specificity, subunit size and composition, and secondary structure. The isolectins, which possessed 6 - 7% covalently bound neutral sugar, exhibited striking similarity in secondary structure and displayed approximately 8 to 10% alpha-helix, 37 - 38% beta-sheet, 20% beta-turns, and 32 - 33% unordered structures. The pH range 7.0 - 8.0 was optimal for the hemagglutinating activity of TCA-I. The activity was thermostable between 0 and 40 degrees C, but it declined abruptly at temperatures higher than 40 degrees C. The association constant for the interaction of TCA-I with lactose was 7,420 M(-1) at 25 degrees C. The exposure and accessibility of its tryptophan residues and the effect of ligand binding were studied by quenching studies using ionic and neutral quenchers (Sultan et al., 2009).

Trichosanthes Cucumerina Seed Lectin

The hemagglutinating activity of the lectin was maximal at pH 8 to pH 11 but plummeted below pH 7, in keeping with the role of imidazole side chains of His residues in the saccharide-binding activity of the lectin. The hemagglutinating activity was stable between 0°C and 60°C, but fell abruptly when the temperature was raised futher. There were three lectin isoforms with pI values of 5.3, 6.2, and 7.1, respectively. The isoform with pI 6.2 was the most abundant. The lectin has very little alpha-helix, 28.4% beta-sheet, 10.6% beta-turns, 7% polyproline type 2 structure, and 54% unordered structure.

4-methylumbelliferyl-beta-D-galactopyranoside binding to the lectin, with an association constant of 6.9 x 10(4)M(-1) at 25°C, enhanced the intensity of fluorescence of the ligand (Kenoth et al., 2003).

Steady-state and time-resolved fluorescence spectroscopic studies on the lectin revealed that its fluorescence emission maximum in the native state and also in the presence of 0.1 M lactose occurred near the wavelength of 331 nm, which moved to 347 nm in the presence of the denaturating agent 8 M urea, signifying the presence of all its tryptophan residues in the native state in a hydrophobic environment. Lactose (0.1 M) enhanced quenching with acrylamide and anionic quencher iodide by a small extent, but in contrast had little effect on quenching with succinimide and cationic quencher cesium ion. The Stern-Volmer plots with cesium ions produced quenching profiles that were biphasic suggesting two or more groups of tryptophan residues with substantial differences in accessibility and/or environment (Kenoth and Swamy, 2003).

TRICHOSANTHES DIOICA SEED LECTIN

The lectin has been crystallized with unit-cell parameters a = b = 167.54, and c = 77.42 A and diffraction to a Bragg spacing of 2.8 A. The protein remained stable from pH 6 to pH 12 and up to 363 K. Total unfolding of the lectin occurred following exposure to 6 M guanidinium hydrochloride. Complete inactivation of the lectin was brought about by urea at higher concentrations. The protein remained stable from pH 6 to pH 12 and up to 363 K. The stability of the lectin is characteristic of type II ribosome-inactivating proteins (Dharkar et al., 2006).

A new galactose-specific 55-kDa lectin with a 24- kDa subunit and a 37- kDa subunit and 4.9% covalently bound neutral sugar has been purified from *Trichosanthes dioica* seeds. It possesses 13.3% alpha-helix, 36.7% beta-sheet, 19.4% beta-turns, and 31.6% unordered structure. The tyrosine residues are crucial to the saccharide binding and hemagglutinating activities of the lectin. The tryptophan residues are hidden deep in the lectin molecule and can only be chemically modified using N-bromosuccinimide under denaturing conditions. The subunit composition and size, secondary structure, and sugar specificity of this lectin are typical of those of type 2 ribosome inactivating proteins (Sultan et al., 2004).

TRICHOSANTHES JAPONICA LECTIN

In histochemical investigations employing biotinylated *Trichosanthes japonica* agglutinin I, staining was not found in normal mucosa, benign adenomas mucinous specimens, and poorly differentiated colon adenocarcinomas. However, 83% of well differentiated and moderately differentiated colon adenocarcinomas demonstrated reaction with *T. japonica* agglutinin I. Because over 90% of colon adenocarcinomas can be differentiated, staining with the lectin may be used in early diagnosis of colon cancers (Yamashita et al., 1995).

The 70-kDa heterodimeric *T. japonica* agglutinin-I is composed of 38-kDa and 32-kDa subunits with interconnecting disulfide bonds. The agglutinin recognizes an N-acetyllactosamine residue. The binding became stronger upon substitution of the beta-galactosyl residue at the C-6 position with a sialic acid or sulfate (Yamashita et al., 1992b).

The 64- kDa *T. japonica* agglutinin II, composed of 29- kDa and 33- kDa subunits, is a glycoprotein possessing N-acetylglucosamine, fucose, mannose and xylose. The strength of lectin binding to beta-galactosyl residue was enhanced as the hydroxyl group at the C-2 position was replaced with an acetylamino or a fucosyl group. A *Trichosanthes japonica* agglutinin II immobilized column was useful for fractionating oligosaccharides or glycoproteins containing blood group type 1H, type 2H, and Sd antigenic determinants (Yamashita et al., 1992).

Trichosanthes Kirilowii Seed Lectin

A 57- kDa galactose- specific lectin composed of a 37- kDa subunit and a 25- kDa subunit with no mitogenic activity toward human lymphocytes was isolated from *T. kirilowii* seeds (Falasca et al., 1989).

Trichosanthes Kirilowii Root Tuber Lectin

Two (TK-I and TK-II) of the three isolectins (TK-I TK-II, and TK-III) resembled each other in blood group and carbohydrate specificities, N-terminal residue, amino acid composition, molecular weight, immunoelectrophoretic and immunodiffusion behavior, insulinomimetic and hemagglutinating activities, and in having subunits with different molecular weights. TK-III differed by having a larger molecular weight, subunits with the same molecular size, a distinctive NH2-terminal residue and isoelectric point and attenuated agglutinating activity toward red blood cells. TK-I, TK-II and TK-III have common antigenic determinants. Their hemagglutinating activity underwent a larger decline in response to beta-linked terminal oligosaccharides containing D-galactose compared with alpha-linked oligosaccharides. The lectins did not have mitogenic activity and had no effect on ConA- induced mitogenic response of lymphocytes (Yeung et al., 1986). The *T. kirilowii* lectin increased lipogenesis in isolated rat epididymal adipocytes but had no effect on corticotropin-induced lipolysis in isolated hamster adipocytes (Ng et al., 1985). *T. kirilowi* lectin exhibited hypoglycemic activity in alloxan-induced diabetic mice (Li et al., 2012) and exerted a protective action against diabetic nephropathy by suppressing the LOX1/NF-κB/caspase-9 signaling pathway (Lu et al., 2018).

Withania Somnifera Leaf Lectin-Like Protein

The 30- kDa monomeric mannose-binding acidic lectin, concanavalin A **like** lectin from *Canavalia ensiformis* Its *in vitro* inhibitory concentration hyphae against main fungal phytopathogens hyphae ranged between 7 and 11 ug. The treated fungal hyphae demonstrated cell wall adhesion (Ghosh, 2009). The lectin displayed insecticidal properties against the hemipteran insect *Probergrothius sanguinolens* and lepidopteran insect *Hyblaea puera* (George et al., 2018).

Discussion

The aforementioned lectins exhibit a diversity of biological activities other than hemagglutinating activity such as anticancer, immunomodulatory, antimicrobial, anti-insect and insulin-like activities (Table 1). They display a variety of sugar specificities, molecular weights (Table 2), number of subunits, presence or absence of carbohydrate moiety, amino acid sequence. They were isolated from seeds, leaves, phloem, fruits, tubers and latex.

Table 1. Selected lectins from medicinal plants and the diversity of their biological activities

Lectin	Activities
Astragalus membranaceus	anticancer and antibacterial (Bai et al., 2018)
Astragalus mongholicus	anticancer and antifungal (Yan et al., 2009 a,b)
Dendrobium findleyanum	antifungal (Sudmoon et al., 2008)
Dendrobium pendulum	no activities shown except hemagglutinating activity
Dioscorea batatas	anti-insect (Ohizumi et al., 2009)
Euphorbia characias	neither mitogenic nor translation inhibitory activity Barbieri et al., 1983)
Euphorbia marginata	immunomodulatory activity (Stirpe et al., 1993)
Euphorbia neriifolia	stimulated the mitogenic response of mouse splenocytes (Seshagirirao and Prasad, 1995)
Euphorbia tirucalli	immunomodulatory activity (Santana et al., 2014)
Hura crepitans	mitogenic and translation inhibitory activities (Barbieri et al., 1983).
Momordica balsamina	bacterium agglutinating activity (Awadallah et al., 2017)
Momordica charantia	bacterium agglutinating, anticancer (Huang et al., 2008; Fang et al., 2012; Kabir et al., 2015; Zhang et al., 2015) immunomodulatory (Huang et al., 2008), antilipolytic, lipogenic (Ng et al., 1989) and HIV-1 reverse transcriptase inhibitory (Wang and Ng, 2001) activities
Trichosanthes anguina	anticancer (Kabir et al., 2012)
Trichosanthes kirilowii	lipogenic (Ng et al., 1985) and antidiabetic (Li et al., 2012; Lu et al., 2018)
Withania somnifera	antifungal (Ghosh, 2009) and anti-insect (George et al., 2018) activities

Table 2. Selected lectins from medicinal plants and the diversity of their sugar binding specificities, molecular weights and number of subunits

Lectin	Sugar Binding Specificity	Molecular Weight (kDa)/no. of Amino Acids (AA)
Astragalus membranaceus	galactose	67
Astragalus Mongholicus	D-galactose & lactose	66.4
Dendrobium findleyanum	mannose	56.2
Dendrobium pendulum	mannose	44-45
Dioscorea batatas	mannose	108 AA
Euphorbia heterophylla	N-acetylgalactosamine	32
Euphorbia marginata	galactose	60
Euphorbia neriifolia	galactose	60
Euphorbia tirucalli	α-lactose	64
Hura crepitans	galactose	140
Momordica balsamina	galactose	56-160
Momordica charantia	galactose	52-129
Sophora japonica	mannose/glucose	32-64
Trichosanthes cordata	galactose	59
Trichosanthes dioica	galactose	55
Trichosanthes kirilowii	galactose	57
Withania somnifera	mannose	30

ACKNOWLEDGMENTS

We gratefully acknowledge the award of Health and Medical Research Fund (no. 12131221) from Food and Health Bureau, The Government of Hong Kong Special Administrative Region, and grants from National Natural Science Foundation of China (no. 81471927).

REFERENCES

Awadallah, AK; Osman, ME; Ibrahim, MA; Bernardes, ES; Dias-Baruffi, M; Konozy, EH. Isolation and partial characterization of 3 nontoxic d-galactose-specific isolectins from seeds of *Momordica balsamina. J Mol Recognit.*, 2017 Feb, 30(2). doi: 10.1002/jmr.2582. Epub 2016 Oct 24.

Baba, K; Ogawa, M; Nagano, A; Kuroda, H; Sumiya, K. Developmental changes in the bark lectin of *Sophora japonica* L. *Planta.*, 1991 Feb, 183(3), 462-70. .

Bai, CZ; Hao, JQ; Hao, XL; Feng, ML. Preparation of *Astragalus membranaceus* lectin and evaluation of its biological function. *Biomed Rep.*, 2018 Oct, 9 (4), 345-349.

Barbieri, L; Falasca, A; Franceschi, C; Licastro, F; Rossi, CA; Stirpe, F. Purification and properties of two lectins from the latex of the euphorbiaceous plants *Hura crepitans* L. (sand-box tree) and *Euphorbia characias* L. (Mediterranean spurge). *Biochem J.*, 1983, Dec 1, 215(3), 433-9.

Barbieri, L; Lorenzoni, E; Stirpe, F. Inhibition of protein synthesis *in vitro* by a lectin from *Momordica charantia* and by other haemagglutinins. *Biochem J.*, 1979, Aug 15, 182(2), 633-5.

Barbieri, L; Zamboni, M; Lorenzoni, E; Montanaro, L; Sperti, S; Stirpe, F. Inhibition of protein synthesis *in vitro* by proteins from the seeds of *Momordica charantia* (bitter pear melon). *Biochem J.*, 1980, Feb 15, 186(2), 443-52.

Chandran, T; Sharma, A; Vijayan, M. Crystallization and preliminary X-ray studies of a galactose- specific lectin frombthe seeds of bitter gourd (*Momordica charantia*). *Acta Crystallogr Sect F Struct Biol Cryst Commun.*, 2010, Sep 1, 66(Pt 9), 1037-40.

Chen, Z; Sun, X; Tang, K. Cloning and expression of a novel cDNA encoding a mannose-binding lectin from *Dendrobium officinale*. *Toxicon.*, 2005, Mar 15, 45(4), 535-40.

Chilson, OP; Kelly-Chilson, AE. Mitogenic lectins bind to the antigen receptor on human lymphocytes. *Eur J Immunol.*, 1989 Feb, 19(2), 389-96.

Dharkar, PD; Anuradha, P; Gaikwad, SM; Suresh, CG. Crystallization and preliminary haracterization of a highly thermostable lectin from *Trichosanthes dioica* and comparison with other *Trichosanthes* lectins. *Acta Crystallogr Sect F Struct Biol Cryst Commun.*, 2006, Mar 1, 62(Pt 3), 205-9.

Dias-Baruffi, M; Sakamoto, M; Rossetto, S; Vozári-Hampe, MM; Roque-Barreira, MC. Neutrophil migration and aggregation induced by euphorbin, a lectin from the latex of *Euphorbia milii, var. milii.*. *Inflamm Res.*, 2000 Dec, 49(12), 732-6.

Falasca, AI; Abbondanza, A; Barbieri, L; Bolognesi, A; Rossi, CA; Stirpe, F. Purification and partial characterization of a lectin from the seeds of *Trichosanthes kirilowii* Maximowicz. *FEBS Lett.*, 1989, Mar 27, 246(1-2), 159-62.

Fang, EF; Zhang, CZ; Ng, TB; Wong, JH; Pan, WL; Ye, XJ; Chan, YS; Fong, WP. *Momordica charantia* lectin, a type II ribosome inactivating protein, exhibits antitumor activity toward human nasopharyngeal carcinoma cells *in vitro* and *in vivo*. *Cancer Prev Res (Phila).*, 2012 Jan, 5(1), 109-21.

Girjal, UV; Neelagund, S; Krishnappa, M. *Ganoderma lucidum*: a source for novel bioactive lectin. *Protein Pept Lett.*, 2011 Nov, 18(11), 1150-7.

George, BS; Silambarasan, S; Senthil, K; Jacob, JP; Ghosh Dasgupta, M. Characterization of an Insecticidal Protein from *Withania somnifera* Against Lepidopteran and Hemipteran Pest. *Mol Biotechnol.*, 2018 Apr, 60(4), 290-301.

Ghosh, M. Purification of a lectin-like antifungal protein from the medicinal herb, *Withania somnifera*. *Fitoterapia.*, 2009 Mar, 80(2), 91-5.

Haak-Frendscho, M; Kino, K; Sone, T; Jardieu, P. Ling Zhi-8: a novel T cell mitogen induces cytokine production and upregulation of ICAM-1 expression. *Cell Immunol.*, 1993 Aug, 150(1), 101-13.

Hankins, CN; Kindinger, J; Shannon, LM. The Lectins of *Sophora japonica*: I. Purification Properties and N-Terminal Amino Acid Sequences of Two Lectins from Leaves. *Plant Physiol.*, 1987 Apr, 83(4), 825-9.

Hankins, CN; Kindinger, JI; Shannon, LM. The Lectins of *Sophora japonica*: II. Purification; Properties, and N-Terminal Amino Acid Sequences of Five Lectins from Bark. *Plant Physiol.*, 1988 Jan, 86(1), 67-70.

Herman, EM; Hankins, CN; Shannon, LM. Bark and Leaf Lectins of *Sophora japonica* Are Sequestered in Protein-Storage Vacuoles. *Plant Physiol.*, 1988 Apr, 86(4), 1027-31.

Horejsí, V; Tichá, M; Novotný, J; Kocourek, J. Studies on lectins. XLVII. Some properties of D-galactose binding lectins isolated from the seeds of *Butea frondosa*, *Erythrina indica* and *Momordica charantia*. *Biochim Biophys Acta.*, 1980, Jun 26, 623(2), 439-48.

Hsu, HC; Hsu, CI; Lin, RH; Kao, CL; Lin, JY. Fip-vvo, a new fungal immunomodulatory protein isolated from *Volvariella volvacea*. *Biochem J.*, 1997, Apr 15, 323, (Pt 2), 557-65.

Huang, L; Adachi, T; Shimizu, Y; Goto, Y; Toyama, J; Tanaka, H; Akashi, R; Sawaguchi, A; Iwata, H; Haga, T. Characterization of lectin isolated from *Momordica charantia* seed as a B cell activator. *Immunol Lett.*, 2008, Dec 22, 121(2), 148-56.

Huang, L; Ikejiri, A; Shimizu, Y; Adachi, T; Goto, Y; Toyama, J; Tanaka, H; Akashi, R; Uchida, K; Miyata, H; Haga, T. Immunoadjuvant activity of crude lectin extracted from *Momordica charantia* seed. *J Vet Med Sci.*, 2008 May, 70(5), 533-5.

Huang, L; Sun, F; Liang, C; He, YX; Bao, R; Liu, L; Zhou, CZ. Crystal structure of LZ-8 from the medicinal fungus *Ganoderma lucidium*. *Proteins.*, 2009, May 1, 75(2), 524-7.

Huang, LH; Yan, QJ; Kopparapu, NK; Jiang, ZQ; Sun, Y. *Astragalus membranaceus* lectin (AML) induces caspase-dependent apoptosis in human leukemia cells. *Cell Prolif.*, 2012 Feb, 45(1), 15-21.

(Huang et al., 2012)

Irazoqui, FJ; Vozari-Hampe, MM; Lardone, RD; Villarreal, MA; Sendra, VG; Montich, GG; Trindade, VM; Clausen, H; Nores, GA. Fine carbohydrate recognition of *Euphorbia milii* lectin. *Biochem Biophys Res Commun.*, 2005, Oct 14, 336 (1), 14-21.

Kabir, SR; Islam, MF; Alom, MJ; Abu Zubair, M; Absar, N. Purification, characterizations of a snake guard seeds lectin with antitumor activity against Ehrlich ascites carcinoma cells *in vivo* in mice. *Protein Pept Lett.*, 2012 Mar, 19(3), 360-8.

Kabir, SR; Nabi, MM; Nurujjaman, M; Abu Reza, M; Alam, AH; Uz Zaman, R; Khalid-Bin-Ferdaus, KM; Amin, R; Khan, MM; Hossain, MA; Uddin, MS; Mahmud, ZH.

Momordica charantia seed lectin: toxicity, bacterial agglutination and antitumor properties. *Appl Biochem Biotechnol.*, 2015 Mar, 175(5), 2616-28.

Kabir, S; Islam, MF; Alom, MJ; Abu Zubair, M; Absar, N. Purification, characterizations of a snake guard seeds lectin with antitumor activity against Ehrlich ascites carcinoma cells *in vivo* in mice. *Protein Pept Lett.*, 2012 Mar, 19(3), 360-8.

Kavitha, M; Sultan, NA; Swamy, MJ. Fluorescence studies on the interaction of hydrophobic ligands with *Momordica charantia* (bitter gourd) seed lectin. *J Photochem Photobiol B.*, 2009, Jan 9, 94(1), 59-64.

Kenoth, R; Komath, SS; Swamy, MJ. Physicochemical and saccharide-binding studies on the galactose-specific seed lectin from *Trichosanthes cucumerina. Arch Biochem Biophys.*, 2003 May 1, 413(1), 131-8.

Kenoth, R; Swamy, MJ. Steady-state and time-resolved fluorescence studies on *Trichosanthes cucumerina* seed lectin. *J Photochem Photobiol B.*, 2003 Mar, 69(3), 193-201.

Kumaran, S; Pandurangan, AK; Shenbhagaraman, R; Esa, NM. Isolation and Characterization of Lectin from the Artist's Conk Medicinal Mushroom, *Ganoderma applanatum* (Agaricomycetes), and Evaluation of Its Antiproliferative Activity in HT-29 Colon Cancer Cells. *Int J Med Mushrooms.*, 2017, 19(8), 675-684.

Kumaran, S; Pandurangan, AK; Shenbhagaraman, R; Esa, NM. Isolation and Characterization of Lectin from the Artist's Conk Medicinal Mushroom, *Ganoderma applanatum* (Agaricomycetes), and Evaluation of Its Antiproliferative Activity in HT-29 Colon Cancer Cells. *Int J Med Mushrooms.*, 2017, 19(8), 675-684.

Li, Q; Ye, XL; Zeng, H; Chen, X; Li, XG. Study on the extraction technology and hypoglycemic activity of lectin from *Trichosanthes kirilowi.* Zhong Yao Cai., 2012 Mar, 35(3), 475-9.

Li, SS. Purification and partial characterization of two lectins from *Momordica charantia. Experientia.*, 1980, May 15, 36(5), 524-7.

Licastro, F; Franceschi, C; Barbieri, L; Stirpe, F. Toxicity of *Momordica charantia* lectin and inhibitor for human normal and leukaemic lymphocytes. *Virchows Arch B Cell Pathol Incl Mol Pathol.*, 1980, 33(3), 257-65.

Lin, WH; Hung, CH; Hsu, CI; Lin, JY. Dimerization of the N-terminal amphipathic alpha-helix domain of the fungal immunomodulatory protein from *Ganoderma tsugae* (Fip-gts) defined by a yeast two-hybrid system and site-directed mutagenesis. *J Biol Chem.*, 1997, Aug 8, 272(32), 20044-8.

Lu, J; Peng, J; Xiang, M; He, L; Wang, D; Xiong, G; Li, S. *Trichosanthes kirilowii* lectin alleviates diabetic nephropathy by inhibiting the LOX1/ NF-κB/caspase-9 signaling pathway. *Biosci Rep.*, 2018, Sep 7, 38(5). pii: BSR20180071.

Mazumder, T; Gaur, N; Surolia, A. The physicochemical properties of the galactose-specific lectin from *Momordica charantia. Eur J Biochem.*, 1981 Jan, 113(3), 463-70.

Manoj, N; Jeyaprakash, AA; Pratap, JV; Komath, SS; Kenoth, R; Swamy, MJ; Vijayan, M. Crystallization and preliminary X-ray studies of snake gourd lectin: homology with type II ribosome-inactivating proteins. *Acta Crystallogr D Biol Crystallogr.*, 2001 Jun, 57(Pt 6), 912-4.

Narahari, A; Nareddy, PK; Swamy, MJ. A new chitooligosaccharide specific lectin from snake gourd (*Trichosanthes anguina*) phloem exudate. Purification, physico-chemical characterization and thermodynamics of saccharide binding. *Biochimie.*, 2011 Oct, 93(10), 1676-84.

Ng, TB; Li, WW; Yeung, HW. Effects of ginsenosides, lectins and *Momordica charantia* insulin-like peptide on corticosterone production by isolated rat adrenal cells. *J Ethnopharmacol.*, 1987 Sep-Oct, 21(1), 21-9.

Ng, TB; Li, WW; Yeung, HW. Effects of lectins with various carbohydrate binding specificities on lipid metabolism in isolated rat and hamster adipocytes. *Int J Biochem.*, 1989, 21(2), 149-55.

Ng, TB; Wong, CM; Li, WW; Yeung, HW. Effect of *Trichosanthes kirilowii* lectin on lipolysis and lipogenesis in isolated rat and hamster adipocytes. *J Ethnopharmacol.*, 1985 Sep, 14(1), 93-8.

Ng, TB; Wong, CM; Li, WW; Yeung, HW. Isolation and characterization of a galactose binding lectin with insulinomimetic activities. From the seeds of the bitter gourd *Momordica charantia* (Family Cucurbitaceae). *Int J Pept Protein Res.*, 1986 Aug, 28(2), 163-72.

Ngai, PH; Ng, TB. A mushroom (*Ganoderma capense*) lectin with spectacular thermostability, potent mitogenic activity on splenocytes, and antiproliferative activity toward tumor cells. *Biochem Biophys Res Commun.*, 2004, Feb 20, 314(4), 988-93.

Nsimba-Lubaki, M; Peumans, WJ; Carlier, AR. Isolation and partial characterization of a lectin from *Euphorbia heterophylla* seeds. *Biochem J.*, 1983, Oct 1, 215(1), 141-5.

Ohizumi, Y; Gaidamashvili, M; Ohwada, S; Matsuda, K; Kominami, J; Nakamura-Tsuruta, S; Hirabayashi, J; Naganuma, T; Ogawa, T; Muramoto, K. Mannose -binding lectin from yam (*Dioscorea batatas*) tubers with insecticidal properties against *Helicoverpa armigera* (Lepidoptera: Noctuidae). *J Agric Food Chem.*, 2009, Apr 8, 57(7), 2896-902.

Pushparajah, V; Fatima, A; Chong, CH; Gambule, TZ; Chan, CJ; Ng, ST; Tan, CS; Fung, SY; Lee, SS; Tan, NH; Lim, RL. Characterisation of a New Fungal Immunomodulatory Protein from Tiger Milk mushroom, *Lignosus rhinocerotis. Sci Rep.*, 2016, Jul 27, 6, 30010.

Santana, SS; Gennari-Cardoso, ML; Carvalho, FC; Roque-Barreira, MC; Santiago Ada, S; Alvim, FC; Pirovani, CP. Eutirucallin, a RIP-2 type lectin from the latex of *Euphorbia tirucalli* L. presents proinflammatory properties *PLoS One.*, 2014, Feb 18, 9(2), e88422

Sattayasai, N. The first trimeric *Galanthus nivalis* agglutinin-related lectin of Orchidaceae was found in *Dendrobium pendulum*: purification, characterization, and effects of stress factors. *Plant Cell Rep.*, 2015 Jul, 34(7), 1253-62.

Sattayasai, N. The first trimeric *Galanthus nivalis* agglutinin-related lectin of Orchidaceae was found in *Dendrobium pendulum*: purification, characterization, and effects of stress factors. *Plant Cell Rep.*, 2015 Jul, 34(7), 1253-62.

Sattayasai, N; Sudmoon, R; Nuchadomrong, S; Chaveerach, A; Kuehnle, AR; Mudalige-Jayawickrama, RG; Bunyatratchata, W. *Dendrobium findleyanum* agglutinin: production, localization, anti-fungal activity and gene characterization. *Plant Cell Rep.*, 2009 Aug, 28(8), 1243-52.

Sattayasai, N; Sudmoon, R; Nuchadomrong, S; Chaveerach, A; Kuehnle, R; Mudalige-Jayawickrama, RG; Bunyatratchata, W. *Dendrobium findleyanum* agglutinin: production, localization, anti-fungal activity and gene characterization. *Plant Cell Rep.*, 2009 Aug, 28(8), 1243-52.

Seshagirirao, K; Prasad, MN. Purification and partial characterization of a lectin from *Euphorbia neriifolia* latex. *Biochem Mol Biol Int.*, 1995 May, 35(6), 1199-204

Siripipatthana, P; Phaonakrop, N; Roytrakul, S; Senawong, G; Mudalige-Jayawickrama, RG; Sattayasai, N. The first trimeric Galanthus nivalis agglutinin-related lectin of Orchidaceae was found in *Dendrobium pendulum*: purification, characterization, and effects of stress factors. *Plant Cell Rep.*, 2015 Jul, 34(7), 1253-62.

Sharma, A; Pohlentz, G; Bobbili, KB; Jeyaprakash, AA; Chandran, T; Mormann, M; Swamy, MJ; Vijayan, M. The sequence and structure of snake gourd (*Trichosanthes anguina*) seed lectin, a three-chain nontoxic homologue of type II RIPs. *Acta Crystallogr D Biol Crystallogr.*, 2013 Aug, 69 (Pt 8), 1493-503.

Siripipatthana, P; Phaonakrop, N; Roytrakul, S; Senawong, G; Mudalige-Jayawickrama, RG; Sudmoon, R; Sattayasai, N; Bunyatratchata, W; Chaveerach, A; Nuchadomrong, S. Thermostable mannose-binding lectin from *Dendrobium findleyanum* with activities dependent on sulfhydryl content. *Acta Biochim Biophys Sin (Shanghai).*, 2008 Sep, 40(9), 811-8.

Stirpe, F; Licastro, F; Morini, MC; Parente, A; Savino, G; Abbondanza, A; Bolognesi, A; Falasca, AI; Rossi, CA. Purification and partial characterization of a mitogenic lectin from the latex of *Euphorbia marginata. Biochim Biophys Acta.*, 1993, Aug 20, 1158(1), 33-9.

Sultan, NA; Kavitha, M; Swamy, MJ. Purification and physicochemical characterization of two galactose-specific isolectins from the seeds of *Trichosanthes cordata. IUBMB Life.*, 2009 Apr, 61(4), 457-69.

Sultan, NA; Kenoth, R; Swamy, MJ. Purification, physicochemical characterization, saccharide specificity, and chemical modification of a Gal/GalNAc specific lectin from the seeds of *Trichosanthes dioica.*, *Arch Biochem Biophys.*, 2004, Dec 15, 432(2), 212-21.

Sultan, NA; Maiya, BG; Swamy, MJ. Thermodynamic analysis of porphyrin binding to *Momordica charantia* (bitter gourd) lectin. *Eur J Biochem.*, 2004 Aug, 271(15), 3274-82.

Sultan, NA; Swamy, MJ. Energetics of carbohydrate binding to *Momordica charantia* (bitter gourd) lectin: an isothermal titration calorimetric study. *Arch Biochem Biophys.*, 2005, May 1, 437(1), 115-25.

Thakur, A; Rana, M; Lakhanpal, TN; Ahmad, A; Khan, MI. Purification and characterization of lectin from fruiting body of *Ganoderma lucidum*: lectin from Ganoderma lucidum. *Biochim Biophys Acta.*, 2007 Sep, 1770(9), 1404-12.

Thakur, A; Pal, L; Ahmad, A; Khan, MI. Complex carbohydrate specificity of lectin from fruiting body of *Ganoderma lucidum*. A surface plasmon resonance study. *IUBMB Life.*, 2007 Dec, 59 (12), 758-64.

Ueda, H; Fukushima, H; Hatanaka, Y; Ogawa, H. Solubility-insolubility interconversion of sophoragrin, a mannose/glucose-specific lectin in *Sophora japonica* (Japanese pagoda tree) bark, regulated by the sugar-specific interaction. *Biochem J.*, 2004, Sep 15, 382(Pt 3), 821-9.

Ueno, M; Ogawa, H; Matsumoto, I; Seno, N. A novel mannose-specific and sugar specifically aggregatable lectin from the bark of the Japanese pagoda tree (*Sophora japonica*). *J Biol Chem.*, 1991, Feb 15, 266(5), 3146-53.

Ueno, M; Arai, T; Kojima, K; Ogawa, HK; Matsumoto, I; Seno, N. Improved affinity chromatographic purification of D-mannose-N- acetyl-D-glucosamine-specific lectin from the bark of *Sophora japonica* eliminating the loss by sugar specific self-aggregation. *J Chromatogr.*, 1992, Apr 24, 597(1-2), 197-200.

Van Damme, EJ; Barre, A; Rouge, P; Peumans, WJ. Molecular cloning of the bark and seed lectins from the Japanese pagoda tree (*Sophora japonica*). *Plant Mol Biol.*, 1997 Feb, 33(3), 523-36.

Wang, H; Ng, TB. Ribosome inactivating protein and lectin from bitter melon (*Momordica charantia*) seeds: sequence comparison with related proteins. *Biochem Biophys Res Commun.*, 1998, Dec 9, 253(1), 143-6.

Wang, HX; Ng, TB. Examination of lectins, polysaccharopeptide, polysaccharide, alkaloid, coumarin and trypsin inhibitors for inhibitory activity against human immunodeficiency virus reverse transcriptase and glycohydrolases. *Planta Med.*, 2001 Oct, 67(7), 669-72.

Yamashita, K; Fukushima, K; Sakiyama, T; Murata, F; Kuroki, M; Matsuoka, Y. Expression of Sia alpha 2--,6Gal beta 1--,4GlcNAc residues on sugar chains of glycoproteins including carcino-embryonic antigens in human colon adenocarcinoma: applications of *Trichosanthes japonica* agglutinin I for early diagnosis. *Cancer Res.*, 1995, Apr 15, 55(8), 1675-9.

Yamashita, K; Ohkura, T; Umetsu, K; Suzuki, T. Purification and characterization of a Fuc alpha 1--,2Gal beta 1--, and GalNAc beta 1--,-specific lectin in root tubers of Trichosanthes japonica. *J Biol Chem.*, 1992b, Dec 15, 267(35), 25414-22.

Yamashita, K; Umetsu, K; Suzuki, T; Ohkura, T. Purification and characterization of a Neu5Ac alpha 2--,6Gal beta 1--,4GlcNAc and HSO3(-)--, 6Gal beta 1--, GlcNAc specific lectin in tuberous roots of *Trichosanthes japonica. Biochemistry.*, 1992b, Nov 24, 31(46), 11647-50

Yan, Q; Jiang, Z; Yang, S; Deng, W; Han, L. A novel homodimeric lectin from *Astragalus mongholicus* with antifungal activity. *Arch Biochem Biophys.*, 2005, Oct 1, 442(1), 72-81.

Yan, Q; Li, Y; Jiang, Z; Sun, Y; Zhu, L; Ding, Z. Antiproliferation and apoptosis of human tumor cell lines by a lectin (AMML) of *Astragalus mongholicus. Phytomedicine.*, 2009 Jun, 16(6-7), 586-93.

Yeung, HW; Ng, TB; Wong, DM; Wong, CM; Li, WW. Chemical and biological characterization of the galactose binding lectins from *Trichosanthes kirilowii* root tubers. *Int J Pept Protein Res.*, 1986 Feb, 27(2), 208-20.

Zhang, CZ; Fang, EF; Zhang, HT; Liu, LL; Yun, JP. *Momordica charantia* lectin exhibits antitumor activity towards hepatocellular carcinoma. *Invest New Drugs.*, 2015 Feb, 33(1), 1-11.

Zhao, S; Guo, YX; Liu, QH; Wang, HX; Ng, TB. Lectins but not antifungal proteins exhibit anti-nematode activity. *Environ Toxicol Pharmacol.*, 2009 Sep, 28(2), 265-8.

In: Hemagglutinins
Editors: T. Bun Ng, J. Wong, R. Tse et al.

ISBN: 978-1-53615-708-6
© 2019 Nova Science Publishers, Inc.

Chapter 8

ACTIONS OF TYPE II RIBOSOME INACTIVATING PROTEINS WITH AN EMPHASIS ON ANTICANCER ACTIVITY

Jack Ho Wong[1,2], Tzi Bun Ng[3,*], Charlene Cheuk Wing Ng[4], Chit Tam[3], Yau Sang Chan[5], Ryno Naude[6], Ryan Tse[1], Tak Fu Tse[1] and Helen Chan[1,†]

[1]Vita Green Pharmaceutical (Hong Kong Limited), Genning Partners Company Limited, and Institute of Medical Research, Hong Kong, China
[2]The Third Affiliated Hospital of Shenzhen University, Shenzhen, China
[3]School of Biomedical Sciences, Faculty of Medicine, The Chinese University of Hong Kong, China
[4]GKT School of Medical Education, King's College London, United Kingdom
[5]Department of Obstetrics and Gynecology, LKS Faculty of Medicine, University of Hong Kong, China
[6]Department of Biochemistry and Microbiology, Nelson Mandela Metropolitan University, Port Elizabeth, South Africa

* Corresponding Author's Email: tzibunng.cuhk.edu.hk.
† Corresponding Author's Email: Helen.chan@vitagreen.com.

ABSTRACT

Ribosome inactivating proteins (RIPs) are proteins capable of interrupting protein synthesis by virtue of their RNA N-glycosidase activity. These proteins are produced mostly by plants, but are now known to be produced also by fungi, bacteria and insects. Type 1 RIPs are composed of a single chain with RNA N-glycosidase activity. Type 2 RIPs, the focus of this review, have a A chain corresponding to the single chain in type 1 RIPs and in addition, a B chain with hemagglutinating activity. Type 2 RIPs are further divided into toxic and nontoxic type 2 RIPs.

The very toxic ones have been used as bioweapons. Immunotoxins based on the type 2 RIPs are potential anticancer agents. Some type 2 RIPs exhibit antiviral, anti-nematode and anti-insect activities. Mutants of the bacterial type 2 RIP Shiga toxin exhibit anticancer and antiangiogenic actions.

Keywords: type 2 ribosome inactivating proteins, hemagglutinating activity, anticancer activity, plants

INTRODUCTION

Ribosome inactivating proteins (RIPs) are plant toxins that permanently damage ribosomes causing cell death. RIPs depurinate a specific adenine residue (A_{4324}) in the conserved sarcin/ricin loop of ribosomal RNA.This change incapacitates the ribosome which can no longer bind the elongation factors properly, thus impeding protein synthesis.

The traditional concept is that there are two main types. Plant type 1 ribosome inactivating proteins are single-chain enzymatic proteins with N- glycosidase and cell-free translation inhibitory activities. Plant type 2 ribosome inactivating proteins are composed of an active A chain corresponding to a type 1 ribosome inactivating protein connected to a B chain with agglutinating activity toward red cells. The B chain binds to the plasma membrane of cancer cells and the A chain exerts cytotoxicity (Walsh et al., 2013; Shrot et al., 2015; Barbier and Gillet, 2018).

There are new members of the RIP family. Shiga and shiga-like toxins are bacterial type 2RIPs which are composed of a catalytic A chain with N- glycosidase activity, and five identical B chains and recognize and bind to the target cells with specific carbohydrate moieties (Chan and Ng, 2013; Walsh et al., 2013). Maize RIP, a type 3 RIP, is distinct from other RIPs due to its synthesis as a precursor with a 25-residue internal inactivation region, which is eliminated during protein activation (Yang et al.,2010).Small RIPs (Wang and Ng, 2000, 2002; Ng et al., 2002, 2003, 2013; Ng and Parkash, 2002; Ng and Wang, 2004; Parkash et al., 2002a, 2002b; Wong et al., 2008) with a molecular mass well below 30 kDa constitute another type of RIPs (Schrot et al., 2015).Mosquito RIPs (Lapadula et al., 2017) have been reported.

The type 1 RIP from *Saponaria officinalis* seeds, saporin-S6, has been employed to prepare antineoplastic conjugates due to its stability and high enzymatic activity(Polito et al.,2011).Evidence suggests that ribosome inactivating proteins are defense proteins protecting plants from micrbial pathogens and predators (Jiménez et al., 2015).

The present review focuses on the anticancer action of type 2 RIPs.

Toxic and Nontoxic Type 2 Ribosome Inactivating Proteins

Some type 2 RIPs such as ricin are extremely toxic to mammalian cells and animals whilst other type 2 RIPs (non-toxic type 2 RIPs) display three to four logs less toxicity. All type 2 RIPs differ in their depurinating activity on the different substrates with differences of up to one to two logs. The toxicity of type 2 RIPs is independent of their enzymatic activity on nucleic acids or on ribosomes (Barbieri et al., 2004).

Abrin A Chain-Vascular Endothelial Growth Factor Isoform 121 Fusion Protein Used in Conjunction with the Cytokine Interleukin-12

ABRaA-vascular endothelial growth factor isoform 121 (VEGF121) is a fusion protein composed of VEGF121, and A chain of the type 2 ribosome inactivating protein abrin from seeds of the rosary pea (or jequirity pea), *Abrus precatorius.* The fusion protein exerts a deleterious effect on blood vessels supplying the cancer cells and elicits necrosis in cells located near the injured blood vessels. Interleukin-12 is a cytokine that promotes specific as well as non-specific immune responses.

The antivascular action of ABRaA-VEGF121 in conjunction with the immunoenhancing action of interleukin-12 yielded a synergistic effect and about one-fifth of the treated tumor bearing mice were completely cured. The tumors from these mice had less blood microvessels and regulatory T lymphocytes and more CD4(+) and CD8(+) cells than the control animals (Ciomber et al., 2014).

Aralin from *Aralia Elata*

Aralin from *Aralia elata* triggers apoptosis in tumor cells. The aralin receptor (110-kDa) is a binding protein (BP) for high-density lipoprotein protein (HDL) and is a HDL receptor. It is referred to as the 110-kDa HDLBP. The sensitivity of tumor cell lines to aralin is governed by the level of expression of aralin. Forced expression of aralin in Huh7 cells resistant to aralin rendered the cells aralin sensitive. HeLa cells with HDLBP knockdown are aralin-resistant. Ectopic expression of the 150-kDa HDLBP brought about elevated expression of the 110-kDa HDLBP and augmented aralin sensitivity. Hence aralin may serve as an anticancer drug for cancers overexpressing HDLBP (Otsuka et al., 2014).

Ca-aralin, an aralin synthesized in the callus, differs from aralin only in glycosylation of B chain. Ca-aralin had a 2-fold higher selective cytotoxic index (ratio of IC50 value against human lung normal cells versus virus-transformed cells) compared with aralin, indicating the greater utility of ca-aralin as an anticancer agent (Tomatsu et al., 2004).

Articulatin-D from *Viscum Articulatum*

Articulatin-D suppressed the viability of leukemia cells. In Jurkat cells articulatin-D altered mitochondrial membrane potential, induced phosphatidylserine exposure, upregulated caspase-8 in extrinsic apoptotic pathway, and stimulated caspase-3-dependent DNA fragmentation of (Mishra et al., 2017).

Cinnamomin from Camphor Tree (*Cinnamomum Camphora*)

Cinnamomin A-chain brought about cleavage of supercoiled double-stranded DNA. The N- and C-terminal domains of cinnamomin A were important for deadenylation of supercoiled double-stranded circular DNA and rRNA (Liu, 2017). Although ricin and cinnamomin A-chain had approximately the same protein synthesis inhibitory activity *in vitro*, the former exerted considerably stronger cytotoxicity on BA/F3beta cells. A hybrid RIP composed of cinnamomin A-chain and ricin B-chain expressed higher and similar cytotoxicity toward BA/F3beta cells compared with cinnamomin and ricin, respectively. On the other hand, the hybrid RIP made up of ricin A-chain and cinnamomin B-chain manifested attenuated and similar cytotoxicity toward BA/F3beta cells compared with ricin and cinnamomin, respectively. Ricin and cinnamomin possess related binding sites on the cell surface albeit with dissimilar affinities (Wang et al., 2006).

Cinnamomin produced leaf necrosis in cranberry (*Vaccinum macrocarpon*) plants and a weaker effect on the foliage of *Capsicum annuum, Lactuca sativa, Lycopersicon esculentum, Nicotiana tabacum* and *Phaseolus vulgaris* (Ivanova et al., 2002).

Cinnamomin inhibited protein synthesis in bollworm larvae in an *in vitro* translation system. The LC50 of dietary cinnamomin was 1839 ppm on bollworm (*Helicoverpa armigera*) larvae and the LC50 of cinnamomin on mosquito (*Culex pipines pallens*) larvae was 168 ppm. (Zhou et al., 2000).

The IC50 of cinnamomin on melanoma M21 cells and human hepatocarcinoma 7721 cells were 11.7 nM and 18.8 nM, respectively. In comparison, the type 1 ribosome-inactivating protein camphorin from *Cinnamomum camphora* seeds inhibited human hepatocarcinoma 7721 cells with a weaker potency i.e., with an IC50 of 59 nM, and did not inhibit melanoma M21cells at all. Cinnamomin markedly repressed the growth of solid melanoma in nude mice. The isolation of an R-fragment from ribosomes of cancer cells exposed to the *C. camphora* ribosome inactivating proteins following treatment with acidic aniline signifies that the proteins act on the ribosomes of the cancer cells (Ling and Liu, 1996).

Ebulin L and Nigrin B (Non-Toxic)

Ebulin l- and nigrin b-transferrin conjugates directed to transferrin receptor overexpressed in cancer cells was constructed with a considerable decline in IC50 compared with those of ebulin l and nigrin b (Citores et al., 2002). An immunotoxin was constructed by conjugating the non-toxic type II ribosome-inactivating protein ebulin l with the mouse anti-human CD105 monoclonal antibody 44G4. The immunotoxin exerted cytotoxicity with an IC50 in the nanomolar concentrations toward human CD105+ cells like L6E9 rat myoblasts transfected with long human CD105 and the L929 mouse fibroblasts transfected with short human CD105. Cells devoid of human CD105 had a substantially lower sensitivity to the immunotoxin. Since ebulin l manifests a much reduced toxicity *in vivo* compared with ricin and other ribosome inactivating proteins employed in the construction of immunotoxins, its immunotoxin may be useful for the treatment of cancer (Benítez et al., 2005).

Eranthis Hyemalis Lectin

N-acetylgalactosamine specific and blood-group-O erythrocyte specific *Eranthis hyemalis* lectin is a dimeric protein (Mr 62000) composed of a 30-kDa subunit and a 32-kDa subunit, bound together by disulphide linkages. There is an abundance of glutamic acid/glutamine, leucine, and aspartic acid/asparagine residues (Cammue et al., 1985). The lectin manifested translation-inhibiting activity and antiviral activity against alfalfa mosaic virus, and larvicidal activity against the southern corn rootworm *Diabrotica undecimpunctata howardii* which infests the maize plant (Kumar et al., 1993).

The lectin by itself, and also the lectin conjugated with citrate gold nanoparticles, delayed the commencement of reproduction and impaired the fecundity of nematodes (Djafari et al., 2018).

Arrested first-stage larvae exposed to the lectin manifested direct mortality, and surviving larvae exhibited food avoidance, lowered fecundity, developmental retardation and constitutive dauer larvae formation. Both inappropriate dauer larval development and difficulty to find the source of food supply were due to the attenuated chemosensory capability and destruction of amphid neurons of *C. elegans* (McConnell et al., 2015).

Eutirucallin from Latex of *Euphorbia Tirucalli*

The hemagglutinating activity of the immunoenhancing lectin toward blood group A, B and O human red blood cells was calcium ion-independent, stable up till 55°C, and inhibited by D-galactose. The partial amino acid sequences exhibited homology to the lectin domains of type-2 ribosome inactivating proteins (Santana et al., 2014).

Foetidissimin II from *Cucurbita Foetidissima* Roots

The 61- kDa foetidissimin II is composed of two chains, a 29-kDa A chain and a 32-kDa B chain. It demonstated RNA-N-glycosidase activity and inhibited cell-free protein-synthesis activity with an IC50 of 0.251 uM and erythroleukemia and adenocarcinoma cells with an IC50 of about 70 nM (Zhang and Halaweish, 2007).

Iris Agglutinin B and Iris Agglutinin R from *Iris Holland*

Iris agglutinin b and iris agglutinin r manifested N-glycosidase activity and exhibited sequence homology to the type 2 ribosome-inactivating proteins reported earlier. All of these hemagglutinins exhibited binding toward mannose and Gal/N-acetylgalactosamine and displayed only about 60% sequence identity to type 1 iris ribosome-inactivating protein. Iris agglutinin b and Iris agglutinin r exhibited low cytotoxicity toward CEM/0, FM3A/0, L1210/0, and Molt4/C8 cells. The cytotoxicity was higher than that of type 1 ribosome-inactivating proteins, but lower than that of mistletoe type 2 ribosome-inactivating protein (Hao et al., 2001).

Lanceolin and Stenodactylin from Caudices of *Adenia Lanceolata* and *A. Stenodactyla*

The 61-kDa lanceolin and 63-kDa stenodactylin are glycoproteins with N-terminal amino acid sequences resembling that of volkensin from *A. volkensii.* They depurinated ribosomes and DNA, but not tRNA or poly (A), suppressed cell-free translation and translation in cells, and exhibited hemagglutinating activity. They evoked apoptosis in cells. Lanceolin and stenodactylin were toxic to mice, with LD50 values of 8.16 ug/kg and 2.76 ug/kg at 48 h, respectively (Stirpe et al., 2007).

Stenodactylin was more toxic than lanceolin and volkensin toward HeLa cells. The uptake, exocytosis and re-uptake of non-degraded stenodactylin occurred to a greater extent than those of volkensin and lanceolin. The potent cytotoxicity of the Adenia RIPs may be attributed to high affinity binding to cells and efficient endocytosis via a clathrin-independent pathway, routing inside the cells analogous to that of other ricin-like toxic ribosome-inactivating proteins such as modeccin and volkensin, and partial resistance to proteases (Battelli et al., 2011).

Stenodactylin structurally and enzymatically resembles its renowned counterpart ricin. Nevertheless, unlike ricin, transport of stenodactylin in the central nervous system and in peripheral nerves occurs in a retrograde manner. Stenodactylin at 1 pM decreased viability of neuroblastoma cells and protein synthesis after incubation for 24 h. It evoked apoptosis and free radical generation. Caspase inhibition using Z-VAD-fmk, together with hydrogen peroxide scavenging using catalase or necroptotic inhibition using necrostatin-1, totally abolished the effects of stenodactylin (Polito et al., 2016).

Modeccin from *Adenia Digitata*

Diphtheria toxin and *Pseudomonas aeruginosa* exotoxin A demonstrated specific binding to the basolateral side but not to the apical side of MDCK-I kidney cells. Their protein synthesis inhibitory action was most pronounced when they were applied to the basolateral side of the MDCK-I, MDBK and Pt K2 kidney cells, and the colon carcinoma Caco-2 cells. A similar phenomenon was observed regarding the action of modeccin and volkensin on MDCK-I cells. In contrast, the cytotoxicity of modeccin and *P. aeruginosa* exotoxin A on the basolateral side in the trophoblastic BeWo cells did not differ from their toxicity applied to the apical side. The protein synthesis inhibitory action of abrin and the bacterial type II ribosome inactivating protein Shigella toxin on the various cell lines examined was the same regardless of whether the ribosome inactivating protein was applied to the basolateral side or to the apical side. Hence entry to the cells was possible in the presence of receptors (Melby et al., 1993).

Momordica *Charantia Lectin*

Momordica charantia lectin had little effect on normal nasopharyngeal 69 cells, but exhibited a cytotoxic action on nasopharyngeal carcinoma CNE-1 and CNE-2 cells, as evidenced by DNA fragmentation, arrest in G1-phase of the cell cycle, mitochondrial injury and apoptosis. The downregulated expression of cyclin D1 and phosphoretinoblastoma protein brought about cell cycle arrest at G1-phase, accompanied by regulation of mitogen-activated protein kinases (MAPK, such as p38 MAPK, JNK, and ERK) phosphorylation and enhanced downstream nitric oxide formation. Simultaneous treatment with SB-203580, a p38 MAPK inhibitor, curtailed nitric oxide generation and action of *M.charantia* lectin on nasopharyngeal carcinoma cells. *M. charantia* lectin augmented cytochrome c release into the cytoplasm, activated caspases-3, -8, and -9, and escalated cleaved PARP production, culminating in DNA fragmentation and apoptosis. *M. charantia* lectin when intraperitoneally injected (1.0 mg/kg daily) produced about 50% remission of the tumors subcutaneously inoculated into nude mice (Fang et al., 2012).

Pulchellin from *Abrus Pulchellus* Seeds

Abrus pulchellus seeds have, at the minimum, seven similar and very toxic type 2 ribosome inactivating proteins referred to as pulchellins. Isoforms P I and P II were similar in toxicity and displayed elevated toxicity on HeLa cells *in vitro* and in mice *in vivo* compared with isoforms P III and P IV. Based on secondary structural characteristics and sugar binding properties the isoforms can be separated into a P I/P II subgroup and a P III/P IV subgroup. The variation in toxicity may be ascribed to differences in cell surface interactions brought about by the B-chain that lead to toxin uptake (Castilho et al., 2008).

The deduced amino acid sequence of recombinant pulchellin A-chain (rPAC) had about 85% identity with abrin-c A-chain. The rPAC exhibited rRNA depurinating activity on yeast ribosomes *in vitro*.

Incubation of rPAC and recombinant pulchellin binding chain (rPBC) in the presence of a reduced/oxidized system gave rise to an 60- kDa active heterodimer (rPAB) with an LD50 of 45 ug/kg in mice upon intraperitoneal administration (Silva et al., 2005). Pulchellin is promising as an adjuvant in breast cancer treatment (de Matos et al., 2012).

Both native pulchellin and rPBC employ the endosomal internalization pathway and retrograde transport through the Golgi apparatus (Goto et al., 2007). Intratumoral injection of mice with pulchellin at 0.75 μg/kg upregulated interferon-γ, interleukin-4 and interleukin-10 and downregulated interleukin -6 and transforming growth factor-β. Macrophages activation occurred as witnessed in the elevated percentage of macrophages expressing adhesion CD54 and costimulatory molecule CD80 (de Matos et al., 2012).

Recombinant ricin A chain (RAC) and recombinant pulchellin (PAC) could be coupled to two different anti-HIV monoclonal antibodies (MAbs), anti-gp120 MAb 924 or anti-gp41 MAb 7B2. The immunotoxins specifically bound native antigens on persistently HIV-infected cells and recombinant antigens on Env-transfected cells. RAC was more cytotoxic than PAC (Sadraeian et al., 2017).

Riproximin from *Ximenia Americana*

Riproximin is the main component in *Ximenia americana* powder used by African healers for cancer therapy. Riproximin inhibited proliferation of human and rodent cancer cells and exhibited anticancer activity in rat models of colorectal and pancreatic cancer with metastasis to the liver. The mechanism of the action of riproximin involves binding of B-chain to cell surface glycans, cellular uptake of riproximin, depurination of the 28S ribosomal RNA due to catalytic action of the A-chain and induction of unfolded protein response (Adwan et al., 2014).

The type II RIPs ricin, riproximin and volkensin elicited unfolded protein response in MDA-MB-231 cells and HCT116 cells, in response to endoplasmic reticulum stress (Horrix et al., 2011).

Riproximin inhibited pancreatic cancer ASML cell proliferation with low IC50 values of 0.8-172 pM, decreased tumor size by 90% and prolonged survival from 17.6 days to 21.8 days after thrice weekly intraperitoneal administration at doses of 1.7-5.4 μg/kg. The standard drug gemcitabine had an attenuated efficacy compared with riproximin (Murtaja et al., 2017).

Immunotoxin Formed from Ricin A Chain RTA and Monoclonal Antibody against Laryngeal Carcinoma Cells

An immunotoxin composed of the toxic ricin A chain conjugated with a monoclonal antibody against laryngeal carcinoma cells selectively killed laryngeal carcinoma cells but not various normal living cells *in vitro* (Zenner, 1986).

Immunotoxin RTA-4D5-KDEL Formed from Ricin A Chain RTA, Anti-HER2 Single-Chain Variable Fragment 4D5 SCFV and Endoplasmic Reticulum Targeting HER2/Neu-Positive Cancer Cells and Locating the Endoplasmic Reticulum

The RTA-4D5-KDEL immunotoxin exhibited a cytotoxic effect on ovarian cancer SKOV-3 cells which was 28-fold and 440-fold higher than that of RTA-4D5 and RTA, respectively. Although RTA- 4D5-KDEL markedly suppressed HER2-overexpressing SKOV-3 cells, it exerted negligible adverse effects on lung cancer H460 cells and normal kidney HEK-293 cells. The immunotoxin RTA-4D5 bound SKOV-3 cells, but not normal HEK-293cells. (Jiao et al., 2018).

Lectin-Rich Mistletoe (*Viscum Album*) Extract

Intratumorally-injected lectin-rich mistletoe extract had higher antitumor efficacy than intravenous gemcitabine on a human pancreatic cancer xenograft (Rostock et al., 2005).

The results of a randomized phase III study indicated that a mistletoe preparation may improve the quality of life in cancer patients (Stauder et al., 2002).

Three Mistletoe Lectins

The three mistletoe lectins ML I, ML II and ML III, augmented by several-fold the secretion of the cytokines interleukin -1 alpha, interleukin-1 beta and interleukin-6 and tumor necrosis factor-alpha by human monocytes at a dosage 100-10,000-fold lower than the toxic dosage (Ribéreau-Gayon et al., 1996).

Viscum Album Agglutinin-1 (VAA-1)

The lectin triggered apoptotic death in leukemic T- cells and B-cells via a receptor-independent mitochondrial apoptotic pathway. Cytochrome c release into the cytosol was enhanced. Apoptosis stimulated by the lectin was nearly abrogated by a peptide caspase inhibitor. Activation of caspases-3, -8, and -9 by proteases in response to the lectin signifies caspase participation. Apoptosis evoked by the lectin was not undermined in CD95 resistant cell clones or in cells made insensitive to other death receptors by overexpressing a dominant- negative FADD mutant (Bantel et al., 1999).

Viscum album agglutinin-1 showed additive anticancer activity in conjunction with cisplatin, doxorubicin, and taxol on human lung cancer A549 cells but the agglutinin synergized with cycloheximide. Arrest of the cancer cells in G1phase of the cell cycle was observed (Siegle et al., 2001). Mistletoe lectin ML-1 synergized with paclitaxel in anticancer action toward cultured human hepatocarcinoma SK-Hep1 cells (Pae et al., 2001).

Dietary inclusion of mistletoe lectin ML-1 suppressed growth of a murine non-Hodgkin lymphoma (NHL) tumour as a consequence of diminished mitotic activity, heightened infiltration of lymphocytes, and increased appearance of apoptotic bodies and impaired angiogenesis. The lectin induced hyperplasia of the small intestine and demonstrated strong binding to lymphoid tissue of Peyer's patches. Viable tumor was undetectable in about one-quarter of the mice receiving the lectin for 11 days (Pryme et al., 2002, 2004).

Nearly one-third of breast cancer cells bound VAA-1 strongly. VAA-1 binding reflects the presence of galactose in plasma membrane or cytoplasmic saccharides, the amount of galactose-containing glycoproteins, glycolipids and mucoproteins in breast cancer cells is related to outcome of the disease. According to Kaplan-Meier statistics, strong binding of VAA-1 to tumor cells is inversely related to prognosis of the disease (Fritz et al., 2004).

The expression of transforming growth factor (TGF)-β, TGF-β receptor II (TGFBR2), TGF-β intracellular signal transducer protein SMAD2, and matrix metalloproteinases (MMP)-2 and MMP-14 was inhibited by Viscumin. Tumor cell motility was impaired (Schötterl et al., 2017).

Mistletoe Lectin II

The beta-galactoside- and N-acetyl-D- galactosamine -specific lectin II (60 kDa), viscotoxin (5 kDa) and polysaccharides in mistletoe extract triggered apoptosis of human acute myeloid leukemia U937 cells. The mistletoe extract upregulated phosphotransferase activity of c-Jun N-terminal kinase 1 (JNK1)/stress-activated protein kinase (SAPK) in U937 cells. Mistletoe lectin II was more active than polysaccharides and viscotoxin in evoking apoptosis and inducing JNK1 activation of U937 cells. The peptide aC-DEVD-CHO, but not aC-YVAD-CHO, suppressed mistletoe lectin II induced catalytic activation of JNK1. Mistletoe lectin II also induced apoptosis in DLD-1 cells, HL-60 cells, Jurkat T cells, primary acute myelocytic leukemic cells, and RAW 264.7 cells (Park et al., 2000).

Recombinant Mistletoe Lectin

Although the E166Q R169Q mutein of recombinant mistletoe lectin (rML) showed a greater diminution of rRNA-N-glycosidase activity (237 fold) than cytotoxic and apoptosis inducing activity (20-22 fold), the consequences of the site-specific mutagenesis indicate that the anticancer activity of recombinant Viscumin (rViscumin) is dependent on its ribosome inactivating protein activity (Langer et al., 1999).

rViscumin upregulated the number of pan-natural killer cells, T-lymphocytes and activated monocytes, decreased the number of tumor colonies, and extended survival time in Balb/c mice bearing RAW -117-P or L-1 sarcoma cells after subcutaneous treatment (Schaffrath et al., 2001).

Daily intraperitoneal injection of 30 ng rViscumin per kg body weight into severe combined immunodeficiency (SCID) mice bearing xenografts of SoTü 3 human ovarian cancer cells for about 80 days resulted in a larger number of tumor-free survivors (13 compared with 2 in vehicle-treated control mice) (Schumacher et al., 2000).

Trastuzumab (Herceptin), a HER-2 antibody, undermined the sensitivity of cancer cells to rViscumin. rViscumin effected HER-2-dependent downregulation of bcl-2 and upregulation of members of the MAPK family, p38, p42/44, and SAPK/JNK, but not caspases-3 and -7 (Abuharbeid et al., 2004).

rViscumin augmented the cytotoxicity of cisplatin, idarubicin, mafosfamide, and vincristine toward human leukemia KG1a cells and K562 cells *in vitro*. rViscumin prevented colony formation in human marrow progenitor cells but there was no effect on the suppression of clonogenicity brought about by cisplatin. rViscumin did not inhibit cisplatin -induced upregulation of mRNA expression of the nucleotide excision repair gene ERCC-1 in K562 cells. Hence the potentiating action of rViscumin on cytotoxic agents *in vitro* was unrelated to the transcriptional activity of DNA repair relevant genes (Galm et al., 2002).

Human bladder carcinoma 5637 cells and human promyelotic HL-60 cells possess about 1.5 million and 3.7 million receptor ganglioside IV(6)Neu5Ac-nLc4Cer molecules per cell, respectively. rViscumin exhibited an IC50 value of 1.16 pM toward human promyelotic HL-60 cells and 12.1 pM toward human bladder carcinoma 5637 cells. rViscumin resistant CHO-K1 cells devoid of alpha 2-6-sialylated gangliosides acquired sensitivity to rViscumin after exogenous addition of human granulocyte gangliosides. Thus, neolacto-series gangliosides with Neu5Acalpha2-6 Galbeta1-4GlcNAc-terminus are rViscumin receptors (Müthing et al., 2002, 2004). Sialic acid-specific rViscumin binds type II N-glycans with CD75s determinants. The structurally homologous but galactoside-specific ricin does not react with terminally sialylated oligosaccharides e.g., CD75s motifs. The binding specificity of rViscumin and elevated CD75s ganglioside expression in a diversity of cancers makes rViscumin promising as an adjuvant for cancer therapy (Müthing et al., 2005). rViscumin displays cytotoxic and immunomodulatory activities *in vitro* and in animal models. Clinical phase I studies revealed its disease stabilizing effects (Zwierzina et al., 2011).

A sandwich immuno-PCR (IPCR) assay has been made available for the detection of rViscumin in blood plasma with a sensitivity of 50 zeptomol (about 100 fg/ml) (Adler et al., 2003).

Korean Mistletoe Lectin (*Viscum Album* Coloratum Agglutinin)

Korean mistletoe lectin is galactose- and N-acetyl-D-galactosamine-specific. In order to circumvent gastrointestinal instability of the lectin, the lectin is incorporated into alginate/chitosan microcapsules which prevent lectin breakdown at low pH values. Coating the lectin with polyacrylic polymers, Eudragit, yielded desirable release profiles with negligible reduction of cytotoxicity. (Lyu et al., 2004).The lectin brought about total regression of xenografts of human colon cancer cells in nude mice. Caspases-2, -3, -8, and -9 were activated and expression of anti-apoptotic molecules including Akt/protein kinase B, nuclear factor-kappaB, receptor interacting protein, and X-linked inhibitor of apoptosis protein was downregulated. Anti-tumor necrosis factor receptor 1 undermined the activation of caspases, especially caspase-8, and suppressed apoptosis evoked by the agglutinin, indicating that tumor necrosis factor receptor 1 was implicated (Khil et al., 2007). Korean mistletoe lectin elicited apoptosis in human leukemia HL-60 cells which was inhibited by dibutyryl cyclic AMP, protein kinase A activator, protein kinase C activator, or 12-O- tertadecanoyl phorbol 13-acetate (TPA). The inhibitory effects of TPA and dibutyryl cyclic AMP were removed by the protein kinase A inhibitor KT5720 and the protein kinase C inhibitor staurospoline. Hence activation of protein kinase A or protein kinase C in leukemia cells may confer protection against apoptosis evoked by Korean mistletoe lectin (Pae et al., 2000).

Korean as well as European mistletoe lectin suppressed tumor-induced angiogenesis and metastasis in mice. The lectin was effective when administered via the oral, subcutaneous and systemic routes. It increased the sensitivity of cancer cells to the triggering of apoptosis by tumor necrosis factor-alpha (Pryme et al., 2006).

Korean mistletoe lectin exerted antiproliferative activity toward B16-BL6 melanoma cells and arrested the cells in sub-G1 phase of the cell cycle and induced apoptosis. The lectin inhibited angiogenesis as indicated by the choriallantoic membrane assay. It prolonged survival of C57BL6 mice inoculated with B16-BL6 melanoma cells, and prevented pulmonary metastasis (Park et al., 2001).

Korean mistletoe lectin induced apoptosis in both SK-Hep-1 (p53-positive) and Hep 3B (p53-negative) hepatocarcinoma cells by upregulation of Bax, downregulation of Bcl-2 and telomerase. The lectin sensitized cancer cells to chemotherapeutic drugs (Lyu et al., 2002).

Chinese Mistletoe Lectin

The lectin inhibited colorectal cancer HT-29 cells and CLY cells *in vitro* as well as *in vivo*. It down-regulated miR-135a & b, and up-regulated expression of their target gene adenomatous polyposis coli and phosphorylation of the related effector β-catenin (Li et al., 2011).

Volkensin from *Adenia Volkensii* Roots

Volkensin from *Adenia volkensii* roots exhibited an LD50 of 50-60 ng/kg in rats. The hemagglutinating activity and carbohydrate-binding specificity of native volkensin as well as recombinant volkensin has been examined to correlate with the activity of volkensin on axonal transport along neurons (Chambery et al., 2007).

Attenuated Mutants of Shiga-Like Toxin I

Two attenuated mutants of Shiga-like toxin I (Stx1), Stx1(W203F) and Stx1(R170H), inhibited proliferation of human umbilical vein endothelial cells (HUVECs) and human ovarian cancer SKOV3 cells *in vitro.* The mutants brought about cell cycle arrest in G(2)-M or S phase and necrosis of SKOV3 cells. In addition, they impeded migration as well as capillary tube formation of HUVECs. The mutants decreased microvessel densities of SKOV3 xenografts in nude mice. Stx1(W203F) suppressed tumor growth in nude mice (Wei et al., 2008).

DISCUSSION AND CONCLUSION

The following list encompasses most of the plant type 2 RIPs that have been reported to date : aralin from *Aralia elata,* articulatin-D from *Viscum articulatum,* cinnamomin from camphor tree (*Cinnamomum camphora*), ebulin l from *Sambucus ebulus* and nigrin b from *Sambucus nigra* (non-toxic), eutirucallin from latex of *Euphorbia tirucalli,* foetidissimin II from *Cucurbita foetidissima* roots, Iris agglutinin b and Iris agglutinin r from *Iris holland,* modeccin from *Adenia digitata,* lanceolin and stenodactylin from caudices of *Adenia lanceolata* and *A. stenodactyla,* bitter gourd (*Momordica charantia*) lectin, pulchellin from *Abrus pulchellus* seeds, mistletoe (*Viscum album*) agglutinin, and volkensin from *Adenia volkensii* roots. Mistletoe agglutinin has undergone clinical trials and may have a disease stabilizing effect on cancer patients and improve their quality of life. In addition, two attenuated mutants of Shiga-like toxin I (Stx1), Stx1 (W203F) and Stx1 (R170H) have anticancer and antiangiogenic effects on ovarian cancer.

Type 2 RIPs have been reported to have anticancer activity against a diversity of cancers including melanoma, breast cancer, ovarian cancer, cervical cancer, leukemia, liver cancer, colorectal cancer, nasopharyngeal cancer, laryngeal cancer, and lung cancer (Table 1). The mechanism of anticancer action has been elucidated in some of them such as articulatin -D and *Momordica charantia* lectin. The mechanisms on cancer cells comprise cell cycle arrest, alteration of mitochondrial membrane potential, upregulation of pro-apoptotic proteins, inhibition of cancer cell proliferation, and suppression of angiogenesis and metastasis.

The following represent examples of anticancer immunotoxins based on type 2 RIPs:(i) abrin A chain-vascular endothelial growth factor isoform 121 fusion protein used in conjunction with the interleukin-12, (ii) immunotoxin RTA-4D5-KDEL formed from ricin A chain RTA, anti-HER2 single-chain variable fragment 4D5 scFv and endoplasmic reticulum targeting HER2/ neu- positive cancer cells and locating the endoplasmic reticulum, and (iii) immunotoxin formed from ricin A chain RTA and monoclonal antibody against laryngeal carcinoma cells. However, RIP immunotoxins may have the drawbacks of instability and immunogenicity. PEGylation using polyethylene glycol polymers to prolong half-life and lower immunogeicity may help (Li et al., 2009).

Besides anticancer action, some type 2 RIPs are known to have antiviral and anti-insect activites which may be exploitable in agriculture.

All in all, some of the type 2 RIPs may be abused as devastating bioweapons while others can be used to the welfare of mankind such as cancer therapy.

Table 1. Anticancer and other actions of Type II ribosome inactivating proteins

Abrin from jequirity bean (*Abrus precatorius*) Abrin A chain-vascular endothelial growth factor isoform 121 fusion protein used in conjunction with the cytokine interleukin-12 for anticancer action (Ciomber et al., 2014).
Aralin from *Aralia elata* cytotoxicity toward tumor cells overexpressing binding protein for high-density lipoprotein protein (Otsuka et al., 2014) e.g., HeLa cells.
Articulatin-D from *Viscum articulatum* cytotoxicity toward leukemia cells (Mishra et al., 2017).
Cinnamomin from camphor tree (*Cinnamomum camphora*) cytotoxicity toward BA/F3beta cells (Wang et al., 2006), melanoma M21 cells and human hepatocarcinoma 7721 cells (Ling and Liu,1996); ani-insect activity (Zhou et al., 2000).
Ebulin l (non-toxic) immunotoxin constructed by conjugating non-toxic type II ribosome-inactivating protein ebulin l with mouse anti-human CD105 monoclonal antibody 44G4 toward human CD105+ cells like L6E9 rat myoblasts transfected with long human CD105 and the L929 mouse fibroblasts transfected with short human CD105(Benítez et al., 2005).
***Eranthis hyemalis* lectin** antiviral activity against alfalfa mosaic virus, and larvicidal activity against southern corn rootworm which infests the maize plant (Kumar et al., 1993), anti-nematode activity (Djafari et al., 2018).
Foetidissimin II from *Cucurbita foetidissima* roots cytotoxicity toward erythroleukemia and adenocarcinoma cells (Zhang and Halaweish, 2007).
Iris agglutinin r from *Iris holland* cytotoxicity toward CEM/0, FM3A/0, L1210/0, and Molt4/C8 cells (Hao et al., 2001).
Lanceolin and stenodactylin from caudices of *Adenia lanceolata* and *A. stenodactyla* Stenodactylin was more toxic than lanceolin and volkensin toward HeLa cells (Battelli et al., 2011).
***Momordica charantia* lectin** nasopharyngeal carcinoma CNE-1 and CNE-2 cells (Fang et al., 2012)
Pulchellin from *Abrus pulchellus* seeds cytotoxicity toward HeLa cells (Castilho et al., 2008); upregulated interferon-γ, interleukin-4 and interleukin-10 and downregulated interleukin -6 and transforming growth factor-β; increased population of macrophages expressing adhesion CD54 and costimulatory molecule CD80 (de Matos et al., 2012).
Riproximin from *Ximenia americana* anticancer activity in rat models of colorectal and pancreatic cancer with hepatic metastasis (Adwan et al., 2014); inhibited pancreatic cancer ASML cell proliferation, reduced tumor growth by 90% and prolonged survival (Murtaja et al., 2017); elicited unfolded protein response in breast cancer MDA-MB-231 cells and human colon cancer HCT116 cells in response to endoplasmic reticulum stress (Horrix et al., 2011)

Table 1. (Continued)

Immunotoxin formed from ricin A chain RTA and monoclonal antibody against laryngeal carcinoma cells cytotoxicity toward laryngeal carcinoma cells (Zenner, 1986).
Immunotoxin RTA-4D5-KDEL formed from ricin A chain RTA, anti-HER2 single-chain variable fragment 4D5 scFv and endoplasmic reticulum targeting HER2/neu-positive cancer cells and locating the endoplasmic reticulum cytotoxic effect on ovarian cancer SKOV-3 cells (Jiao et al., 2018)
Recombinant mistletoe lectin rViscumin upregulated the number of pan-natural killer cells, T-lymphocytes and activated monocytes, decreased the number of tumor colonies, and extended survival time in Balb/c mice bearing RAW -117-P or L-1 sarcoma cells after subcutaneous treatment (Schaffrath et al., 2001);SCID mice bearing xenografts of SoTü 3 human ovarian cancer cells(Schumacher et al., 2000);augmented the cytotoxicity of cisplatin, idarubicin, mafosfamide, and vincristine toward human leukemia KG1a cells and K562 cells *in vitro* (Galm et al., 2002)
Korean mistletoe lectin (*Viscum album* coloratum agglutinin) total regression of xenografts of human colon cancer cells in nude mice(Khil et al., 2007), elicited apoptosis in human leukemia HL-60 cells (Pae et al., 2000), antiproliferative activity toward B16-BL6 melanoma cells, prolonged survival of C57BL6 mice inoculated with B16-BL6 melanoma cells, and prevented pulmonary metastasis (Park et al.,2001), induced apoptosis in both SK-Hep-1 (p53-positive) and Hep 3B (p53-negative) hepatocarcinoma cells by upregulation of Bax, downregulation of Bcl-2 and telomerase. The lectin sensitized cancer cells to chemotherapeutic drugs (Lyu et al., 2002).
Chinese mistletoe lectin inhibited colorectal cancer human colorectal cancer HT-29 cells and CLY cells *in vitro* as well as *in vivo* (Li et al., 2011)
Attenuated mutants of Shiga-like toxin I cytotoxicity toward human ovarian cancer SKOV3 cells suppressed tumor growth in nude mice (Wei et al., 2008).

ACKNOWLEDGMENTS

We gratefully acknowledge the award of Health and Medical Research Fund (no. 12131221) from Food and Health Bureau, The Government of Hong Kong Special Administrative Region, and grants from National Natural Science Foundation of China (no. 81417927).

REFERENCES

Abuharbeid S, Apel J, Sander M, Fiedler B, Langer M, Zuzarte ML, Czubayko F,Aigner A. Cytotoxicity of the novel anti-cancer drug rViscumin depends on HER-2 levels in SKOV-3 cells. *Biochem Biophys Res Commun.* 2004 Aug 20;321(2):403-12.

Adler M, Langer M, Witthohn K, Eck J, Blohm D, Niemeyer CM. Detection of rViscumin in plasma samples by immuno-PCR. *Biochem Biophys Res Commun.* 2003 Jan 17;300(3):757-63.

Adwan H, Bayer H, Pervaiz A, Sagini M, Berger MR. Riproximin is a recently discovered type II ribosome inactivating protein with potential for treating cancer. *Biotechnol Adv.* 2014 Nov 1;32(6):1077-90.

Bantel H, Engels IH, Voelter W, Schulze-Osthoff K, Wesselborg S. Mistletoe lectin activates caspase-8/FLICE independently of death receptor signaling and enhances anticancer drug-induced apoptosis. *Cancer Res.* 1999 May 1;59(9):2083-90.

Barbier J, Gillet D. Ribosome inactivating proteins: From plant defense to treatments against human misuse or diseases. *Toxins* (Basel). 2018 Apr 18;10(4).pii: E160.

Barbieri L, Ciani M, Girbés T, Liu WY, Van Damme EJ, Peumans WJ, Stirpe F.Enzymatic activity of toxic and non-toxic type 2 ribosome-inactivating proteins. *FEBS Lett.* 2004 Apr 9;563(1-3):219-22.

Battelli MG, Scicchitano V, Polito L, Farini V, Barbieri L, Bolognesi A. Binding and intracellular routing of the plant-toxic lectins, lanceolin and stenodactylin. *Biochim Biophys Acta.* 2010 Dec;1800(12):1276-82.

Benítez J, Ferreras JM, Muñoz R, Arias Y, Iglesias R, Córdoba-Díaz M, del Villar R, Girbés T. Cytotoxicity of an ebulin l-anti-human CD105 immunotoxin on mouse fibroblasts (L929) and rat myoblasts (L6E9) cells expressing human CD105. *Med Chem.* 2005 Jan;1(1):65-70.

Cammue BP, Peeters B, Peumans WJ. Isolation and partial characterization of an N-acetyl galactosamine - specific lectin from winter-aconite (*Eranthis hyemalis*) root tubers. *Biochem J.* 1985 May 1;227(3):949-55.

Castilho PV, Goto LS, Roberts LM, Araújo AP. Isolation and characterization of four type 2 ribosome inactivating pulchellin isoforms from *Abrus pulchellus* seeds. *FEBS J.* 2008 Mar;275(5):948-59.

Chambery A, Severino V, Stirpe F, Parente A. Cloning and expression of the B chain of volkensin, type 2 ribosome inactivating protein from *Adenia volkensii harms*: co-folding with the A chain for heterodimer reconstitution. *Protein Expr Purif.* 2007 Feb;51(2):209-15.

Chan YS, Ng TB. Shiga toxins: from structure and mechanism to applications. *Appl Microbiol Biotechnol.* 2016 Feb;100(4):1597-610

Ciomber A, Smagur A, Mitrus I, Cichoń T, Smolarczyk R, Sochanik A, Szala S, Jarosz M. Antitumor effects of recombinant antivascular protein ABRaA-VEGF121 combined with IL-12 gene therapy. *Arch Immunol Ther Exp* (Warsz). 2014 Apr;62(2):161-8.

Citores L, Ferreras JM, Muñoz R, Benítez J, Jiménez P, Girbés T. Targeting cancer cells with transferrin conjugates containing the non-toxic type 2 ribosome-inactivating proteins nigrin b or ebulin l. *Cancer Lett.* 2002 Oct 8;184(1):29-35.

de Matos DC, de Ribeiro LC, Tansini A, Ferreira LS, Placeres MC, Colombo LL, Carlos IZ. Immunological response in mice bearing LM3 breast tumor undergoing pulchellin treatment. *BMC Complement Altern Med.* 2012 Jul 24;12:107.

Djafari J, McConnell MT, Santos HM, Capelo JL, Bertolo E, Harvey SC, Lodeiro C, Fernández-Lodeiro J. Synthesis of gold functionalised nanoparticles with the *Eranthis hyemalis* lectin and preliminary toxicological studies on *Caenorhabditis elegans. Materials* (Basel). 2018 Aug 6;11(8). pii: E1363.

Fang EF, Zhang CZ, Ng TB, Wong JH, Pan WL, Ye XJ, Chan YS, Fong WP. *Momordica charantia* lectin, a type II ribosome inactivating protein, exhibits antitumor activity toward human nasopharyngeal carcinoma cells *in vitro* and *in vivo. Cancer Prev Res* (Phila). 2012 Jan;5(1):109-21.

Fritz P, Dippon J, Kierschke T, Siegle I, Möhring A, Moisa A, Mürdter TE. Impact of mistletoe lectin binding in breast cancer. *Anticancer Res.* 2004 Mar-Apr;24(2C):1187-92.

Galm O, Fabry U, Efferth T, Osieka R. Synergism between rViscumin and cisplatin is not dependent on ERCC-1 expression. *Cancer Lett.* 2002 Dec 10;187(1-2):143-51.

Goto LS, Castilho PV, Cominetti MR, Selistre-Araújo HS, Ulian Araújo AP.Endocytosis of pulchellin and its recombinant B-chain into K-562 cells: binding and uptake studies. *Biochim Biophys Acta.* 2007 Dec;1770(12):1660-6.

Hao Q, Van Damme EJ, Hause B, Barre A, Chen Y, Rougé P, Peumans WJ. Iris bulbs express type 1 and type 2 ribosome-inactivating proteins with unusual properties. *Plant Physiol.* 2001 Feb; 125 (2) :866-76.

Horrix C, Raviv Z, Flescher E, Voss C, Berger MR. Plant ribosome-inactivating proteins type II induce the unfolded protein response in human cancer cells. *Cell Mol Life Sci.* 2011 Apr;68(7):1269-81.

Ivanova DG, Sarkar HK, Singh BR. Cloning, expression, and biological activity of recombinant alpha-cinnamomin: toxicity to cranberry and other plant species. *J Nat Toxins.* 2002 May;11(2):95-102.

Jiao P, Zhang J, Dong Y, Wei D, Ren Y. Construction and characterization of the recombinant immunotoxin RTA-4D5-KDEL targeting HER2/neu-positive cancer cells and locating the endoplasmic reticulum. *Appl Microbiol Biotechnol.* 2018 Aug 24.doi: 10.1007/s00253-018-9291-z.

Jiménez P, Tejero J, Cordoba-Diaz D, Quinto EJ, Garrosa M, Gayoso MJ, Girbés T. Ebulin from dwarf elder (*Sambucus ebulus* L.): a mini-review. *Toxins* (Basel).2015 Feb 25;7(3):648-58.

Khil LY, Kim W, Lyu S, Park WB, Yoon JW, Jun HS. Mechanisms involved in Korean mistletoe lectin-induced apoptosis of cancer cells. *World J Gastroenterol.* 2007 May 28;13(20):2811-8.

Kumar MA, Timm DE, Neet KE, Owen WG, Peumans WJ, Rao AG. Characterization of the lectin from the bulbs of *Eranthis hyemalis* (winter aconite) as an inhibitor of protein synthesis. *J Biol Chem.* 1993 Nov 25; 268(33):25176-83.

Langer M, Möckel B, Eck J, Zinke H, Lentzen H. Site -specific mutagenesis of mistletoe lectin: the role of RIP activity in apoptosis. *Biochem Biophys Res Commun.* 1999 Nov 2;264(3):944-8.

Lapadula WJ, Marcet PL, Mascotti ML, Sanchez-Puerta MV, Juri Ayub M. Metazoan ribosome inactivating protein encoding genes acquired by horizontal gene transfer. *Sci Rep.* 2017 May 12;7(1):1863.

Li LN, Zhang HD, Zhi R, Yuan SJ. Down-regulation of some miRNAs by degrading their precursors contributes to anti-cancer effect of mistletoe lectin-I. *Br J Pharmacol.* 2011 Jan;162(2):349-64.

Li M, Chen Y, Liu Z, Shen F, Bian X, Meng Y. Anti-tumor activity and immunological modification of ribosome-inactivating protein (RIP) from *Momordica charantia* by covalent attachment of polyethylene glycol. *Acta Biochim Biophys Sin* (Shanghai) 2009;41:792–9.

Ling J, Liu WY. Cytotoxicity of two new ribosome - inactivating proteins, cinnamomin and camphorin, to carcinoma cells. *Cell Biochem Funct.* 1996 Sep;14(3):157-61.

Liu WY. Research on ribosome-inactivating proteins from angiospermae to gymnospermae and cryptogamia. *Am J Transl Res.* 2017 Dec 15;9(12):5719-5742.

Lyu SY, Choi SH, Park WB. Korean mistletoe lectin-induced apoptosis in hepatocarcinoma cells is associated with inhibition of telomerase via mitochondrial controlled pathway independent of p53. *Arch Pharm Res.* 2002;25(1):93-101.

Lyu SY, Kwon YJ, Joo HJ, Park WB. Preparation of alginate/chitosan microcapsules and enteric coated granules of mistletoe lectin. *Arch Pharm Res.* 2004 Jan;27(1):118-26.

McConnell MT, Lisgarten DR, Byrne LJ, Harvey SC, Bertolo E. Winter aconite (*Eranthis hyemalis*) lectin as a cytotoxic effector in the lifecycle of *Caenorhabditis elegans*. *Peer J.* 2015 Aug 20;3:e1206.

Mishra R, Das MK, Singh S, Sharma RS, Sharma S, Mishra V. Articulatin-D induces apoptosis via activation of caspase-8 in acute T-cell leukemia cell line. *Mol Cell Biochem.* 2017 Feb;426(1-2):87-99.

Murtaja A, Eyol E, Xiaoqi J, Berger MR, Adwan H. The ribosome inhibiting protein riproximin shows antineoplastic activity in experimental pancreatic cancer liver metastasis. *Oncol Lett.* 2018 Feb;15(2):1441-1448.

Müthing J, Meisen I, Bulau P, Langer M, Witthohn K, Lentzen H, Neumann U, Peter-Katalinić J. Mistletoe lectin I is a sialic acid-specific lectin with strict preference to gangliosides and glycoproteins with terminal Neu5Ac alpha 2-6Gal beta 1-4GlcNAc residues. *Biochemistry.* 2004 Mar 23;43(11):2996-3007.

Müthing J, Meisen I, Kniep B, Haier J, Senninger N, Neumann U, Langer M, Witthohn K, Milosević J, Peter-Katalinić J. Tumor-associated CD75s gangliosides and CD75s-bearing glycoproteins with Neu5Acalpha2-6Galbeta1- 4GlcNAc-residues are receptors for the anticancer drug rViscumin. *FASEB J.* 2005 Jan;19(1):103-5.

Melby EL, Jacobsen J, Olsnes S, Sandvig K. Entry of protein toxins in polarized epithelial cells. *Cancer Res.* 1993 Apr 15;53(8):1755-60.

Ng TB, Parkash A, Tso WW. Purification and characterization of alpha- and beta benincasins, arginine /glutamate-rich peptides with translation inhibiting activity from wax gourd seeds. *Peptides.* 2003 Jan;24(1):11-6.

Ng TB, Wang HX. Flammin and velin: new ribosome inactivating polypeptides from the mushroom *Flammulina velutipes*. *Peptides.* 2004 Jun;25(6):929-33.

Ng TB, Parkash A. Hispin, a novel ribosome inactivating protein with antifungal activity from hairy melon seeds. *Protein Expr Purif.* 2002 Nov;26(2):211-7.

Ng TB, Parkash A, Tso WW. Purification and characterization of moschins, arginine glutamate rich proteins with translation-inhibiting activity from brown pumpkin (*Cucurbita moschata*) seeds. *Protein Expr Purif.* 2002 Oct;26(1):9-13.

Otsuka H, Gotoh Y, Komeno T, Ono T, Kawasaki Y, Iida N, Shibagaki Y, Hattori S, Tomatsu M, Akiyama H, Tashiro F. Aralin, a type II ribosome-inactivating protein from *Aralia elata*, exhibits selective anticancer activity through the processed form of a 110-kDa high-density lipoprotein-binding protein: a promising anticancer drug. *Biochem Biophys Res Commun.* 2014 Oct 10;453(1):117-23.

Pae HO, Oh GS, Seo WG, Shin MK, Hong SG, Lee HS, Chung HT. Mistletoe lectin synergizes with paclitaxel in human SK-hep1 hepatocarcinoma cells. *Immunopharmacol Immunotoxicol.* 2001 Nov;23(4):531-40.

Pae HO, Seo WG, Shin M, Lee HS, Lee HS, Kim SB, Chung HT. Protein kinase A or C modulates the apoptosis induced by lectin II isolated from Korean mistletoe, *Viscum album* var. Coloratum, in the human leukemic HL-60 cells. *Immunopharmacol Immunotoxicol.* 2000 May;22(2):279-95.

Park R, Kim MS, So HS, Jung BH, Moon SR, Chung SY, Ko CB, Kim BR, Chung HT. Activation of c-Jun N-terminal kinase 1 (JNK1) in mistletoe lectin II-induced apoptosis of human myeloleukemic U937 cells. *Biochem Pharmacol.* 2000 Dec 1;60(11):1685-91.

Park WB, Lyu SY, Kim JH, Choi SH, Chung HK, Ahn SH, Hong SY, Yoon TJ, Choi MJ.Inhibition of tumor growth and metastasis by Korean mistletoe lectin is associated with apoptosis and antiangiogenesis. *Cancer Biother Radiopharm.* 2001 Oct;16(5):439-47.

Polito L, Bortolotti M, Pedrazzi M, Mercatelli D, Battelli MG, Bolognesi A. Apoptosis and necroptosis induced by stenodactylin in neuroblastoma cells can be completely prevented through caspase inhibition plus catalase or necrostatin-1. *Phytomedicine.* 2016;23(1):32-41.

Polito L, Bortolotti M, Pedrazzi M, Bolognesi A. Immunotoxins and other conjugates containing saporin-s6 for cancer therapy. *Toxins* (Basel). 2011 Jun;3(6):697-720.

Pryme IF, Bardocz S, Pusztai A, Ewen SW. Dietary mistletoe lectin supplementation and reduced growth of a murine non-Hodgkin lymphoma. *Histol Histopathol.* 2002;17(1):261-71.

Pryme IF, Bardocz S, Pusztai A, Ewen SW, Pfüller U. A mistletoe lectin (ML-1)-containing diet reduces the viability of a murine non-Hodgkin lymphoma tumor. *Cancer Detect Prev.* 2004;28(1):52-6.

Pryme IF, Bardocz S, Pusztai A, Ewen SW. Suppression of growth of tumour cell lines in vitro and tumours *in vivo* by mistletoe lectins. *Histol Histopathol.* 2006 Mar;21(3):285-99.

Ribéreau-Gayon G, Dumont S, Muller C, Jung ML, Poindron P, Anton R. Mistletoe lectins I, II and III induce the production of cytokines by cultured human monocytes. *Cancer Lett.* 1996 Dec 3;109 (1-2):33-8.

Rostock M, Huber R, Greiner T, Fritz P, Scheer R, Schueler J, Fiebig HH. Anticancer activity of a lectin-rich mistletoe extract injected intratumorally into human pancreatic cancer xenografts. *Anticancer Res.* 2005 May-Jun;25(3B):1969-75.

Santana SS, Gennari-Cardoso ML, Carvalho FC, Roque-Barreira MC, Santiago Ada S, Alvim FC, Pirovani CP. Eutirucallin, a RIP-2 type lectin from the latex of *Euphorbia tirucalli* L. presents pro-inflammatory properties. *PLoS One*. 2014 Feb18; 9(2):e88422.

Sadraeian M, Guimarães FEG, Araújo APU, Worthylake DK, LeCour LJ, Pincus SH.Selective cytotoxicity of a novel immunotoxin based on pulchellin A chain for cells expressing HIV envelope. *Sci Rep*. 2017 Aug 8;7(1):7579.

Schötterl S, Hübner M, Armento A, Veninga V, Wirsik NM, Bernatz S, Lentzen H, Mittelbronn M, Naumann U. Viscumins functionally modulate cell motility-associated gene expression. *Int J Oncol*. 2017 Feb;50(2):684-696.

Schaffrath B, Mengs U, Schwarz T, Hilgers RD, Beuth J, Möckel B, Lentzen H, Gerstmayer B. Anticancer activity of rViscumin (recombinant mistletoe lectin) in tumor colonization models with immunocompetent mice. *Anticancer Res.* 2001 Nov-Dec;21(6A):3981-7.

Schumacher U, Feldhaus S, Mengs U. Recombinant mistletoe lectin (rML) is successful in treating human ovarian cancer cells transplanted into severe combined immunodeficient (SCID) mice. *Cancer Lett.* 2000 Mar 31;150(2):171-5.

Siegle I, Fritz P, McClellan M, Gutzeit S, Mürdter TE. Combined cytotoxic action of *Viscum album* agglutinin-1 and anticancer agents against human A549 lung cancer cells. *Anticancer Res.* 2001 Jul-Aug;21(4A):2687-91.

Siegle I, Fritz P, McClellan M, Gutzeit S, Müthing J, Burg M, Möckel B, Langer M, Metelmann-Strupat W, Werner A, Neumann U, Peter-Katalinic J, Eck J. Preferential binding of the anticancer drug rViscumin (recombinant mistletoe lectin) to terminally alpha2-6-sialylated neolacto-series gangliosides. *Glycobiology*. 2002 Aug; 12 (8) :485-97.

Silva AL, Goto LS, Dinarte AR, Hansen D, Moreira RA, Beltramini LM, Araújo AP.Pulchellin, a highly toxic type 2 ribosome-inactivating protein from *Abrus pulchellus*. Cloning heterologous expression of A-chain and structural studies. *FEBS J*. 2005 Mar;272(5):1201-10.

Stauder H, Kreuser ED. Mistletoe extracts standardised in terms of mistletoe lectins (ML I) in oncology:current state of clinical research. *Onkologie*. 2002 Aug;25(4):374-80.

Stirpe F. Ribosome-inactivating proteins. *Toxicon*. 2004;44:371–83.

Stirpe F, Bolognesi A, Bortolotti M, Farini V, Lubelli C, Pelosi E, Polito L, Dozza B, Strocchi P, Chambery A, Parente A, Barbieri L. Characterization of highly toxic type 2 ribosome-inactivating proteins from *Adenia lanceolata* and *Adenia stenodactyla* (Passifloraceae). *Toxicon*. 2007 Jul;50(1):94-105.

Tomatsu M, Mujin T, Shibamoto N, Tashiro F, Ikuta A. Production of aralin, a selective cytotoxic lectin against human transformed cells, in callus culture of Aralia elata. *Planta Med*. 2004 May;70(5):469-71.

Walsh MJ, Dodd JE, Hautbergue GM. Ribosome inactivating proteins: potent poisons and molecular tools. *Virulence*. 2013 Nov 15;4(8):774-84

Wang BZ, Zou WG, Liu WY, Liu XY. The lower cytotoxicity of cinnamomin (a type II RIP) is due to its B-chain. *Arch Biochem Biophys*. 2006 Jul 1;451(1):91-6.

Wang H, Ng TB. Luffangulin, a novel ribosome inactivating peptide from ridge gourd (*Luffa acutangula*) seeds. *Life Sci*. 2002 Jan;70(8): 899-906.

Wang HX, Ng TB. Lagenin, a novel ribosome inactivating protein with ribonucleolytic activity from bottle gourd (*Lagenaria siceraria*) seeds. *Life Sci*. 2000 Oct 13;67(21):2631-8.

Wei F, Cao S, Ren X, Liu H, Yu J, Li H, Hao X. Efficient antiproliferative and antiangiogenic effects on human ovarian cancer growth by gene transfer of attenuated mutants of Shiga-like toxin I. *Int J Gynecol Cancer*. 2008 Jul-Aug;18(4):677-91.

Wong JH, Wang HX, Ng TB. Marmorin, a new ribosome inactivating protein with antiproliferative and HIV-1 reverse transcriptase inhibitory activities from the mushroom *Hypsizigus marmoreus*. *Appl Microbiol Biotechnol*. 2008 Dec;81(4):669-74.

Xu H, Liu WY.Cinnamomin--a versatile type II ribosome-inactivating protein. *Acta Biochim Biophys Sin* (Shanghai). 2004 Mar;36(3):169-76.

Yang Y, Mak AN, Shaw PC, Sze KH. Solution structure of an active mutant of maize ribosome inactivating protein (MOD) and its interaction with the ribosomal stalk protein P2. *J Mol Biol*. 2010 Feb 5;395(5):897-907.

Zenner HP. Selective killing of laryngeal carcinoma cells by a monoclonal immunotoxin. *Ann Otol Rhinol Laryngol*. 1986 Mar-Apr;95(2 Pt 1):115-20.

Zhang D, Halaweish FT.Isolation and characterization of ribosome-inactivating proteins from Cucurbitaceae. *Chem Biodivers*. 2007 Mar;4(3):431-42.

Zhou X, Li XD, Yuan JZ, Tang ZH, Liu WY. Toxicity of cinnamomin--a new type II ribosome -inactivating protein to bollworm and mosquito. *Insect Biochem Mol Biol.* 2000 Mar;30(3):259-64.

Zwierzina H, Bergmann L, Fiebig H, Aamdal S, Schöffski P, Witthohn K, Lentzen H. The preclinical and clinical activity of aviscumine: a potential anticancer drug. *Eur J Cancer.* 2011 Jul;47(10):1450-7.

In: Hemagglutinins
Editors: T. Bun Ng, J. Wong, R. Tse et al.
ISBN: 978-1-53615-708-6
© 2019 Nova Science Publishers, Inc.

Chapter 9

RECENT RESEARCH ON ANTICANCER, ANTIMICROBIAL, ANTIPARASITIC AND IMMUNOENHANCING ACTIVITIES OF MUSHROOM HEMAGGLUTININS

Jack Ho Wong[1,2], Tzi Bun Ng[3,*], Charlene Cheuk Wing Ng[4], Chit Tam[3], Yau Sang Chan[5], Ryno Naude[6], Ryan Tse[1], Tak Fu Tse[1], Helen Chan[1,†], Ajay Kumar Gautam[7], Neha Gupta[7], Dakshita Tanaji Narvekar[7], Amita Bhadkari[7], Shubhlukshmi Kumari[7], Nidhi Srivastava[8] and Sameer Suresh Bhagyawant[7]

[1]Vita Green Pharmaceutical (Hong Kong Limited),
Genning Partners Company Limited, and
Institute of Medical Research, Hong Kong, China
[2]The Third Affiliated Hospital of Shenzhen University, Shenzhen, China
[3]School of Biomedical Sciences, Faculty of Medicine,
The Chinese University of Hong Kong, China
[4]GKT School of Medical Education, King's College London, United Kingdom
[5]Department of Obstetrics and Gynecology,
LKS Faculty of Medicine, University of Hong Kong, China
[6]Department of Biochemistry and Microbiology,
Nelson Mandela Metropolitan University, Port Elizabeth, South Africa

[*] Corresponding Author's E-mail: tzibunng@cuhk.edu.hk.
[†] Corresponding Author's E-mail: Helen.chan@vitagreen.com.

[7]School of Studies in Biotechnology, Jiwaji University, Gwalior, India
[8]Department of Bioscience and Biotechnology, Banasthali Vidhyapeeth, Banasthali, India

Abstract

This review reports the various potentially exploitable biological activities observed in recently reported mushroom hemagglutinins comprising anticancer, immunostimulatory, antifungal, antibacterial, anti-HIV reverse transcriptase, antinematode and anti-insect activities.

Keywords: mushroom, hemagglutinins, anticancer, immunostimulatory, antifungal, antibacterial, anti-HIV reverse transcriptase, antinematode, anti-insect, activities

Introduction

Mushrooms produce an abundance of proteins including deoxyribonucleases (Zhang et al. 201 4), amylases (Du et al. 2013), feruloyl esterases (Wang et al. 2014), laccases (Xu et al. 2013), α-galactosidases (Zhang et al. 2015), nitrite reductases (Zhang et al. 2015), ribosome inactivating proteins (Pan et al. 2013), phytases (Li et al. 2013, Zhang et al. 2013), metalloproteases (Wu et al. 2011), antibacterial proteins (Zheng et al. 2011), angiotensin I-converting enzyme inhibitory proteins (Geng et al. 2015), and antiviral proteins (Ma et al. 2013).

These proteins exhibit a spectrum of exploitable activities. The aim of this review is to cover recent research on mushroom hemagglutinins.

Agaricus Bisporus Lectin

Agaricus bisporus lectin did not inhibit arginase activity but suppressed nitric oxide production by mouse peritoneal macrophages in response to the lipopolysaccharide *in vitro*. Oral administration of the lectin to BALB/c mice induced a marked inhibition of nitric oxide production by peritoneal macrophages after the lipopolysaccharide stimulus.

The lectin exerted an antiproliferative action on Con A-stimulated mononuclear cells, and reduced the incidence of autoimmune encephalomyelitis and mitigated the symptoms (Ditamo et al. 2016).

AGROCYBE AEGERITA LECTIN

Cancer cell lines exposed to *Agrocybe aegerita* lectin displayed excessive accumulation of microtubule-associated protein light chain 3 II (LC3-II), formation of EGFP-LC3 puncta and acidic autophagic vacuoles, and the formation of autophagosomes. Inhibition of autophagy with chloroquine potentiated *Agrocybe aegerita-* lectin elicited apoptosis in liver cancer cells.

Inhibition of autophagy may minimize the dose of lectin for antitumor treatment. Inhibitors of autophagy may potentiate the action of anticancer drugs that work by stimulating both apoptosis and autophagy (Liu et al. 2017).

The lectin, which exhibited hemagglutinating activity toward rabbit red cells at 12.5 μg/ml, bound to four white spot syndrome virus WSSV structural proteins, including VP39B, VP41B, VP53A and VP216, and protected shrimps from infection by the virus.

This characteristic activity of the lectin may be exploited in shrimp culture (Chang et al. 2018).

ALEURIA AURANTIA LECTIN, COPRINOPSIS CINEREA LECTIN AND MARASMIUS OREADES AGGLUTININ

Aleuria aurantia lectin, *Coprinopsis cinerea* lectin and *Marasmius oreades* agglutinin retarded nematode *(Haemonchus contortus)* larval development. Only *M. oreades* agglutinin had a killing action on the larvae. The remaining aforementioned lectins exerted a larvistatic action arresting larval growth at the L1 stage. The lectins showed binding to the larval gut and interacted with glycans in the adult digestive tract. Some of these glycan structures are hidden antigens not exposed to the immune system of the host and thus have a potential for development of drugs and vaccines (Heim et al. 2015).

CERRENA UNICOLOR LECTIN

The galactose-specific lectins isolated from *Cerrena unicolor* mycelia using affinity chromatography displayed different molecular weights ranging from 15 to 116kDa and also distinct kinetic parameters (Davitashvili et al. 2015).

Coprinopsis Cinerea Lectin

Coprinopsis cinerea lectin was cytotoxic to the fruitfly *Drosophila melanogaster* and. the nematode *Caenorhabditis elegans.* The entomotoxic and nematotoxic actions are attributed to a high-affinity GlcNAc (Fucα1,3)β1,4GlcNAc binding site on the protein surface. The lectin dimerizes in solution and in crystals. Substitution of two amino acids at the interface of the dimer, R18A and F133A, affected lectin dimerization and undermined toxicity without impairing sugar binding. Crosslinking of surface N-glycoproteins on the intestinal cells of fruitfly and the nematode plays an important role in the mechanism of toxicity of the lectin to fruitfly and the nematode (Bleuler-Martinez et al. 2017).

Ganoderma Applanatum Lectin

The the hemagglutinating activity of the 35- kDa. *G. applanatum* lectin was inhibited by a number of sugars including fructose, galactose, glucose, maltose, mannitol, mannose, sorbose, and sucrose. The lectin induced apoptosis in HT-29 colon adenocarcinoma cells (Kumaran et al. 2017).

Hericium Erinaceus Lectin

The lectin exhibited potent mitogenic activity toward mouse splenocytes, and demonstrated antiproliferative activity toward hepatoma (HepG2) and breast cancer (MCF7) cells with an IC(50) of 56.1 microM and 76.5 microM, respectively. It manifested HIV-1 reverse transcriptase inhibitory activity with an IC50 of 31.7 microM. The lectin exhibited potent mitogenic activity toward murine splenocytes but was devoid of antifungal activity (Li et al. 2010).

Recombinant *Hericium erinaceus* lectins rHEL2a and rHEL2b (both about 120 kDa in molecular weight) but not 15-kDa rHEL1 agglutinated animal erythrocytes. The hemagglutinating activity of the lectins was attenuated by glycoproteins with mucin-type O-glycans. rHEL2 isolectins showed interactions with O-glycans with the core 1 O-glycan motif, Galβ(1,3) GalNAc.

Like peanut agglutinin, they recognize O-glycans attached to leukosialin as tumor-associated carbohydrate antigens on the surface of K562 human leukemia cells (Kim et al. 2018).

Hohenbuehelia Serotina Lectin

The heterodimeric 31.2-kDa *Hohenbuehelia serotina* lectin exhibited hemagglutinating activity which was thermolabile and remained stable up to 50°C for only half an hour with only about half of the activity remaining. Stability of the activity in NaOH solution up to 12.5mmol/L and in HCl solution up to 6 mmol/L was observed. The activity was enhanced in the presence of Ca2+ and Mn2+ ions and inhibited by galactose and lactose. Half of the hemagglutinating activity remained at 60°C and activity was undetectable beyond 90°C. When co-inoculated with the proVAX/S2 vaccine, the lectin upregulated the level of hepatitis B surface antigen and CD4+ cell expression of interferon, augmented expression of major histocompatibility complex II in C57BL/6 mice, exerted proliferative activity toward T cells, and promoted maturation of dendritic cell. The lectin may find application in prevention of HBV infection (Xu et al. 2017).

Macrolepiota Procera Lectin

The lectin from the parasol mushroom *Macrolepiota procera* is a ricin B-like lectin with a β-trefoil fold. It is specific for terminal N-acetyllactosamine and other β-galactosides. Besides carbohydrate- binding site at the α-repeat, there is a carbohydrate-binding site at the γ-repeat.with a low affinity for galactose. A hydrophobic binding site with specificity for non- carbohydrate molecules is also present. The lectin showed toxicity toward the nematode *Caenorhabditis elegans* (Žurga et al. 2014).

Marasmius Oreades Agglutinin

It is a Galα1-3Gal specific lectin from the fairy ring mushroom. It possesses an N-terminal carbohydrate-binding domain and a C-terminal dimerization domain with calcium-dependent cysteine protease activity. The protease activity brought about, in MDCKII cells following internalization, changes including cytoskeletal rearrangement, cell detachment and cell death, and β1-integrin breakdown. The lectin protects against predators (Juillot et al. 2016).

Paecilomyces Japonica Agglutinin

Paecilomyces japonica agglutinin is a 16- kDa acidic protein with cytotoxicity on human breast cancer MDA-MB-2319 cells, human pancreas cancer AsPc-1 cells, and human stomach cancer SNU-1 cells (Park et al. 2004).

PHOLIOTA ADIPOSA LECTIN

Pholiota adiposa lectin is a homodimeric 32- kDa protein which inhibited proliferation of breast cancer MCF7 cells and hepatoma Hep G2 cells and the activity of HIV-1 reverse transcriptase with an IC50 of 3.2 microM, 2.1 microM and 1.9 microM, respectively (Zhang et al. 2009).

PHOLIOTA SQUARROSA LECTIN

The core α1-6 fucosylation-specific *Pholiota squarrosa* lectin can be produced by chemical synthesis due to its short length consisting of only 40 residues. The lectin displays a β-prism motif with a three-fold rotational symmetry.

The three antiparallel β-sheets are intimately linked by swapping of β-strands, resulting in a novel structure.

Three tryptophan residues occupy the structural core, and their indole rings demonstrate NH-π electrostatic interactions. NMR studies employing an excess of fucose disclosed that fucose gains entry to a pocket located where the β-sheet edges meet, with the methyl group found at the bottom, producing hydrophobic and hydrogen- bonding interactions with the lectin (Yamasaki et al. 2018).

PLEUROCYBELLA PORRIGENS LECTIN

Suzuki et al. (2014) reported the heterologous expression of recombinant lectin from the toxic mushroom *Pleurocybella porrigens* by the basidiomycete *Phanerochaete sordida* YK-624 with the help of glyceraldehyde 3-phosphate dehydrogenase gene promoter. The expression of the recombinant protein was confirmed by assay of hemagglutinating activity, and molecular mass determination by MALDI-TOF and SDS-PAGE9 (Suzuki et al. 2014).

PLEUROTUS FERULAE LECTIN

Pleurotus ferulae lectin is dimeric and composed of two 17.5 kDa subunits. The lectin was stable at 60°C. Its hemagglutinating activity began to decline with a further increase in temperature, and was reduced in the presence of calcium and magnesium ions and a variety of sugars including galactose, galactosamine, glucose, and lactose. The activity reached a maximum at pH 6-9.

The lectin exhibited mitogenic activity toward murine spleen cells (Xu et al. 2014).

PLEUROTUS FLORIDA LECTIN

Pleurotus florida lectin brought about the following changes in hepatocytes and splenocytes of arsenic-treated rats. It reduced apoptosis in both types of cells, restored the suppressed activities of the antioxidative enzymes catalase and superoxide dismutase, and lowered the escalated levels of lipid peroxidation and protein carbonyl in red blood cells (Bera et al. 2011a; Rana et al. 2012, 2014, 2015).

The lectin normalized sodium arsenite-induced alterations of cell proliferation index, phagocytic activity, superoxide dismutase activity, and nitric oxide production in hepatocytes (Rana et al. 2011). Similar changes in the kidneys of arsenic-treated rats were observed in the kidneys in response to the lectin (Bera et al. 2011b).

PLEUROTUS OSTREATUS LECTIN

Pleurotus ostreatus lectin brought about activation of the Toll-like receptor 6 signal pathway in dendritic cells, boosted the titer of HBV-specific antibodies and augmented the helper T cell response to surmount HBV tolerance in transgenic mice (He et al. 2017).

POLYPORUS SQUAMOSUS LECTIN 1A

Polyporus squamosus lectin 1a binds to sialylated glycans and glycoconjugates with high specificity and selectivity.

It displays an N-terminal carbohydrate-binding domain and a C-terminal domain with Ca2+-dependent proteolytic activity which contributes to its cytotoxicity on mammalian cancer cells including HeLa cells (Manna et al. 2017).

PSATHYRELLA ASPEROSPORA LECTIN

Psathyrella asperospora lectin manifested potent cytotoxicity on monkey kidney VERO cells and human colon cancer HT29 cells. It is essentially N-acetylglucosamine specific (with submicromolar affinity) with some affinity toward sialic acid (millimolar affinity) (Ribeiro et al. 2017).

PYCNOPORUS SANGUINEUS LECTIN

Pycnoporus sanguineus lectin is composed of 68.7 and 55.2 kDa subunits with isoelectric points at 5.5 and 5.2. The lectin-induced hemagglutination of rat red cells was inhibited by galactose, other monosaccharides, and glycoproteins including bovine lactoferrin, bovine transferrin, fetal calf fetuin and horseradish peroxidase (Albores et al. 2016).

RUSSULA DELICA LECTIN

It was a homodimeric 60-kDa protein which inhibited proliferation of breast cancer MCF7 cells and hepatoma Hep G2 cells and activity of HIV-1 reverse transcriptase with an IC50 of 0.52 microM, 0.88 microM, and 0.26 microM, respectively (Zhao et al. 2010).

RUSSULA LEPIDA LECTIN

The 32-kDa lectin inhibited proliferation of breast cancer MCF7 cells and hepatoma Hep G2 cells with an IC50 of 0.9 microM and 1.6 microM, respectively.

The lectin, when administered by intraperitoneal injections (5.0mg/kg body weight/day for 20 days), caused the weight of S-180 tumor to shrink to about one-third of the size of the negative control group (Zhang et al. 2010).

SCLEROTIUM ROLFSII LECTIN

Sclerotium rolfsii lectin diaplayed high binding specificity for (Galβ1-3GalNAc-α-O-Ser/Thr, T or TF, the oncofetal Thomsen-Friedenreich carbohydrate antigen) found to be expressed in over 90% of human cancers.

The lectin triggered apoptosis in human breast, colon and ovarian cancer cells. It elicited changes in MAPK and c-JUN- associated signaling pathways in 2 hours; change of cell miRNA expression in 12 hours; and cell cycle changes, DNA replication and apoptosis pathways in 24 hours. The lectin exerted an inhibitory action on tumor growth *in vivo* (Barkeer et al. 2015).

Sparassis Latifolia Lectin

Sparassis latifolia lectin demonstrated activity against the fungi *Candida* and *Fusarium* species as well as activity against bacteria including *E. coli* and resistant strains of *Staphylococcus aureus* and *Pseudomonas aeruginosa* by interacting with the surfaces of the microorganisms (Chandrasekaran et al. 2016).

Xylaria Hypoxylon Lectin

Xylaria hypoxylon lectin together with inactivated foot-and-mouth disease vaccine brought about, in immunized mice, a strong activation of Th1 and Tc1 cell responses, especially Tfh cell responses, dendritic cell maturation via augmenting the expression of major histocompatibility complex II molecules and co-stimulatory molecules CD40 and CD86. *X. hypoxylo* lectin-specific IgG or inflammatory factors were indiscernible, signifying the safety of *X. hypoxylon* lectin as an adjuvant for inactivated foot-and-mouth disease vaccine (Kang et al. 2015).

Fungal Immunomodulatory Proteins from Flammulina Velutipes, Ganoderma Lucidum, Ganoderma Sinense and Volvariella Volvacea

Fungal immunomodulatory proteins from *Flammulina velutipes, Ganoderma lucidum, Ganoderma sinense* and *Volvariella volvacea* cells displayed hemagglutinating activities and upregulated cytokine production by murine spleen cells (Wang et al. 2013).

The amino acids L10, W12, and D45 are the key residues instrumental to the potent interleukin 2 stimulatory activity of the fungal immunomodulatory protein LZ-8 isolated from *Ganoderma lucidum* (Bao et al. 2018).

Discussion and Conclusion

The account above highlights the wide range of biological activities observed in mushroom lectins encompassing anticancer activity toward different cancer cell lines such as breast cancer cells and hepatoma cells, mitogenic activity toward splenocytes, cytokine production, antifungal, antibacterial, antiviral, anti-HIV reverse transcriptase, antioxidant, renoprotective, hepatoprotective, entomotoxic and nematotoxic activities (Table 1).

More activities of mushroom lectins will be uncovered as research goes on.

Table 1. Diverse activities of mushroom hemagglutinins

Anticancer activity
Agrocybe aegerita lectin (Liu et al. 2017), *Ganoderma applanatum* lectin (Kumaran et al. 2017), *Hericium erinaceus* lectin (Li et al. 2010), *Marasmius oreades* agglutinin (Juillot et al. 2016), *Paecilomyces japonica* agglutinin (Park et al. 2004), *Pholiota adiposa* lectin (Zhang et al. 2009), *Polyporus squamosus* lectin 1a (Manna et al. 2017), *Psathyrella asperospora* lectin (Ribeiro et al. 2017). *Russula delica* lectin (Zhao et al. 2010), *Russula lepida* lectin (Zhang et al. 2010), *Sclerotium rolfsii* lectin (Barkeer et al. 2015).
Immunomodulatory activity
Agaricus bisporus lectin (Ditamo et al. 2016), *Pleurotus ferulae* lectin (Xu et al. 2014), *Xylaria hypoxylon* lectin (Kang et al. 2015), Fungal immunomodulatory proteins from *Flammulina velutipes, Ganoderma lucidum, Ganoderma sinense* and Volvariella volvacea (Wang et al. 2013;Ditamo et al. 2016; Bao et al. 2018).
Antifungal activity
Sparassis latifolia lectin (Chandrasekaran et al. 2016).
Antibacterial activity
Sparassis latifolia lectin (Chandrasekaran et al. 2016).
Anti-HIV reverse transcriptase activity
Pholiota adiposa lectin (Zhang et al. 2009), *Russula lepida* lectin (Zhang et al. 2010)
Anti- white spot syndrome virus activity
Agrocybe aegerita lectin (Chang et al. 2018),
Anti-hepatitis B virus activity
Hohenbuehelia serotina lectin (Xu et al. 2017), *Pleurotus ostreatus* lectin (He et al. 2017)
Anti-nematode activity
Aleuria aurantia lectin, *Coprinopsis cinerea* lectin and *Marasmius oreades* agglutinin (Heim et al. 2015), Coprinopsis cinerea lectin (Bleuler-Martinez et al. 2017), *Macrolepiota procera* lectin (Žurga et al. 2014)
Anti-insect activity
Coprinopsis cinerea lectin (Bleuler-Martinez et al. 2017)
Hepatoprotective, renoprotective and spleno-protective activities
Pleurotus florida lectin (Bera et al. 2011a, b; Rana et al. 2011, 2012, 2014, 2015).

ACKNOWLEDGMENT

We gratefully acknowledge the award of HMRF research grant (no.12131221) from Food and Health Bureau, The Government of Hong Kong Special Administrative Region, and a grant from National Natural Science Foundation of China (no. 81471927).

REFERENCES

Albores, S., Moros, M., Cerdeiras, M. P., de la Fuente, J. M., Grazu, V., Fraguas, L. F. A lectin purified from blood red bracket Mushroom, *Pycnoporus sanguineus* (Agaricomycetidae) mycelium displayed affinity toward bovine transferrin. *Int. J. Med. Mushrooms,* 2016; 18(1): 67 - 74.

Barkeer, S., Guha, N., Hothpet, V., Saligrama Adavigowda, D., Hegde, P., Padmanaban, A., Yu, L. G., Swamy, B. M., Inamdar, S. R. Molecular mechanism of anticancer effect of *Sclerotium rolfsii* lectin in HT29 cells involves differential expression of genes associated with multiple signaling pathways: A microarray analysis. *Glycobiology,* 2015; 25 (12):1375 - 91.

Bera, A. K., Rana, T., Bhattacharya, D., Das, S., Pan, D., Das, S. K. Sodium arsenite-induced alteration in hepatocyte function of rat with special emphasis on superoxide dismutase expression pathway and its prevention by mushroom lectin. *Basic Clin. Pharmacol. Toxicol.,* 2011a; 109 (4):240 - 4.

Bera, A. K., Rana, T., Das, S., Bhattacharya, D., Pan, D., Bandyopadhyay, S., Das, S. K. Mitigation of arsenic-mediated renal oxidative stress in rat by *Pleurotus florida* lectin. *Hum. Exp. Toxicol.,* 2011b; 30(8): 940 - 51.

Bleuler-Martinez, S., Stutz, K., Sieber, R., Collot, M., Mallet, J. M., Hengartner, M., Schubert, M., Varrot, A., Künzler, M. Dimerization of the fungal defense lectin CCL2 is essential for its toxicity against nematodes. *Glycobiology*, 2017; 27(5):486 - 500.

Chandrasekaran, G., Lee, Y. C., Park, H., Wu, Y., Shin, H. J. Antibacterial and antifungal activities of lectin extracted from fruiting bodies of the Korean cauliflower medicinal mushroom, *Sparassis latifolia* (Agaricomycetes). *Int. J. Med. Mushrooms,* 2016; 18(4):291 - 9.

Chang, Y. S., Chen, J. N., Chang, K. H., Chang, Y. M., Lai, Y. J., Liu, W. J. Cloning and expression of the lectin gene from the mushroom *Agrocybe aegerita* and the activities of recombinant lectin in the resistance of shrimp white spot syndrome virus infection. *Dev. Comp. Immunol.*, 2018 Jul. 19. pii: S0145-305X(18) 30297-0.

Cordara, G., Manna, D., Krengel, U. Family of papain-like fungal chimerolectins with distinct Ca2+-dependent activation mechanism. *Biochemistry,* 2017; 56(35):4689 - 4700.

Davitashvili, E., Kapanadze, E., Kachlishvili, E., Mikiashvili, N. A., Elisashvili, V. Isolation and Characterization of lectins formed by *Cerrena unicolor* (Higher Basidiomycetes) in solid-state fermentation of sorghum and wheat straw. *Int. J. Med. Mushrooms*, 2015; 17(5):427 - 34.

Geng, X., Tian, G., Zhang, W., Zhao, Y., Zhao, L., Ryu, M., Wang, H., Ng, T. B. Isolation of an angiotensin I-converting enzyme inhibitory protein with antihypertensive effect in spontaneously hypertensive rats from the edible wild mushroom *Leucopaxillus tricolor*. *Molecules,* 2015; 20(6): 10141 - 53.

He, M., Su, D., Liu, Q., Gao, W., Kang, Y. Mushroom lectin overcomes hepatitis B virus tolerance via TLR6 signaling. *Sci. Rep.,* 2017; 7(1):5814.

Heim, C., Hertzberg, H., Butschi, A., Bleuler-Martinez, S., Aebi, M., Deplazes, P., Künzler, M., Štefanić, S. Inhibition of *Haemonchus contortus* larval development by fungal lectins. *Parasit. Vectors,* 2015; 8:425.

Juillot, S., Cott, C., Madl, J., Claudinon, J., van der Velden, N. S., Künzler, M., Thuenauer, R., Römer, W. Uptake of *Marasmius oreades* agglutinin disrupts integrin-dependent cell adhesion. *Biochim. Biophys. Acta,* 2016; 1860(2):392 - 401.

Kang, J., Zuo, Y., Guo, Q., Wang, H., Liu, Q., Liu, Q(2)., Xia, G., Kang, Y. *Xylaria hypoxylon* lectin as adjuvant elicited Tfh cell responses. *Scand. J. Immunol.,* 2015; 82(5):436 - 42.

Kim, S. *Hericium erinaceus* isolectins recognize mucin-type O-glycans as tumor-associated carbohydrate antigens on the surface of K562 human leukemia cells. *Int. J. Biol. Macromol.,* 2018 pii: S0141-8130. [Epub ahead of print].

Kumaran, S., Pandurangan, A. K., Shenbhagaraman, R. Esa, N. M. Isolation and characterization of lectin from the artist's conk medicinal mushroom, *Ganoderma applanatum* (Agaricomycetes), and evaluation of its antiproliferative activity in HT-29 colon cancer cells. *Int. J. Med. Mushrooms,* 2017; 19(8):675 - 684.

Li, Zhang, G., Ng, T. B., Wang, H. A novel lectin with antiproliferative and HIV-1 reverse transcriptase inhibitory activities from dried fruiting bodies of the monkey head mushroom *Hericium erinaceum. J. Biomed. Biotechnol.,* 2010; 2010:716515.

Liu, Q., Chen, H., Wang, H., Ng, T. B. Isolation and Characterization of a ubiquitin-like ribonuclease from the cultured deep root mushroom, *Oudemansiella radicata* (Higher Basidiomycetes). *Int. J. Med. Mushrooms,* 2015;17(11):1037 - 45.

Manna, D., Pust, S., Torgersen, M. L., Cordara, G., Künzler, M., Krengel, U., Sandvig, K. *Polyporus squamosus* lectin 1a (PSL1a) exhibits cytotoxicity in mammalian cells by disruption of focal adhesions, inhibition of protein synthesis and Induction of Apoptosis. *PLoS One*, 2017 Jan. 23; 12(1):e0170716.

Rana, T., Bera, A. K., Bhattacharya, D., Das, S., Pan, D., Das, S. K. Evidence of antiapoptotic properties of *Pleurotus florida* lectin against chronic arsenic toxicity in renal cells of rats. *J. Environ. Pathol. Toxicol. Oncol.,* 2012; 31(1):39 - 48.

Rana, T., Bera, A. K., Bhattacharya, D., Das, S., Pan, D., Das, S. K. Effect of *Pleurotus florida* lectin (PFL) on arsenic-induced activities of splenocytes in rat. *Toxicol. Ind. Health*, 2014; 30(4):336 - 46.

Rana, T., Bera, A. K., Bhattacharya, D., Das, S., Pan, D., Das, S. K. Characterization of arsenic-induced cytotoxicity in liver with stress in erythrocytes and its reversibility with *Pleurotus florida* lectin. *Toxicol. Ind. Health,* 2015; 31(2):108 - 22.

Rana, T., Bera, A. K., Das, S., Bhattacharya, D., Pan, D., Bandyopadhyay, S., De, S., Das, S. K. Mushroom lectin protects arsenic induced apoptosis in hepatocytes of rodents. *Hum. Exp. Toxicol.,* 2011; 30(4):307 - 17.

Rana, T., Bera, A. K., Das, S., Bhattacharya, D., Pan, D., Das, S. K. Inhibition of oxidative stress and enhancement of cellular activity by mushroom lectins in arsenic induced carcinogenesis. *Asian Pac. J. Cancer Prev.*, 2016; 17(9):4185 - 4197.

Rana, T., Bera, A. K., Das, S., Bhattacharya, D., Pan, D., Bandyopadhyay, S., Mondal, D. K., Samanta, S., Bandyopadhyay, S., Das, S. K. *Pleurotus florida* lectin normalizes duration dependent hepatic oxidative stress responses caused by arsenic in rat. *Exp. Toxicol. Pathol.*, 2012; 64(7-8):665 - 71.

Ribeiro, J. P., Ali Abol Hassan, M., Rouf, R., Tiralongo, E., May, T. W., Day, C. J., Imberty, A., Tiralongo, J., Varrot, A. Biophysical characterization and structural determination of the potent cytotoxic *Psathyrella asperospora* lectin. *Proteins,* 2017; 85(5): 969 - 975.

Singh, R. S., Kaur, H. P., Kennedy, J. F. Modulation of immunocyte functions by a mucin-specific lectin from *Aspergillus gorakhpurensis*. *Int. J. Biol. Macromol.*, 2017 Aug.; 101:172 - 178.

Singh, R. S., Kaur, H. P., Singh, J. Purification and characterization of a mucin specific mycelial lectin from *Aspergillus gorakhpurensis*: application for mitogenic and antimicrobial activity. *PLoS One,* 2014 Oct. 6; 9(10):e109265.

Singh, S. S., Wang, H., Chan, Y. S., Pan, W., Dan, X., Yin, C. M., Akkouh, O., Ng, T. B. Lectins from edible mushrooms. *Molecules,* 2014; 20(1):446 - 69.

Suzuki, T., Dohra, H., Omae, S., Takeshima, Y., Choi, J. H., Hirai, H., Kawagishi, H. Heterologous expression of a lectin from *Pleurocybella porrigens* (PPL) in *Phanerochaete sordida* YK-624. *J. Microbiol. Methods,* 2014; 100: 70 - 6.

Wang, X. F., Li, Q. Z., Bao, T. W., Cong, W. R., Song, W. X., Zhou, X. W. *In vitro* rapid evolution of fungal immunomodulatory proteins by DNA family shuffling. *Appl. Microbiol. Biotechnol.,* 2013; 97(6):2455 – 65.

Xu, C. J., Wang, Y. X., Niu, B. N., Liu, B., Li, Y. B., Wang, X. M., Lu, S. L. Isolation and characterization of a novel lectin with mitogenic activity from *Pleurotus ferulae*. *Pak. J. Pharm. Sci.,* 2014; 27(4 Suppl.): 983 - 9.

Xu, Y., Chen, S., Liu, Q. Lectin from the late oyster mushroom, *Hohenbuehelia serotina* Agaricomycetes), and its novel effect as an adjuvant of the HBV DNA vaccine. *Int. J. Med. Mushrooms,* 2017; 19(12):1123 - 1133.

Yamasaki, K., Yamasaki, T., Tateno, H. The trimeric solution structure and fucose-binding mechanism of the core fucosylation-specific lectin PhoSL. *Sci. Rep.,* 2018 May 17; 8(1):7740.

Zhang, G., Sun, J., Wang, H., Ng, T. B. First isolation and characterization of a novel lectin with potent antitumor activity from a Russula mushroom. *Phytomedicine,* 2010; 17(10):775 - 81.

Zhang, G. Q., Sun, J., Wang, H. X., Ng, T.B. A novel lectin with antiproliferative activity from the medicinal mushroom *Pholiota adiposa*. *Acta Biochim. Pol.,* 2009; 56(3):415 - 21.

Zhang, W., Du, F., Wang, L., Zhao, L., Wang, H., Ng, T. B. Hydrolysis of oligosaccharides by a thermostable α-galactosidase from *Termitomyces eurrhizus*. *Int. J. Mol. Sci.,* 2015; 16(12): 29226 – 35.

Zhang, W., Tian, G., Feng, S., Wong, J. H., Zhao, Y., Chen, X., Wang, H., Ng, T. B. *Boletus edulis* nitrite reductase reduces nitrite content of pickles and mitigates Intoxication in nitrite-intoxicated mice. *Sci. Rep.,* 2015; 5:14907.

Zhang, W., Tian, G., Geng, X., Zhao, Y., Ng, T. B., Zhao, L., Wang, H. Isolation and characterization of a novel lectin from the edible mushroom *Stropharia rugosoannulata*. *Molecules,* 2014; 19(12): 19880 - 91.

Zhao, S., Zhao, Y., Li, S., Zhao, J., Zhang, G., Wang, H., Ng, T. B. A novel lectin with highly potent antiproliferative and HIV-1 reverse transcriptase inhibitory activities from the edible wild mushroom *Russula delica*. *Glycoconj J.,* 2010; 27(2):259 - 65.

Žurga, S., Pohleven, J., Renko, M., Bleuler-Martinez, S., Sosnowski, P., Turk, D., Künzler, M., Kos, J., Sabotič, J. A novel β-trefoil lectin from the parasol mushroom (*Macrolepiota procera*) is nematotoxic. *FEBS J.*, 2014; 281(15):3489 - 506.

In: Hemagglutinins
Editors: T. Bun Ng, J. Wong, R. Tse et al.
ISBN: 978-1-53615-708-6
© 2019 Nova Science Publishers, Inc.

Chapter 10

C TYPE LECTINS IN INVERTEBRATES AND VERTEBRATES

***Jack Ho Wong*[1,2], *Tzi Bun Ng*[3,*], *Charlene Cheuk Wing Ng*[4], *Chit Tam*[3], *Yau Sang Chan*[5], *Ryno Naude*[6], *Ryan Tse*[1], *Tak Fu Tse*[1] *and Helen Chan*[1,†]**

[1]Vita Green Pharmaceutical (Hong Kong Limited),
Genning Partners Company Limited, and Institute of Medical Research,
Hong Kong, China
[2]The Third Affiliated Hospital of Shenzhen University, Shenzhen, China
[3]School of Biomedical Sciences, Faculty of Medicine,
The Chinese University of Hong Kong, China
[4]GKT School of Medical Education,
King's College London, United Kingdom
[5]Department of Obstetrics and Gynecology, LKS Faculty of Medicine,
University of Hong Kong, China
[6]Department of Biochemistry and Microbiology,
Nelson Mandela Metropolitan University, Port Elizabeth, South Africa

ABSTRACT

C-type lectins are Ca++-dependent carbohydrate-binding proteins that resemble each other in primary and secondary structures of their carbohydrate- recognition domains,

---

* Corresponding Author's E-mail: tzibunng.cuhk.edu.hk.
† Corresponding Author's E-mail; Helen.chan@vitagreen.com.

alternatively called C-type lectin domains. Different invertebrates (including arthropods, mollusks and echinoderms) and vertebrates produce C-type lectins. C-type lectins protect these organisms against pathogens including viruses and bacteria by interacting with the pathogen-associated molecular patterns on the surfaces of infectious microbes. C-type lectins may have other functions such as playing a role in regeneration in Hydra, food sorting in oysters, and as a receptor for vitellogenin transport in the hemolymph during yolk formation in crustaceans. The C type lectin in venom of the snake *Bothrops pauloensis* may have therapeutic value due to its action on inhibition of collagen deposition and secretion of inflammatory cytokines. C type lectins in venoms of snakes manifest inhibitory activity on breast cancer cells.

Keywords: C-type lectins, invertebrates, vertebrates, carbohydrate- recognition domains, viruses, bacteria

INTRODUCTION

C-type lectins are Ca++-dependent carbohydrate-binding proteins that resemble each other in primary and secondary structures of their carbohydrate- recognition domains, alternatively called C-type lectin domain. Some C-type lectins are secreted proteins whereas other C-type lectins are transmembrane proteins. They play a role in homeostasis, innate immunity, inflammatory responses, and leukocyte and platelet trafficking (Cummings and McEver, 2017). The intent of this review is to examine C-type lectins produced by different organisms.

C-Type Lectins Produced by Different Organisms

Metazoans

The C-type lectin in *Hydra magnipapillata* is a receptor for polyfucosylated glycans. Disruption of interactions between the lectin and glycan stopped regeneration in *Hydra magnipapillata* (Sahadevan et al., 2014).

Genome sequencing of the cnidarian *Nematostella vectensis* disclosed 67 predicted C-type lectins, with 92 C-type lectin domains. Conserved residues in the glycan-binding pocket indicate that about 50% of the C-type lectin-like domains have glycan-binding function. The cnidarian C-type lectin domain sequences are more similar to those of the vertebrates (Wood-Charlson and Weis, 2009).

CRUSTACEANS

C-Type Lectin FmLC3 from the Shrimp *Fenneropenaeus merguiensis*

The mannose-specific C-type lectin FmLC3 exhibited a molecular mass of 17.96 kDa and a pI of 4.92. The predicted primary structure of the lectin is characterized by a signal peptide, a carbohydrate recognition domain with an EPN motif and one Ca2+ binding site-2, encompassing a double-loop region with two conserved disulfide linkages. In normal or unchallenged shrimps, the FmLC3 mRNA expression was discernible only in the hepatopancreas whereas FmLC3 was observed in the heart, hemolymph, intestine and stomach but undetectable in the expressed tissue, suggesting that following protein synthesis in the hepatopancreas, secretion to the other tissues occurred. FmLC3 expression in the shrimp was up-regulated following exposure to the white spot syndrome virus or *Vibrio harveyi*, but attenuated subsequent to knockdown using gene-specific double-stranded RNA in spite of inoculation of the pathogenic microbes, which was accompanied by a rise in cumulative mortality and decrease in median lethal time (Runsaeng et al., 2017).

CTL FmLdlr with Low-Density Lipoprotein Receptor (LDLR) Domain from the Shrimp *Fenneropenaeus merguiensis*

The lectin sequence possesses a LDLR domain and a carbohydrate recognition domain with a QAP motif putative for galactose-specific binding. Expression of the lectin was observed only in healthy shrimp hemocytes, and enhanced after exposure to the white spot syndrome virus or the bacterium *V. parahaemolyticus*. Gene silencing together with administration of the pathogen decreased the median lethal time and elevated the cumulative mortality. Although the recombinant proteins of FmLdlr, LDLR domain and carbohydrate recognition domain agglutinated various bacteria, the LDLR domain manifested the weakest activity. Only the carbohydrate recognition domain and FmLdlr enhanced phagocytosis and encapsulation by hemocytes. Both FmLdlr and carbohydrate recognition domain displayed antibacterial activity toward *V. parahaemolyticus*. Only FmLdlr and carbohydrate recognition domain bound to white spot syndrome virus proteins, capsid VP15, envelope VP28, and tegument VP39A, with the binding affinity of FmLdlr being stronger than that of CRD. Thus FmLdlr plays a role in shrimp immune defense, via the carbohydrate recognition domain, in agglutinating bacteria, facilitating phagocytosis and encapsulation, and in antimicrobial action and binding to viral proteins. Since the LDLR domain displays greater binding affinity to vitellogenin than the entire lectin molecule and the carbohydrate recognition domain, it is likely that FmLdlr may serve as a receptor for vitellogenin transport in the hemolymph during vitellogenesis in shrimps (Kwankaew et al., 2018).

C-Type Lectin from Fenneropenaeus merguiensis hemocytes

The lectin has 284 amino acids and an isoelectric point of 4.35, and two similar carbohydrate recognition domains, each with a Ca2+ binding site-2 and a galactose-specific QPD motif. Expression of the FmLC5 transcripts, which was detectable only in hemocytes, was heightened following exposure to lipopolysaccharide, white spot syndrome virus or *Vibrio harveyi*, and downregulated after RNAi- based silencing despite co-inoculation of the bacterium or the virus, resulting in a shortened median lethal time and a higher mortality (Senghoi et al., 2017).

C-Type Lectin MnCTL from the Freshwater Shrimp *Macrobrachium nipponense*

The lectin has 257 amino acids and a single carbohydrate recognition domain with an EPN (Glu-Pro-Asn) motif in calcium-binding site 2. Interestingly, the lectin bears a closer evolutionary relationship to vertebrate lectins than invertebrate lectins. Its expression in the stomach and gills is elevated following exposure to white spot syndrome virus or bacteria. Knock-down of the lectin gene in *M. nipponense* exposed to white spot syndrome virus caused a downregulation of transcript levels of anti-lipopolysaccharide factors. The recombinant lectin demonstrated agglutinating activity toward the bacteria *Staphylococcus aureus* and *Vibrio parahaemolytic*. Bacterial clearance brought about by the lectin was attenuated by silencing using dsRNA interference (Huang et al., 2017).

C-Type Lectin (MjCC-CL) from the Kuruma Shrimp (*Marsupenaeus japonicus*)

Recognition of bacterial glycans by the C-type carbohydrate recognition domain of MjCC-CL directly activates the JAK/STAT pathway via interaction of the coiled-coil domain with the surface receptor Domeless, and upregulation of antimicrobial peptides expression. In contrast, in vertebrates, the JAK/STAT pathway is indirectly activated by cell death and stress signals through cytokines or growth factors (Sun et al., 2017).

C-Type Lectin MjGCTL from *Marsupenaeus japonicus*

The lectin owns a CRD with a QAP motif. The recombinant lectin, which is neither galactose-specific nor mannose-specific, agglutinated Gram-positive and Gram-negative bacteria in a calcium-dependent way. The structure of QAP motif and carbohydrate-specificity of rMjGCTL was different from both EPN and QPD, which bind to mannose and galactose, respectively, indicating that QAP is a new motif (Alenton et al., 2017).

C-Type Lectin PtCTL1 from the Swimming Crab *Portunus* trituberculatus

PtCTL1 has 166 amino acids, a C-type lectin carbohydrate-recognition domain and four conserved Cys residues. The recombinant PtCTL1 agglutinated erythrocytes, Gram-positive and Gram-negative bacteria, and exhibited Ca2+-dependent lipopolysaccharide-binding and agglutinating activities. Its mRNA transcripts were found in hemocytes and hepatopancreas and transcripts in the latter were upregulated after exposure to *V. alginolyticus* (Lu et al., 2017).

C-Type Lectins PtCTL-2 and PtCTL-3 from *Portunus trituberculatus*

The two lectins differ in several aspects although there are similarities. PtCTL-2 possesses 485 amino acids, a signal peptide and a single carbohydrate-recognition domain .PtCTL-3 possesses 241 amino acids with a coiled-coil region and a single- carbohydrate-recognition domain. The motifs governing carbohydrate binding specificity in PtCTL-2 and PtCTL-3 are Glu-Pro-Arg and Gln-Pro-Asp. Recombinant PtCTL-2 but not recombinant PtCTL-3 bound glucan, lipopolysaccharide and peptidoglycan with relatively high affinity. Recombinant PtCTL-2 bound seven and recombinant PtCTL-3 six kinds in the microbe binding assay. Recombinant forms of PtCTL-2 and PtCTL-3 showed resemblance in their agglutinating activity toward Gram-positive bacteria, Gram-negative bacteria and fungi. PtCTL-2 and PtCTL-3 possess a novel and a typical key motif, respectively. PtCTL-3 displays a nonclassical carbohydrate binding specificity different from that of PtCTL-2 (Huang et al., 2017).

C-Type Lectin PtCTL4 from *Portunus trituberculatus*

The full length of the nucleic acid sequence of the lectin gene has 654 bp and a 480-bp open reading frame. The lectin has a CRD domain with Ca2+ binding site 2 and 6 conserved Cys residues. The hepatopancreas demonstrated the highest expression level of the lectin whereas eyestalk, gonad hemocytes, and muscle expressed low levels. Following exposure to *V. alginolyticus,* expression of the lectin attained peak levels three hours later. Basal levels were resumed at 48 h. The data indicate that the lectin serves as a pattern recognition receptor in the innate immune response of the crab. The recombinant lectin agglutinated rabbit red cells, and in addition Gram-positive bacteria (*Bacillus aquimaris, Staphylococcus aureus,* and *Micrococcus lysodeik*) and Gram-negative bacteria (*Aeromonas hydrophila, Chryseobacterium indologenes* and *V. alginolyticus*) in the presence of calcium ions. PtCTL4 acts as a pattern recognition receptor in the innate immune response of *P. trituberculatus* (Zhang et al., 2018).

C-Type Lectin (PcLec6) from the Red Swamp Crayfish *Procambarus clarkii*

C-type lectins are important immune molecules that participate in crustacean defense response. The lectin has 385 amino acids, with a C-type lectin domain (CTLD) and a serine-rich region. Expression of the lectin in the gills and lymph organ was augmented following exposure to white spot syndrome virus or the bacterium *Vibrio alginolyticus*. Both recombinant lectin and its CTLD protein demonstrated binding to bacteria, and promoted clearance of *V. alginolyticus*, but only recombinant lectin protected red swamp crayfish from white spot syndrome virus infection (Zhang et al., 2018).

SpCTL-B from Hemocytes of the Mud Crab *Scylla paramamosain*

The full-length of SpCTL-B cDNA is 1278 bp with a 348 bp open reading frame. The predicted protein contains a single carbohydrate-recognition domain. The hepatopancreas had the highest level of lectin expression. The mRNA levels of the lectin in hepatopancreas and hemocytes increased following exposure to white spot syndrome virus, *Vibrio parahaemolyticus*, polyI:C and lipopolysacharide. The binding of recombinant lectin to *Aeromonas hydrophila, Escherichia coli, Staphylococcus aureus,* β-hemolytic *Streptococcus,* and *V. alginolyticus* was stronger than its binding with *V. parahaemolyticus* and *Saccharomyces cerevisiae*. The lectin agglutinated, in a Ca2+-dependent fashion, Gram-positive β-hemolytic *Streptococcus* and *S. aureus* as well as Gram-negative *A. hydrophila, E. coli, V. alginolyticus* and *V. parahaemolyticus,* and retarded their growth. The lectin-induced agglutination was abolished in the presence of lipopolysaccharide and mannose, but not galactose and fructose. RNAi silencing of the lectin downregulated expression of three phagocytosis-related genes and reduced the antibacterial capability of the mud crab. The hemocyte levels of transcripts of the antimicrobial peptides SpALF4 (anti-lipopolysaccharide factor), SpALF5, SpALF6, SpHistin, and SpCrustin underwent a decline (Wei et al., 2018).

INSECTS

Insect C-Type Lectins

Insect C-type lectins play a role in agglutination, encapsulation, opsonization, nodule formation, melanization, and prophenol oxidase activation, as well as in maintaining gut microbiome homeostasis (Xia et al., 2017).

C-Type Lectin from Silkworm *Bombyx mori*. BmCTL-S2

The lectin, which has 221 amino acids, is expressed in hemocytes and the fat body. The level of its transcript in the midgut and the fat body was upregulated after exposure to bacteria. The recombinant lectin displayed binding and antibacterial activities toward *Bacillus subtilis* and *Staphylococcus aureus* (Shahzad et al., 2017).

C-Type Lectin Mosquito Galactose-Specific C-Type Lectin 7 Mos GCTL-7

The lectin was upregulated by Japanese encephalitis virus infection and facilitated infection *in vivo* and *in vitro*. Ca2+ dependent binding of mosGCTL-7 to N-glycan at N154 on the viral envelope protein. is crucial to entry of the Japanese encephalitis virus. Antibodies against mosGCTL-7 minimized the incidence of infections (Liu et al., 2017).

C-Type Lectins Produced by the Beet Armyworm *Spodoptera exigua*

Gasmi et al., (2018) discovered two groups of C-type lectins produced by *Spodoptera exigua* which are distinct in phylogeny, domain architecture, and expression pattern. One group of C-type lectins resembles other lepidopteran lectins. The other group of C-type lectins are bracoviral-related lectins (bracovirus-like lectins, Se-BLLs). The bracovirus-like lectin Se-BLL3 mitigated the mortality of *Spodoptera frugiperda* larvae brought about by *Junonia coenia* densovirus infection but not the mortality of *S. exigua* larvae. Another bracovirus-like lectin Se-BLL2 made the *S. exigua* larvae more tolerant to baculovirus infection (Gasmi et al., 2017).

MOLLUSKS

Mucosal C-Type Lectins (CvML3912 and CvML3914) from the Oyster *Crassostrea virginica*

The expression of the transcripts of the two oyster mucosal C-type lectins in the feeding organs was enhanced by starvation. Oysters silenced by means of Short Dicer-substrate small interfering RNA (DsiRNA) targeting the two lectins exhibited a reduced level of the two transcripts and also attenuated food sorting capability (Pales Espinosa and Allam, 2018).

ECHINODERMS

C-Type Lectin SJL-1 from the Body Surface Layer of the Japanese Sea Cucumber, *Apostichopus japonica*

A 15-kDa C-type lectin SJL-1 from the body surface layer of the Japanese sea cucumber, *Apostichopus japonica* SJL-1 agglutinated yeast as well as Gram-positive and Gram-negative bacteria and plays a role in surface defense. SJL-1-positive cells were dispersed below the outermost layer of the sea cucumber integument (Ono et al., 2018).

FISH

Nattectin-Like Protein (CaNTC) in the Qihe Crucian Carp *Carassius auratus*

The lectin possesses an EPN/WND motif and four conserved disulfide-bonded residues (Cys57-Cys150, Cys126-Cys142) necessary for carbohydrate-binding specificity. Enhancement of lectin expression in the head-kidney, spleen and liver was observed following exposure to the bacterium *Aeromonas hydrophila*. The recombinant lectin binds lipopolysaacharide and exhibits Ca2+-dependent agglutination of rabbit red cells and the bacteria *Aeromonas hydrophila, Escherichia coli* and *Staphylococcus aureus* (Wang et al., 2017).

C-Type Lectin Gene Controlling Resistance to Benedenia Disease in the Japanese Yellowtail *Seriola quinqueradiata*

Expression of the mRNA of this C-type lectin was detected in skin tissue of the Japanese yellowtails *Seriola quinqueradiata* infected by the monogenean parasite *Benedenia seriolae*. Scanning for single nucleotide polymorphisms (SNPs) led to the discovery of a SNP in the C-type lectin/C-type lectin-like domain associated with *B. seriolae* infection levels. The novel C-type lectin gene controls resistance to Benedenia disease in Japanese yellowtails (Nakamoto et al., 2017).

AMPHIBIANS

Expression of gene of the *Xenopus* C type lectin XCL-2 was upregulated in tails undergoing regeneration, indicating the importance of XCL-2 in tail regeneration (Ishino et al., 2007).

SNAKES

C-Type Lectin BpLec Isolated from Venom of the Snake *Bothrops pauloensis*

BpLec inhibited collagen deposition and downregulated the secretion of inflammatory cytokines without affecting the activities of N-acetyl glucosaminidase and myeloperoxidase (Castanheira et al., 2017).

C-Type Lectin Macrovipecetin from Venom of the Viper *Macrovipera lebetina*

Macrovipecetin decreased the viability of melanoma cells 100 fold more than cisplatin. In conjunction with cisplatin, macrovipecetin elevated the sensitivity of SK-MEL-28 melanoma cells by increasing their apoptosis through upregulated expression of the apoptosis inducing factor and activation of AKT, ERK1/2, p38, and NF-κB. Macrovipecetin by itself, or in conjunction with cisplatin, upregulated the expression of Bax, Bad, Bim, p53 and TRADD, inhibited the expression of Bcl-2, and suppressed levels of reactive oxygen species in SK-MEL-28 cells. In addition, the following processes were downregulated: melanoma cell adhesion, migration and invasion, via modulating the expression of β-catenin, E-cadherin, αvβ3 integrin, RhoA, c-Src and vimentin (Hammouda et al., 2017).

C-Type Lectin-Like Protein Lebecin from Venom of the Viper *Macrovipera lebetina*

The 30-kDa lebecin, which is comprised of a -residue α subunit and a 131-residue β subunit, exerted antiproliferative activity toward MDA-MB231 human breast cancer cells, prevented MDA-MB231 cell adhesion to a variety of adhesion substrata mediated by integrin, and suppressed migration of MDA-MB231 cells toward fibronectin and fibrinogen in haptotaxis assays (Jebali et al., 2014).

C-type Lectin-Like Protein Vixapatin from Venom of Israeli Viper *Vipera xantina palestinae*

The protein is a selective α2β1 integrin inhibitor which prevented metastasis of melanoma. It suppressed the α2 mediated adhesion of K562 over-expressing cells (IC50 = 3 nM). At a dose of 3 nM it exerted antiproliferative activity toward human dermal microvascular endothelial cells. At a dose of 25 nM, collagen I- stimulated human fibrosarcoma HT-1080 cell migration was suppressed. At a dose of 50 nM, rat C6 glioma and human breast carcinoma MDA-MB-231 cells were inhibited. At a dose of 1 μM, human dermal microvascular endothelial cells tube formation decreased by 75% in a Matrigel assay. At 1 μM vixapatin, physiological angiogenesis stimulated by bovine fibroblast growth factor was diminished by 70%, and pathological angiogenesis elicited by C6 glioma undermined by 94% in an assay using shell-less embryonic quail chorioallantoic membrane (Momic et al., 2012).

Birds

C-Type Lectin-Like Proteins Ovocleidin-17 and Ansocalcin from Decalcified Chicken and Goose Eggshell Protein Extracts

Ovocleidin-17 and ansocalcin bound bacterial polysaccharides, and exerted killing activity on *Bacillus subtilis, Pseudomona aeruginosa* and *Staphylococcus aureus* (Wellman-Labadie et al., 2008).

C-Type Lectin-Like Proteins Struthiocalcin -1 and Struthiocalcin-2 from Ostrich Calcified Eggshell Layer Matrix

Struthiocalcin-1 and struthiocalcin-2 showed 65% and 46% sequence identity with goose eggshell protein compared to 41% and 36%with chicken eggshell ovocleidin-17, respectively (Mann et al., 2004)

MAMMALS

Mammalian Macrophage C-Type Lectin

MCL (macrophage C-type lectin) helps to destroy bacteria during peritonitis caused by *Escherichia coli*. Mice lacking in MCL, when subjected to *E. coli*-induced sepsis, displayed poorer bacterial clearing ability and survival than control mice, although the two groups of mice did not differ in phagocytic activity and production of generation of inflammatory cytokines and reactive oxygen species. Macrophages of mice lacking in MCL, when the animals were subjected to *E. coli*-induced sepsis, exhibited impaired phagosome maturation and phagosomal acidification. MCL facilitates phagosome maturation and acidification during sepsis and may be a therapeutic candidate (Lee et al., 2017).

Macrophage-Inducible C-Type Lectin (Mincle)

Mincle is an immune receptor on myeloid cells for the mycobacterial cell wall glycolipid trehalose- 6,6'-dimycolate and the mammalian necrotic cell-derived alarmin histone deacetylase complex unit Sin3-associated protein 130. When bound to its ligands, Mincle stimulates secretion of innate cytokines and other immune mediators. Mincle is associated with tumor progression, autoimmunity, and sterile inflammation (Patin et al., 2017).

Binding between Mouse Colonic Mucins and the C-Type Lectin Receptors

Intestinal mucins elicit immune responses when they are recognized by dendritic cells through protein-carbohydrate interactions. Binding between mouse colonic mucins and the C-type lectin receptors, mouse dendritic cell-associated lectin (mDectin)-2 and specific intercellular adhesion molecule-grabbing nonintegrin-related-1 (SIGN-R1), was achieved through high-mannose N-glycans (Leclaire et al., 2018).

C-Type Lectin Receptors

Dectin-1 is a member of the family of C-type lectin receptors. No differences in binding or phagocytosis were exhibited by phagocytes deficient in Mannose Receptor, Mincle, Card-9, or FcRγ. Dectin-1 and other C-type lectins do not play a major role in phagocytosis and innate defense of phagocytes in response to *Cryptococcus neoformans* spores (Walsh et al., 2017).

Discussion and Conclusion

The foregoing account disclosed that a variety of invertebrates including prawns, shrimps, insects, mollusks and echinoderms, and vertebrates including fish, amphibians, snakes, birds and mammals, produce C-type lectins. In most of the cases it has been shown that C-type lectins protect the organisms against pathogens including viruses and bacteria (Tables 1-4). In other cases, C-type lectins play a role in regeneration in Hydra (Sahadevan et al., 2014), food sorting in oysters (Pales Espinosa and Allam, 2018), and as a receptor for vitellogenin transport in the hemolymph during vitellogenesis in shrimps (Kwankaew et al., 2018). The C type lectin in venom of the snake *Bothrops pauloensis* may have therapeutic value due to its action on inhibition of collagen deposition and secretion of inflammatory cytokines (Castanheira et al., 2017). C type lectins in venoms of other snakes manifest inhibitory activity on breast cancer cells. C-type lectins regulate innate and adaptive antimicrobial immune responses, and also play a central role in autoimmune diseases (Table 5). Defects in C-type lectins produce derangements in physiology and development, and heightened susceptibility to infections and non-infectious diseases (Brown et al., 2018). Continuing research on C-type lectins will lead to a better understanding of their roles in different organisms and design of products or medications for the welfare of mankind.

Table 1. Antibacterial activity of C-type lectins

FmLdlr from shrimp (*Fenneropenaeus merguiensis*) with low-density lipoprotein receptor (LDLR) domain (Kwankaew et al., 2018)
MnCTL from freshwater shrimp *Macrobrachium nipponense* (Huang et al., 2017).
C-type lectin from silkworm *Bombyx mori*. BmCTL-S2 (Shahzad et al., 2017).
C-type lectin SJL-1from body surface layer of Japanese sea cucumber *Apostichopus japonica* (Ono et al., 2018)
Nattectin-like protein (CaNTC) in Qihe crucian carp *Carassius auratus* (Wang et al., 2017)
C-type lectin-like proteins ovocleidin-17 and ansocalcin from decalcified chicken and goose eggshell protein extracts (Wellman-Labadie et al., 2008)
Mammalian macrophage C-type lectin (Lee et al., 2017)

Table 2. Up-regulation of crustacean C-type lectins following exposure to white spot syndrome virus, *Vibrio harveyi*, *V. alginolyticus*, and *V. parahaemolyticus*

FmLC3 from shrimp Fenneropenaeus merguiensis (Runsaeng et al., 2017).
FmLdlr from shrimp (*Fenneropenaeus merguiensis*) with low-density lipoprotein receptor (LDLR) domain (Senghoi et al., 2017; Kwankaew et al., 2018)
MnCTL from freshwater shrimp *Macrobrachium nipponense* (Huang et al., 2017).
PtCTL1 from the swimming crab *Portunus trituberculatus* (Lu et al., 2017)
PtCTL4 from *Portunus trituberculatus* (Zhang et al., 2018)
PcLec6 from the red swamp crayfish *Procambarus clarkii* (Zhang et al., 2018)
SpCTL-B from hemocytes of mud crab *Scylla paramamosain* (Wei et al., 2018)

Table 3. Up-regulation of insect and fish C-type lectins following exposure to bacteria

C-type lectin from silkworm *Bombyx mori*. BmCTL-S2 (Shahzad et al., 2017).
Nattectin-like protein (CaNTC) in Qihe crucian carp *Carassius auratus* (Wang et al., 2017)

Table 4. Antiviral and proviral activities of C-type lectins

Antiviral bracovirus-like lectin Se-BLL3 produced by beet armyworm *Spodoptera exigua* mortality of *Spodoptera frugiperda* larvae brought about by *Junonia coenia* densovirus infection(Gasmi et al.,2017)
Proviral C-type lectin mosquito galactose-specific C-type lectin 7 Mos GCTL-7 facilitated Japanese encephalitis virus infection and facilitated infection *in vivo* and *in vitro* (Liu et al., 2017).

Table 5. Other activities of C-type lectins

Anticancer
C-type lectin macrovipecetin from viper (*Macrovipera lebetina*) venom (Hammouda et al., 2017)
C-type lectin like protein lebecin from viper (*Macrovipera lebetina*) venom (Jebali et al., 2014)
C-type lectin-like protein vixapatin from Israeli viper Vipera xantina palestinae snake venom (Momic et al., 2012)
Anti-inflammatory
C-type lectin BpLec isolated from snake *Bothrops pauloensis* venom(Castanheira et al., 2017)
Antiparasitic
C-type lectin gene controlling resistance to infection by monogenean parasite *Benedenia seriolae* in Japanese yellowtail *Seriola quinqueradiata* (Nakamoto et al., 2017)
Food sorting
Mucosal C-type lectins (CvML3912 and CvML3914) from oyster *Crassostrea virginica* (Pales Espinosa and Allam, 2018).
Vitellogenin transport
FmLdlr from shrimp (Fenneropenaeus merguiensis) with low-density lipoprotein receptor (LDLR) domain serving as vitellogenin transport receptor for vitellogenin transport in the hemolymph during vitellogenesis in shrimps (Kwankaew et al., 2018)
Tail regeneration
Xenopus C type lectin XCL-2 (Ishino et al., 2007).

Acknowledgments

We gratefully acknowledge the award of HMRF research grant (no. 12131221) from Food and Health Bureau, The Government of Hong Kong Special Administrative Region, and a grant from National Natural Science Foundation of China (no. 81471927).

References

Alenton RR, Koiwai K, Miyaguchi K, Kondo H, Hirono I. Pathogen recognition of a novel C-type lectin from *Marsupenaeus japonicus* reveals the divergent sugar-binding specificity of QAP motif. *Sci Rep.* 2017;7:45818.

Brown GD, Willment JA, Whitehead L. C-type lectins in immunity and homeostasis. *Nat Rev Immunol.* 2018;18(6):374-389.

Castanheira LE, Lopes DS, Gimenes SNC, Deconte SR, Ferreira BA, Alves PT, Filho LRG, Tomiosso TC, Rodrigues RS, Yoneyama KAG, Araújo FA, Rodrigues VM. Angiogenenic effects of BpLec, a C-type lectin isolated from *Bothrops pauloensis* snake venom. *Int J Biol Macromol.* 2017 Sep;102:153-161.

Cummings RD, McEver RP. In: Varki A, Cummings RD, Esko JD, Stanley P, Hart GW, Aebi M, Darvill AG, Kinoshita T, Packer NH, Prestegard JH, Schnaar RL, Seeberger PH, editors. *Essentials of Glycobiology [Internet].* 3rd edition. Cold Spring Harbor (NY): Cold Spring Harbor Laboratory Press; 2015-2017. Chapter 34. 2017.

Gasmi L, Jakubowska AK, Ferré J, Ogliastro M, Herrero S. Characterization of two groups of *Spodoptera exigua* Hübner (Lepidoptera: Noctuidae) C-type lectins and insights into their role in defense against the densovirus JcDV. *Arch Insect Biochem Physiol.* 2018 Jan;97(1). doi: 10.1002/arch.21432. (Gasmi et al., 2017).

Hammouda MB, Riahi-Chebbi I, Souid S, Othman H, Aloui Z, Srairi-Abid N, Karoui H, Gasmi A, Magnenat EM, Wells TNC, Clemetson KJ, Rodríguez-López JN, Essafi-Benkhadir K. Macrovipecetin, a C-type lectin from *Macrovipera lebetina* venom, inhibits proliferation migration and invasion of SK-MEL-28 human melanoma cells and enhances their sensitivity to cisplatin. *Biochim Biophys Acta.* 2018;1862(3):600-614.

Huang M, Mu C, Wu Y, Ye F, Wang D, Sun C, Lv Z, Han B, Wang C, Xu XW. The functional characterization and comparison of two single CRD containing C-type lectins with novel and typical key motifs from *Portunus trituberculatus. Fish Shellfish Immunol.* 2017;70:398-407.

Huang X, Li T, Jin M, Yin S, Wang W, Ren Q. Identification of a *Macrobrachium nipponense* C-type lectin with a close evolutionary relationship to vertebrate lectins. *Mol Immunol.* 2017;87:141-151.

Ishino T, Kunieda T, Natori S, Sekimizu K, Kubo T. Identification of novel members of the *Xenopus* Ca2+ -dependent lectin family and analysis of their gene expression during tail regeneration and development. *J Biochem.* 2007;141(4):479-88.

Khan KA, Naylor AJ, Khan A, Noy PJ, Mambretti M, Lodhia P, Athwal J, Korzystka A, Buckley CD, Willcox BE, Mohammed F, Bicknell R. Multimerin - 2 is a ligand for group 14 family C-type lectins CLEC14A, CD93 and CD248 spanning the endothelial pericyte interface. *Oncogene.* 2017 Nov 2;36(44):6097-6108.

Kwankaew P, Praparatana R, Runsaeng P, Utarabhand P. An alternative function of C-type lectin comprising low-density lipoprotein receptor domain from *Fenneropenaeus merguiensis* to act as a binding receptor for viral protein and vitellogenin. *Fish Shellfish Immunol.* 2018;74:295-308.

Leclaire C, Lecointe K, Gunning PA, Tribolo S, Kavanaugh DW, Wittmann A, Latousakis D, MacKenzie DA, Kawasaki N, Juge N. Molecular basis for intestinal mucin recognition by galectin-3 and C-type lectins. *FASEB J* 2018:32(6); :3301-3320.

Lee WB, Yan JJ, Kang JS, Kim LK, Kim YJ. Macrophage C-type lectin is essential for phagosome maturation and acidification during *Escherichia coli*-induced peritonitis. *Biochem Biophys Res Commun.* 2017 Dec 2;493(4):1491-1497.

Liu K, Qian Y, Jung Y, Zhou B, Cao R, Shen T, Shao D, Wei J, Ma Z, Chen P, Zhu H, Qiu Y. Mos GCTL-7, a C-type lectin protein, mediates Japanese encephalitis virus infection in mosquitoes. *J Virol.* 2017; 91(10). pii: e01348-16.

Lu J, Yu Z, Mu C, Li R, Song W, Wang C. Characterization and functional analysis of a novel C-type lectin from the swimming crab *Portunus trituberculatus. Fish Shellfish Immunol. 2017* May; 64:185-192.

Mann K, Siedler F. Ostrich (*Struthio camelus*) eggshell matrix contains two different C-type lectin-like proteins. Isolation, amino acid sequence, and posttranslational modifications. *Biochim Biophys Acta.* 2004;1696(1):41-50.

Mu L, Yin X, Xiao Y, Bian X, Yang Y, Wu L, Ye J. A C-type lectin (CL11X1-like) from Nile tilapia (*Oreochromis niloticus*) is involved in host defense against bacterial infection. *Dev Comp Immunol.* 2018; 84: 230-240.

Nakamoto M, Takeuchi Y, Akita K, Kumagai R, Suzuki J, Koyama T, Noda T, Yoshida K, Ozaki A, Araki K, Sakamoto T. A novel C-type lectin gene is a strong candidate gene for Benedenia disease resistance in Japanese yellowtail, *Seriola quinqueradiata. Dev Comp Immunol.* 2017 Nov; 76:361-369.

Ono K, Suzuki TA, Toyoshima Y, Suzuki T, Tsutsui S, Odaka T, Miyadai T, Nakamura O(3). SJL-1, a C-type lectin, acts as a surface defense molecule in Japanese sea cucumber, *Apostichopus japonicus. Mol Immunol.* 2018; 97: 63-70.

Pales Espinosa E, Allam B. Reverse genetics demonstrate the role of mucosal C-type lectins in food particle selection in the oyster *Crassostrea virginica. Exp Biol.* 2018 Mar 14;221(Pt 6). pii: jeb174094.

Patin EC, Orr SJ, Schaible UE. Macrophage Inducible C-Type Lectin as a multifunctional player in immunity. *Front Immunol.* 2017 Jul 25; 8:861.

Runsaeng P, Puengyam P, Utarabhand P. A mannose-specific C-type lectin from *Fenneropenaeus merguiensis* exhibited antimicrobial activity to mediate shrimp innate immunity. *Mol Immunol.* 2017 Dec;92:87-98.

Sahadevan S, Antonopoulos A, Haslam SM, Dell A, Ramaswamy S, Babu P. Unique, polyfucosylated glycan-receptor interactions are essential for regeneration of *Hydra magnipapillata. ACS Chem Biol.* 2014;9(1):147-55.

Senghoi W, Runsaeng P, Utarabhand P. FmLC5, a putative galactose-binding C-type lectin with two QPD motifs from hemocytes of *Fenneropenaeu s merguiensis* participates in shrimp immune defense. *J Invertebr Pathol.* 2017;150:136-144.

Shahzad T, Zhan MY, Yang PJ, Yu XQ, Rao XJ. Molecular cloning and analysis of a C-type lectin from silkworm *Bombyx mori. Arch Insect Biochem Physiol.* 2017 Jul;95(3). doi: 10.1002/arch.21391. Epub 2017 Jun 15.

Sun JJ, Lan JF, Zhao XF, Vasta GR, Wang JX. Binding of a C-type lectin's coiled-coil domain to the Domeless receptor directly activates the JAK/STAT pathway in the shrimp immune response to bacterial infection. *PLoS Pathog.* 2017 Sep 20;13(9):e1006626.

Tsunezumi J, Higashi S, Miyazaki K. Matrilysin (MMP-7) cleaves C-type lectin domain family 3 member A (CLEC3A) on tumor cell surface and modulates its cell adhesion activity. *J Cell Biochem.* 2009 Mar 1;106(4):693-702

Walsh NM, Wuthrich M, Wang H, Klein B, Hull CM. Characterization of C-type lectins reveals an unexpectedly limited interaction between *Cryptococcus neoformans* spores and Dectin-1. *PLoS One.* 2017; 12(3):e0173866.

Wang L, Zhang J, Kong X, Zhao X, Pei C, Li L. A C-type lectin, Nattectin-like protein (CaNTC) in Qihe crucian carp *Carassius auratus*: Binding ability with LPS, PGN and various bacteria, and agglutinating activity against bacteria. *Fish Shellfish Immunol.* 2017: 67: 382-392.

Wang P, Zhuo XR, Tang L, Liu XS, Wang YF, Wang GX, Yu XQ, Wang JL. C-type lectin interacting with β-integrin enhances hemocytic encapsulation in the cotton bollworm, *Helicoverpa armigera. Insect Biochem Mol Biol.* 2017; 86:29-40.

Wei X, Wang L, Sun W, Zhang M, Ma H, Zhang Y, Zhang X, Li S. C-type lectin B (SpCTL-B) regulates the expression of antimicrobial peptides and promotes phagocytosis in mud crab *Scylla paramamosain. Dev Comp Immunol.* 2018;84:213-229.

Wellman-Labadie O, Lakshminarayanan R, Hincke MT. Antimicrobial properties of avian eggshell-specific C-type lectin-like proteins. *FEBS Lett.* 2008;582(5):699-704.

Wood-Charlson EM, Weis VM. The diversity of C-type lectins in the genome of a basal metazoan, *Nematostella vectensis. Dev Comp Immunol.* 2009 Aug;33(8):881-9.

Xia X, You M, Rao XJ, Yu XQ. Insect C-type lectins in innate immunity. *Dev Comp Immunol*. 2017 Dec 1. pii: S0145-305X (17)30576-1.

Zhang X, Lu J, Mu C, Li R, Song W, Ye Y, Shi C, Liu L, Wang C. Molecular cloning of a C-type lectin from *Portunus trituberculatus*, which might be involved in the innate immune response. *Fish Shellfish Immunol.* 2018 Mar 1;76:216-223.

Zhang XW, Man X, Huang X, Wang Y, Song QS, Hui KM, Zhang HW. Identification of a C-type lectin possessing both antibacterial and antiviral activities from red swamp crayfish. *Fish Shellfish Immunol*. 2018 Mar 10;77:22-30.

In: Hemagglutinins
Editors: T. Bun Ng, J. Wong, R. Tse et al.

ISBN: 978-1-53615-708-6
© 2019 Nova Science Publishers, Inc.

Chapter 11

FISH LECTINS WITH ANTICANCER, ANTIBACTERIAL, BACTERIAL AGGLUTINATING, FUNGAL CELL AGGLUTINATING, IMMUNOENHANCING, ANTIVIRAL, AND ANTIPARASITIC ACTIVITIES

Jack Ho Wong[1,2], Tzi Bun Ng[3,*], Charlene Cheuk Wing Ng[4], Chit Tam[3], Yau Sang Chan[5], Ryno Naude[6], Ryan Tse[1], Tak Fu Tse[1] and Helen Chan[1,†]

[1] Vita Green Pharmaceutical (Hong Kong Limited), Genning Partners Company Limited, and Institute of Medical Research, Hong Kong, China
[2] The Third Affiliated Hospital of Shenzhen University, Shenzhen, China
[3] School of Biomedical Sciences, Faculty of Medicine, The Chinese University of Hong Kong, China
[4] GKT School of Medical Education, King's College London, UK
[5] Department of Obstetrics and Gynecology, LKS Faculty of Medicine, University of Hong Kong, China
[6] Department of Biochemistry and Microbiology, Nelson Mandela Metropolitan University, Port Elizabeth, South Africa

ABSTRACT

The fish lectins discussed in this review were reported from many different species and isolated from tissues as diverse as serum, gills, liver, eggs and skin mucus. These lectins can be classified under different categories consisting of galectins, intelectins, lily-

* Corresponding Author's Email: tzibunng@cuhk.edu.hk.
† Corresponding Author's Email: helen.chan@vitagreen.com

type lectin, mannose binding lectins and C-type lectins. Most of the aforementioned fish lectins manifest antibacterial or bacterial cell agglutinating activity. Lectins from bighead carp, black rockfish, cobia, pearl spot, striped murrel, tongue sole and zebrafish exhibit inhibitory activity toward bacteria. Black rockfish, blunt snout bream, Qihe crucian carp, rock bream and turbot lectins induce bacterial cell agglutination. Very few fish lectins have been demonstrated to have antiviral or antifungal activity. Pearl spot and rock bream lectins elicit antiviral activity. Pearl spot lectin has fungal cell agglutinating activity. Several fish lectins show anticancer, immunoenhancing or antiparasite/parasite cell agglutinating activity. The blunt snout bream, catfish, chinook salmon and pearl spot lectins exhibit inhibitory activity toward cancer cells. Black rockfish, chinook salmon, cobia, tongue sole and zebrafish lectins exert an immunoenhancing action. Turbot lectin demonstrates antiparasite activity. Rock bream and turbot lectins have parasite cell agglutinating activity. The antibacterial, bacterial cell agglutinating, immunoenhancing, antiviral, antiparasite and parasite cell agglutinating activities serve to protect the fish against invading pathogenic organisms. An upsurge in the mRNA transcripts of the fish lectins is discerned to provide additional protection in response to bacterial or viral infection.

Keywords: fish lectins, galectins, intelectins, lily-type lectin, mannose binding lectins, C-type lectins, anticancer, immunoenhancing, antibacterial, antiviral, antiparasite

INTRODUCTION

Lectins have previously been reported from a multitude of fish species such as (i) ayu, (ii) gilt head bream, (iii) carp including grass carp, bighead carp and gibel carp, (iv) catfish including Arabian Gulf catfish, blue catfish, catfish, and channel catfish, (v) white-spotted charr, (vi) cobia fish, (vii) eels including conger eel, electric eel, and Japanese eel, (viii) blue gourami, (x) Spanish mackerel, (xi) toxic moray, (xii) perch and pike perch, (xiii) Potca fish, (xiv) powan, (xv) olive rainbow smelt and rainbow smelt, (xvi) roach, (xvii) rudd, (xviii) Atlantic salmon and chinook salmon, (xix) common skate, (xx) tench, (xxi) tilapia, (xxii) Japanese trout and steelhead trout, and (xxiii) zebrafish, The tissues from which the lectins were isolated included eggs, electric organs, gills, intestines, stomachs, livers, skin, mucus, plasma, and serum. The lectins exhibit a variety of molecular weights, number of subunits, extents of glycosylation, sugar binding specificities and amino acid sequences. The lectins display antitumor, antimicrobial, and immunoregulatory activities and developmental functions (Ng et al., 2015). The review presented in this chapter serves to provide more recent information on fish lectins.

Aristichthys nobilis (Bighead Carp) Gill Lectin with Antibacterial Activity

The lectin was a 220-kDa homomultimeric fucose- binding glycoprotein with a 13.4% carbohydrate content and the N-terminal sequence AGEQGGQCSA. The lectin did not need calcium ions for hemagglutinating activity toward rabbit red cells. The lectin exhibited antibacterial activity toward *Vibrio harveyi* but was devoid of antifungal activity (Pan et al., 2010).

Carassius auratus (Qihe Crucian Carp) C-Type Lectin (Nattectin-Like Protein) Lectin with Bacterial Cell Agglutinating Activity

Following exposure to *Aeromonas hydrophila* and poly I: C, respectively, the temporal expression of the lectin in the head kidney, spleen and liver was enhanced in a time-dependent fashion. The recombinant lectin bound lipopolysaccharide and all bacterial species tested, as well as agglutinated rabbit red cells, Gram-positive *Staphylococcus aureus*, and Gram-negative *Escherichia coli* and *A. hydrophila,* in a Ca2+-dependent fashion (Wang et al., 2017).

Channa striatus (Striped Murrel) Lily Type Lectin-2 Lectin with Antibacterial Activity

The lectin possessed a B-lectin domain and three carbohydrate binding sites, a hallmark of lily type lectin. The expression of the mRNA transcripts of the lectin in the gills were escalated upon exposure to the bacterium *Aeromonas hydrophila* and the fungus *Aphanomyces invadans.* The lectin bound to the membrane of *Escherichia coli,* destroyed integrity of the membrane resulting in a precipitous decline in the bacterial count.

In contrast, the lectin was devoid of any cytotoxicity towards fish peripheral blood leukocytes (Arasu et al., 2017).

Channa striatus Galectin with Antibacterial Activity

Galectins are galactoside-binding lectins which bind to bacterial membrane carbohydrates and play the role of pathogen recognition receptors. The constructed *Channa striatus* cDNA library contained a galectin-4 sequence. The hepatic expression of the galectin was markedly upregulated after exposure to the pathogens *Aeromonas hydrophila* and *Aphanomyces invadans* and to the viral analog poly I:C. Introduction of the pentamer oligotryptophan W5 to the C-terminal of membrane-binding G4 peptide (putative dimerization site of the lectin) considerably augmented its membrane disrupting activity toward the aquaculture bacterial pathogen *Vibrio harveyi* . However, there was neither cytotoxic nor hemolytic activity toward *Channa striatus* peripheral blood cells (Arasu et al., 2017).

Cynoglossus semilaevis (Tongue Sole) CD94 Lectin with Immunoenhancing and Antibacterial Activities

The lectin, with a C-type lectin-like domain, has 209 amino acids and exhibits 43-50% sequence identity with known teleost CD94s. The lectin was expressed in a variety of tissues and the expression was heightened upon exposure to bacteria. The recombinant lectin bound and agglutinated Gram-positive as well as Gram-negative bacteria in a calcium-dependent manner. Exposure of bacteria to the recombinant lectin *in vitro* increased the mortality of the bacteria and promoted phagocytic activity of peripheral blood leukocytes toward the bacteria (Li and Hu, 2016).

Dano rerio (Zebrafish) Egg Lectin with Antibacterial and Immunoenhancing Activities

The lectin interacted with Gram-positive as well as Gram-negative bacteria and stimulated the phagocytic activity of the macrophages toward the bacteria. The lectin microinjected into the zebrafish embryos boosted the embryonic resistance to the bacterial pathogen *Aeromonas hydrophila* (Wang et al., 2016).

Danio rerio Hepatic Lectin with Antibacterial and Immunoenhancing Activities

The lectin was a typical Ca2+-dependent carbohydrate binding protein. Its hepatic expression was enhanced by lipopolysaccharide and lipoteichoic acid. The recombinant C-type lectin domain of the lectin bound to and agglutinated both Gram-positive and Gram-negative bacteria and upregulated the phagocytic activity of the macrophages toward bacteria. Galactose suppressed the binding of recombinant domain bound to insoluble lipopolysaccharide, lipoteichoic acid and peptidoglycan. The production of pre-inflammatory cytokines was attenuated by overexpression of the lectin which was located in the membrane (Yang et al., 2018).

Etroplus suratensis **(Pearl Spot) Serum C-Type Lectin**

Lectin with Antibacterial, Antiviral, Anticancer and Fungal Cell Agglutinating Activities

The 75-kDa C-type lectin showed antibacterial activity against Gram negative *Aeromonas hydrophila* and *Vibrio parahaemolyticus.* It manifested antibiofilm activity against Gram-negative bacteria, antiviral and anticancer activities (Rubeena et al., 2019). The lectin agglutinated the yeast *Saccharomyces cerevisiae* (Rubeena et al., 2019).

Larimichthys crocea **C-Type Lectin with Bacterial Cell Agglutinating Activity**

The lectin displayed 33%-49% homology with other fish C-type lectins. It possessed a single carbohydrate-binding domain with four conserved disulfide-bonded cysteines and EPN/AND motifs in lieu of invariant EPN/WND motifs required for carbohydrate-binding specificity and constructing Ca(2+)-binding sites. The mRNA transcripts of the lectin were present in all tissues studied with the highest expression in the gills.

Following exposure to poly I:C and *Vibrio parahaemolyticus,* the temporal expression of the lectin in the head kidney, spleen and liver was elevated. Transcripts were also induced in the *head kidney*, gills, spleen and skin, after challenge with *Cryptocaryon irritans*. The recombinant lectin exhibited D-mannose-inhibitable hemagglutinating activity toward rabbit, human and large yellow croaker red cells in a Ca(2+)-dependent manner. The lectin agglutinated, in the presence of Ca(2+) ions, all bacterial species tested including Gram-positive bacteria (*Bacillus subtilis, Micrococcus lysoleikticus* and *Staphylococcus aureus*) and Gram-negative bacteria (*Aeromonas hydrophila, E. coli, Vibrio alginolyticus* and *V. parahaemolyticus,*) (Lv et al., 2016).

Megalobrama Amblycephala **(Blunt Snout Bream) Intelectin with Anticancer and Bacterial Cell Agglutinating Activities**

Intelectins are lectins with regulatory roles in early embryogenesis, iron metabolism, and innate immune response. The blunt snout bream intelectin pro-protein has a molecular weight of 34 kDa, and is characterized by a fibrinogen-related domain and eight conserved cysteines. Its mRNA was found at various stages of developmental but underwent a decline after hatching. In the adult, the liver was the organ manifesting the maximal expression level. Following exposure to *Aeromonas hydrophila*, heightened levels of the mRNA of the intelectin were noted in the kidney, gills, liver, intestine, and spleen, and elevated amounts of the intelectin protein were present in the kidney, liver, and spleen, disclosing the role of the intelectin in the immune response. The recombinant intelectin bound and agglutinated *E. coli* in a Ca2+-dependent fashion (Ding et al., 2017).

Intelectins are found in mammals, amphibians, and fish. Recombinant *Megalobrama amblycephala* intelectin bound and agglutinated bacteria. Following exposure to *Aeromonas hydrophila,* an increased level of *M. amblycephala* intelectin was noted in the macrophage-like kupffer cells in the liver. Recombinant *Megalobrama amblycephala* intelectin inhibited proliferation of two types of tumor cells (Ding et al., 2019).

Oncorhynchus tshawytscha (Chinook Salmon) Roe Lectin with Anticancer and Immunoenhancing Activities

The 30-kDa chinook salmon roe lectin with fair thermostability and pH stability inhibited proliferation of human hepatoma Hep G2 cells and human breast cancer MCF-7 cells. The formation of nitric oxide by murine peritoneal macrophages was upregulated. There was no inhibitory effect on fungal growth (Bah et al., 2011).

Oplegnathus fasciatus (Rock Bream) Galectin-1 with Bacterial Cell Agglutinating, Antiviral and Parasite Cell Agglutinating Activities

Galectins play a number of developmental and immunoregulatory roles and act as pattern recognition receptors of the innate immune system for combating microorganisms. The cDNA of *Oplegnathus fasciatus* (rock bream) galectin-1 codes for an amino acid sequence containing135 residues and a single carbohydrate recognition domain. The predicted tertiary structure was a typical β-sandwich structure. The cardiac expression of its mRNA transcripts was the highest among the various tissues. The expression of its mRNA transcripts in the head kidney, gills, and liver was elevated following exposure to rock bream irido virus, the bacteria *Streptococcus iniae* and *Edwardsiella tarda*, and the mitogens lipopolysaccharide and poly I:C. The agglutination of human red cells brought about by the recombinant lectin was inhibitable by D-galactose and lactose. The recombinant lectin exhibited inhibitory activity against rock bream irido virus and bound the ciliate parasite *Miamiensis avidus* and many bacteria including *Edwardsiella tarda, Escherichia coli, Vibrio tapetis, Vibrio harveyi, Vibrio anguillarum S. parauberis,* and *S. iniae* (Thulasitha et al., 2016a).

Oplegnathus fasciatus (Rock Bream) Galectin Related Protein with Bacterial Cell Agglutinating and Parasite Cell Agglutinating Activities

The cDNA of the galectin related protein codes for a sequence of 145 amino acids with a single carbohydrate recognition domain containing only two of the seven residues requisite for carbohydrate recognition. The galectin like B protein does not have a signal peptide. It bears a close relationship to the prototype galectins as revealed by homology assessment, multiple sequence alignment, and phylogenetic analysis. Its mRNA was expressed in many bream tissues and the expression in the spleen and gills was enhanced in the presence of viral or bacterial infection. The recombinant galectin like B protein was devoid of carbohydrate binding activity but demonstrated agglutinating activity toward the ciliate parasite *Miamiensis avidus* and the bacteria *Edwardsiella tarda*, *Escherichia coli, Listeria monocytogenes, Streptococcus iniae,* and *Vibrio tapetis* (Thulasitha et al., 2016b).

Oreochromis niloticus (Nile Tilapia) Mannose-Binding Lectin

Mannose binding lectins, as a soluble pattern recognition receptor, recognize antigens and regulate non-specific innate immunity, such as inflammation, migration, opsonization, phagocytosis and killing. Following exposure to the bacteria *Aeromonas hydrophila* and *Streptococcus agalactiae,* an augmentation of the mRNA expression level of Nile tilapia mannose binding lectin in monocytes/macrophages was observed (Mu et al., 2012).

Rachycentron canadum (Cobia) Ovarian Lectin with Antibacterial and Immunoenhancing Activities

The 180-kDa tetrameric lectin exhibited an antibacterial action toward *Escherichia coli*. However, it was devoid of an inhibitory effect on a number of fungal species including *Mycosphaerella arachidicola*, *Rhizoctonia solani, Fusarium oxysporum,* and *Coprinus comatus*. The lectin demonstrated mitogenic activity toward murine spleen cells (Ngai and Ng, 2007).

Scophthalmus maximus (Turbot) Skin Mucus Lily-Type Lectin with Bacterial Cell Agglutinating and Antiparasite Activities

Elevated expression of the lectin was noted in gills, intestine and skin. Thermal stress, ciliate infection, and salinity modified expression levels in these tissues. The recombinant lectin exhibited by D-mannose inhibitable agglutinating activity toward murine red cells without the presence of calcium ions. The lectin bound the bacteria *Vibrio anguillarum* and *Edwardsiella tarda*, and manifested cytotoxicity toward the ciliate parasite *Philasterides dicentrarchi* (Huang et al., 2016).

Sebastes schlegelii (Black Rockfish) C-Type Lectin with Antibacterial and Immunoenhancing Activities

The lectin is a C-type lectin domain family 4 member which also shows pronounced homology to CD209 antigen-like protein. The lectin was expressed in a variety of tissues and the expression was increased in response to infection by viruses and bacteria. The recombinant lectin bound the infectious spleen and kidney necrosis virus and the bacteria *Vibrio anguillarum* and *Edwardsiella tarda.* The agglutinating activity toward bacteria was Ca2+-dependent and nullified by mannose or in the absence of calcium. The recombinant lectin fortified the killing activity of macrophages toward bacteria, and conferred on fish an augmented resistance against bacterial infection but more susceptibility to viral infections. The lectin serves as a pattern recognition receptor which increases killing activity toward bacteria, but may nevertheless act as a target for viral manipulation of host defense system (Xue et al., 2018).

Sebastes schlegelii (Black Rockfish) Lily-Type Lectin Homolog with Bacterial Cell Agglutinating Activity

Lily-type lectins are soluble pathogen recognition receptors which possess at least one characteristic carbohydrate recognition domain, through which the lectins exhibit reversible and specific binding to cognate sugar moieties located on the cell membranes of pathogens and elicit the host innate immune responses. Three mannose-binding sites are present in the protein sequence of black rockfish lily-type lectin homolog. Two of these sites are the same as those in monocot mannose-binding lectins. A considerable amount of expression of black rockfish lily-type lectin homolog was observed in rockfish mucosal tissues located both externally and internally. Following exposure to both poly I:C and *Streptococcus iniae* an enhancement of mRNA expression of black rockfish lily-type lectin homolog in blood and gills was detected. Surprisingly, the opposite results (i.e., an attenuated mRNA expression) were observed in response to lipopolysaccharide. The recombinant lily-type lectin homolog agglutinated fish red cells as well as a number of bacteria pathogenic to fish. The hemagglutinating activity was augmented in the presence of calcium ions (Kugapreethan et al., 2018).

Silurus asotus (Catfish) Egg Lectin with Anticancer Activity

Catfish egg lectin potentiates the anticancer potency of low dosages of the anticancer drug vincristine by reducing the gene expression of multidrug resistance associate protein-1 to inhibit the multidrug resistance by binding to the target ligand Gb3(Galα1-4Galβ1-4Glc)-glycosphingolipid on Burkitt's lymphoma cells. The multidrug resistance associate protein-1 was more highly expressed in Burkitt's lymphoma cells in comparison with multi-drug resistant protein another member of the ATP-binding cassette family (Fujii et al., 2012).

DISCUSSION

Etroplus suratensis (pearl spot) serum C-type lectin, *Megalobrama amblycephala* (blunt snout bream) intelectin, *Oncorhynchus tshawytscha* (chinook salmon) roe lectin and *Silurus asotus* (catfish) egg lectin manifest anticancer activity. *Aristichthys nobilis* (bighead carp) gill lectin, *Channa striatus* (striped murrel) lily type lectin-2 lectin, *C. striatus* galectin, *Cynoglossus semilaevis* (tongue sole) CD94 lectin, *Dano rerio* (zebrafish) egg lectin, *Etroplus suratensis* (pearl spot) serum C-type lectin, *Rachycentron canadum* (cobia) ovarian lectin, and *Sebastes schlegelii* (black rockfish) C-type lectin exhibit antibacterial activity. *Carassius auratus* (Qihe crucian carp) C-type lectin (nattectin-like protein), *Larimichthys crocea* C-type lectin, *M. amblycephala*) intelectin, *Oplegnathus fasciatus* (rock bream) galectin-1 and galectin related protein, *Scophthalmus maximus* (turbot) skin mucus lily-type lectin, and *Sebastes schlegelii* (black rockfish) lily-type lectin homolog show bacterial cell agglutinating activity. *Etroplus suratensis* (pearl spot) serum C-type lectin and *O. fasciatus* galectin-1 have antiviral activity. *E. suratensis* serum C-type lectin has fungal cell agglutinating activity. *C. semilaevis* CD94 lectin, *Dano rerio* (zebrafish) egg lectin and hepatic lectin, *O. tshawytscha* roe lectin, *R. canadum* ovarian lectin, and *S. schlegelii* C-type lectin have immunoenhancing activity. *S. maximus* skin mucus lily-type lectin has antiparasite activity. *S. maximus* skin mucus lily-type lectin *O. fasciatus* galectin-1 and galectin related protein have parasite cell agglutinating activity.

The aforementioned fish lectins originate from many different species and belong to a few different categories including galectins, intelectins, lily-type lectin, mannose binding lectins and C-type lectins. They were isolated from tissues as diverse as serum, gills, liver, eggs and skin mucus. Most of the aforementioned fish lectins manifest antibacterial or bacterial cell agglutinating activity. Very few have been demonstrated to have antiviral or antifungal activity. Several lectins show anticancer, immunoenhancing or antiparasite/parasite cell agglutinating activity. The antibacterial, bacterial cell agglutinating, immunoenhancing, antiviral, antiparasite, parasite cell agglutinating activities serve to protect fish against invading pathogenic organisms. An upsurge in the mRNA transcripts of the fish lectins is discerned to provide additional protection in response to bacterial or viral infection.

Acknowledgments

We gratefully acknowledge the award of Health and Medical Research Fund (no. 12131221) from Food and Health Bureau, The Government of Hong Kong Special Administrative Region, and grants from National Natural Science Foundation of China (no. 81471927)

References

Arasu A, Kumaresan V, Ganesh MR, Pasupuleti M, Arasu MV, Al-Dhabi NA, Arockiaraj J. Bactericidal activity of fish galectin 4 derived membrane-binding peptide tagged with oligotryptophan. *Dev Comp Immunol.* 2017 Jun;71:37-48.

Arasu A, Kumaresan V, Palanisamy R, Arasu MV, Al-Dhabi NA, Ganesh MR, Arockiaraj J. Bacterial membrane binding and pore formation abilities of carbohydrate recognition domain of fish lectin. *Dev Comp Immunol. 2017* Feb;67:202-212.

Bah CS, Fang EF, Ng TB, Mros S, McConnell M, Bekhit Ael-D. Purification and characterization of a rhamnose-binding chinook salmon roe lectin with antiproliferative activity toward tumor cells and nitric oxide-inducing activity toward murine macrophages. *J Agric Food Chem.* 2011 May 25;59(10):5720-8.

Ding Z, Zhao X, Wang J, Zhang F, Wang W, Liu H. Intelectin mediated phagocytosis and killing activity of macrophages in blunt snout bream (*Megalobrama amblycephala*). *Fish Shellfish Immunol.* 2019 Jan 5;87:129-135.

Ding Z, Zhao X, Zhan Q, Cui L, Sun Q, Lin L, Wang W, Liu H. Characterization and expression analysis of an intelectin gene from *Megalobrama amblycephala* with excellent bacterial binding and agglutination activity. *Fish Shellfish Immunol.* 2017 Feb;61:100-110.

Fujii Y, Sugawara S, Araki D, Kawano T, Tatsuta T, Takahashi K, Kawsar SM, Matsumoto R, Kanaly RA, Yasumitsu H, Ozeki Y, Hosono M, Miyagi T, Hakomori SI, Takayanagi M, Nitta K. MRP1 expressed on Burkitt's lymphoma cells was depleted by catfish egg lectin through Gb3-glycosphingolipid and enhanced cytotoxic effect of drugs. *Protein J.* 2012 Jan;31(1):15-26.

Huang Z, Ma A, Xia D, Wang X, Sun Z, Shang X, Yang Z, Qu J. Immunological characterization and expression of lily-type lectin in response to environmental stress in turbot (*Scophthalmus maximus*). *Fish Shellfish Immunol.* 2016 Nov;58:323-331.

Kugapreethan R, Wan Q, Nilojan J, Lee J. Identification and characterization of a calcium-dependent lily-type lectin from black rockfish (*Sebastes schlegelii*): Molecular antennas are involved in host defense via pathogen recognition. *Dev Comp Immunol.* 2018 Apr;81:54-62.

Li XP, Hu YH. CD94 of tongue sole *Cynoglossus semilaevis* binds a wide arrange of bacteria and possesses antibacterial activity. *Fish Shellfish Immunol.* 2016 Nov;58:641-649.

Lv C, Zhang D, Wang Z. A novel C-type lectin, Nattectin-like protein, with a wide range of bacterial agglutination activity in large yellow croaker *Larimichthys crocea. Fish Shellfish Immunol.* 2016 Mar;50:231-41.

Mu L, Yin, Yang, Wu L, Wu H, Li B, Guo Z, Ye J. Functional characterization of a mannose-binding lectin (MBL) from Nile tilapia (*Oreochromis niloticus*) in non-specific cell immunity and apoptosis in monocytes/ macrophages. *Fish Shellfish Immunol.* 2019 Jan 14;87: 265-274.

Ng TB, Fai Cheung RC, Wing Ng CC, Fang EF, Wong JH. A review of fish lectins. Curr Protein Pept Sci. 2015;16(4):337-51.

Ngai PH, Ng TB. A mannose-specific tetrameric lectin with mitogenic and antibacterial activities from the ovary of a teleost, the cobia (*Rachycentron canadum*). *Appl Microbiol Biotechnol.* 2007 Feb;74(2):433-8.

Pan S, Tang J, Gu X. Isolation and characterization of a novel fucose-binding lectin from the gill of bighead carp (*Aristichthys nobilis*). *Vet Immunol Immunopathol.* 2010 Feb 15;133(2-4):154-64.

Rubeena AS, Divya M, Vaseeharan B, Karthikeyan S, Ringø E, Preetham E. Antimicrobial and biochemical characterization of a C-type lectin isolated from pearl spot (*Etroplus suratensis*). *Fish Shellfish Immunol.* 2019 Jan 10;87:202-211

Thulasitha WS, Umasuthan N, Whang I, Nam BH, Lee J. Antimicrobial response of galectin-1 from rock bream *Oplegnathus fasciatus*:Molecular, transcriptional, and biological characterization. *Fish Shellfish Immunol.* 2016 Mar;50:66-78.

Thulasitha WS, Whang I, Umasuthan N, Kang HS, Mothishri MS, Lee S, Qiang W, Noh JK, Lee J. A galectin related protein from *Oplegnathus fasciatus*: Genomic, molecular, transcriptional features and biological responses against microbial pathogens. *Dev Comp Immunol.* 2016 Mar;56:13-24.

Wang L, Zhang J, Kong X, Zhao X, Pei C, Li L. A C-type lectin, Nattectin-like protein (CaNTC) in Qihe crucian carp *Carassius auratus*: Binding ability with LPS, PGN and various bacteria, and agglutinating activity against bacteria. *Fish Shellfish Immunol.* 2017 Aug;67:382-392.

Wang Y, Bu L, Yang L, Li H, Zhang S. Identification and functional characterization of fish-egg lectin in zebrafish. *Fish Shellfish Immunol.* 2016 May;52:23-30.

Xue D, Guang-Hua W, Yan-Li S, Min Z, Yong-Hua H. Black rockfish C-type lectin, SsCTL4: A pattern recognition receptor that promotes bactericidal activity and virus escape from host immune defense. *Fish Shellfish Immunol.* 2018 Aug;79:340-350.

Yang Q, Wang P, Wang S, Wang Y, Feng S, Zhang S(1), Li H. The hepatic lectin of zebrafish binds a wide range of bacteria and participates in immune defense. *Fish Shellfish Immunol*. 2018 Nov; 82:267-278.

In: Hemagglutinins
Editors: T. Bun Ng, J. Wong, R. Tse et al.
ISBN: 978-1-53615-708-6
© 2019 Nova Science Publishers, Inc.

Chapter 12

ANTICANCER AND OTHER ACTIVITIES OF AMPHIBIAN HEMAGGLUTININS

Jack Ho Wong[1,2], Tzi Bun Ng[3,*], Charlene Cheuk Wing Ng[4], Chit Tam[3], Yau Sang Chan[5], Ryno Naude[6], Ryan Tse[1], Tak Fu Tse[1] and Helen Chan[1,†]

[1]Vita Green Pharmaceutical (Hong Kong Limited), Genning Partners Company Limited, and Institute of Medical Research, Hong Kong, China

[2]The Third Affiliated Hospital of Shenzhen University, Shenzhen, China

[3]School of Biomedical Sciences, Faculty of Medicine, The Chinese University of Hong Kong, China

[4]GKT School of Medical Education, King's College London, United Kingdom

[5]Department of Obstetrics and Gynecology, LKS Faculty of Medicine, University of Hong Kong, China

[6]Department of Biochemistry and Microbiology, Nelson Mandela Metropolitan University, Port Elizabeth, South Africa

ABSTRACT

Bullfrog galectin is abundant in the ovary, but is also present in the kidney, liver, sciatic nerve, skeletal muscle, and skin. Bullfrog galectin exerted a suppressive action on

---

* Corresponding Author's E-mail: tzibunng.cuhk.edu.hk.

† Corresponding Author's E-mail: Helen.chan@vitagreen.com.

all human leukemia cells tested including HL-60, K562, and U937 cells but not human colon cancer Colo 201cells or murine mammary tumor FM3A cells. A sialic acid binding lectin produced from bullfrog eggs demonstrated RNase activity toward poly U and poly C. It acts on the mitochondrial pathway and the endoplasmic reticulum stress-mediated pathway to produce its antitumor action. It inhibits the growth of tumor cells such as P388 and L1210 leukemia cells, human leukemia Jurkat cells, mesothelioma cells and human breast cancer cells. An alpha-N-acetyl galactosamine –binding lectin isolated from bullfrog eggs demonstrated hemagglutinating activity toward blood group A-erythrocytes but not blood group B- or O-erythrocytes. A soluble beta-galactoside- binding S type lectin from unfertilized bullfrog eggs does not depend on Ca++ or SH-reagent for hemagglutinating activity. *Japanese* egg lectin agglutinates tumor cells. *Xenopus laevis* oocyte cortical granule lectin (XL35) plays a role in fertilization and embryonic development. Salamander *Andrias davidianus* skin lectin has antifungal and antibacterial activities.

Keywords: amphibian, anticancer, antifungal and antibacterial activities

INTRODUCTION

Lectins have been reported from bacteria (Podgorskiĭ et al.,2014), viruses (Mair et al., 2014), fungi (Ng and Wong, 2013; Kobayashi and Kawagishi, 2014; Singh et al., 2014), algae (Akkouh et al.,2015 Singh et al., 2015), ferns (Tateno et al., 2003), gymnosperms (Yagi et al., 2002), flowering plants (Rudiger and Gabius, 2001; Van Damme, 2014;Yau et al.,2015), protozoans (Csaba, 2016), corals (Kvennefors et al., 2008), annelids (Hemmi et al., 2013),arthropods (Ng et al., 1985; Grubhoffer et al.,2004; Wang and Wang, 2013; Grubhoffer et al.,2004), mollusks (Ng et al.,1985; Vasta et al., 2015), echinoderms (Bordbar et al., 2011), protochordate (Vasta et al., 1999), tunicates (Wright and Cooper,1984), fish (Ng et al., 2015), amphibians(Lee et al., 2004; Ishino et al., 2007), reptiles (Arlinghau et al; 2013Eble), birds (Laursen and Nielsen, 2000), and mammals (Mason and Tarr, 2015). The intent of this article is to review lectins produced by amphibians. It is known that some amphibian tissues such as frog skin (Zhang et al., 2005; Leroy et al., 2006; Conlon et al., 2014; Kumar et al., 2015;Wang et al., 2015; Xu and Lai, 2015) and frog brain (Danger et al., 1985; Lázár.2001; Ukena et al., 2003) are abundant sources of peptides with a diversity of activities.

Bullfrog Galectin-1

Galectin-1 is a 15- kDa beta-galactoside-binding lectin from bullfrog (*Rana catesbeiana*) oocytes. Bullfrog galectin is abundant in the ovary, but is also present in the kidney, liver, sciatic nerve, skeletal muscle, and skin. It is located in basement membranes, cell-cell junctions, extracellular matrix, or secretory materials in several organs. In the ovary, this lectin is localized in the rim of the yolk platelet, but not in the cortical granule.

Rana oocyte yolk platelets are rich in galectin, unlike *Xenopus laevis* oocyte yolk platelets and cortical granules which possess other classes of lectins (Uchiyama et al., 1997).

The glycan binding of galectin-1 has been reported. Human blood type A-hexasaccharide (GalNAcalpha1-3(Fucalpha1-2 Galbet; 1-4GlcNAc beta1 -4Galbeta1-4Glc) bound to the galectin most strongly followed by Forssman antigen (GalNAcalpha1-3GalNAcbeta1- 3Gal alpha1 -4Galbeta1-4Glc) and type-A-tetra saccharide (GalNAcalpha1-3(Fucalpha1-2 Gal beta1-4GlcNAc beta1-4Glc). The glycan binding profile is unique compared with other galectins. Galectin-1 also bound to human rhabdomyosarcoma cells (Kawsar et al., 2009).

Anticancer Activity of Bullfrog Galectin-1

Bullfrog galectin-1 RCG1 exerted a suppressive action on all human leukemia cells tested including HL-60, K562, and U937 cells but not human colon cancer Colo 201cells or murine mammary tumor FM3A cells. Unlike other galectins which promote apoptosis, RCG1 arrested growth of cancer cells which was inhibited by lactose, but not by melibiose or sucrose. RCG1 did not cause the cancer cells to undergo necrosis or apoptosis (Yasumitsu et al., 2011).

Bullfrog Sialic Acid-Binding lectin (SBL or cSBL)

A 111-residue sialic acid binding lectin from frog (*Rana catesbeiana*) eggs with a unique amino acid sequence was reported by Titani et al. (1987). *Rana catesbeiana* egg lectin with five intrachain disulfide bonds bound and agglutinated human and animal cancer cells but not erythrocytes, fibroblasts lymphocytes. Its potency of agglutination of cancer cells was markedly lowered by submaxillary mucin, fetuin and keratan sulfate, but not by their desialylated forms or glycoproteins with a lower extent of sialylation such as transferrin (Nitta et al., 1987).

Bullfrog sialic acid-binding lectin is a basic protein with amino acid sequence resemblance to pancreatic ribonuclease and human angiogenin. Although the disulfide bond linking Cys-65 and Cys-72 of pancreatic RNase is missing in SBL and angiogenin, there are three other disulfide bonds common to all of them and found in similar locations (Nitta et al., 1993). SBL possesses secondary and tertiary structures similar to those of bovine pancreatic ribonucleases (EC 3.1.27.5) as judged by the conservation of several Cys residues which form disulfide bonds, a group of nonpolar residues, and a number of residues found in the active site of bovine ribonuclease activity((Lewis et al., 1989).

RNase Activity of SBL

SBL demonstrated RNase activity toward poly U and poly C. Human placental RNase inhibitor inhibited the RNase activity of RNase A and angiogenin but not the RNase and cancer cell agglutinating activities of SBL (Nitta et al., 1993).

Anticancer Activity of SBL

Sialic acid-binding lectin (SBL) from *Rana catesbeiana* oocytes is a multi-functional protein with ribonuclease as well as lectin activities. Thus it is referred to as leczyme. SBL-stimulated agglutination and its anticancer action are undermined in the presence of sialomucin but not asialomucin. SBL did not affect sialidase- treated cells. SBL synergized with interferon- γ and tumor necrosis factor-related apoptosis ligand (TRAIL) in its antitumor action. Sialylated glycoconjugates on the tumor cell surface may play a role in the hemagglutinating as well as antitumor activity of the lectin, SBL acts on the mitochondrial pathway as well as the endoplasmic reticulum stress-mediated pathway to produce its antitumor action (Tatsuta et al., 2014 a,b,c).

Anticancer Activity of SBL against Mouse Leukemia P388 Cells and L1210 Leukemia Cells

cSBL-treated P388 cells showed widespread RNA fragmentation after 1 h. However, cSBL-treated SBL-resistant RC-150 cells were unaffected even after 24 h. cSBL downregulated intracellular calcium ion concentration, protein kinase A activity, glutathione levels but upregulated activities of protein kinase G and the detoxifying enzyme glutathione-S- transferase.in P388 cells. Similarly treated cSBL-insensitive RC-150 cells did not show any changes (Nitta et al., 1996).

The distribution of heat shock protein Hsp70 and heat shock cognate protein Hsc70 in P388 cells underwent striking changes before SBL exerted its pro-apoptotic action. Attenuated Hsp70 expression mitigated SBL-elicited apoptosis (Tatsuta et al., 2014).

Treatment of mouse leukemia P388 cells with SBL/RC-RNase brought about apoptosis as evidenced by nuclear condensation and DNA fragmentation but had little effects on fibroblasts umbilical vein endothelial cells. SBL/RC-RNase urpregulated caspase-8 and then apoptosis via the caspase-3 pathway. Removal of sialic acid from P388 cells and/or the co-presence of free bovine submaxillary mucin. reduced the cytotoxicity of SBL/RC-RNase. Following adhesion to cholesterol-rich microdomains SBL/RC-RNase gained entry to the cells. Both methyl-β-cyclodextrin which removes cholesterol, and quercetin which inhibits cell membrane heat shock protein Hsp70 and heat shock cognate protein Hsc 70, attenuated apoptosis triggered by SBL/RC-RNase (Ogawa et al., 2014).

Sialic acid-binding lectin from *Rana catesbeiana* eggs inhibited the growth of tumor cells such as P388 and L1210 leukemia cells (Nitta et al., 1994).

Anticancer Activity of SBL against Human Leukemia Jurkat Cells

SBL was cytotoxic to human leukemia Jurkat cells irrespective of P-glycoprotein expression. It brought about mitochondrial perturbation and caspase- dependent apoptosis (Tatsuta et al., 2013a).

SBL triggered multiple apoptotic pathways. It upregulated the intrinsic pathway of apoptosis and causes rigid mitochondrial perturbation. SBL independently activated caspase-4, which initiates ER stress-mediated apoptosis, activates caspase-8 and caspase-9. Inhibition of caspase-4 led to suppression of SBL-triggered apoptosis (Tatsuta et al., 2013b).

Anticancer Activity of SBL against Human Breast Cancer Cells

SBL induced cell death in a variety of human breast cancer cells including SK-BR-3, MCF-7, and MDA-MB231 cells but did not affect normal cells. The lectin, through its RNase activity, upregulated the p38 mitogen-activated protein kinase (MAPK) signaling pathway. When p38 phosphorylation MAPK was suppressed by a p38 MAPK inhibitor or short interference RNA knockdown of p38 MAPK expression, a greater number of MDA-MB231 cells survived the treatment with SBL. H103A, a mutant of SBL devoid of RNase activity, only weakly elicited cell death. H103A could be internalized into the cells and inhibited MAPK phosphorylation of p38. SBL, but not H103A, activated caspase-3/7 and upregulated poly (ADP-ribose)-polymerase cleavage. The pan- caspase inhibitor zVAD-fmk, and the p38 MAPK inhibitor SB203580, inhibited caspase-3/7 activation brought about by SBL. zVAD-fmk attenuated cell death caused by SBL (Kariya et al., 2016).

Anticancer Activity of SBL against Mesothelioma Cells

Malignant mesothelioma is an aggressive cancer caused by asbestos with few therapeutic choices exposure. Sialic acid-binding ***lectin*** from *Rana catesbeiana* eggs (cSBL) caused morphological and biochemical apoptotic changes in both xenograft-competent H2452 and MSTO human mesothelioma cell lines. cSBL-induced apoptosis via an intrinsic pathway and suppressed tumor growth in two xenograft nude mouse models, without any adverse effects. Although cSBL by itself exerted cytotoxicity toward mesothelioma cell. Exposure of H28 cells to a combination of pemetrexed and cSBL ensued in more potent cytotoxicity than pemetrexed combined with cisplatin, which represents the usual treatment given to mesothelioma. The synergistic action of a combination of pemetrexed and cSBL was attributed to the cytostatic action of pemetrexed and the cytotoxicity of cSBL. It thus appears that cSBL has therapeutic potential for the treatment of mesothelioma (Satoh et al., 2017, Tatsuta et al., 2018). Bid mediated the synergistic actions of SBL and TRAIL on malignant mesothelioma (Tatsuta et al., 2014c).

Effect of Amino Acid Substitution on Anticancer Activity of cSBL

The anticancer activity of cSBL and its uptake into tumor cells increased as the net charge of cSBL was elevated. Substitution of D24 and E88 in cSBL with arginine worked better than that of E97. The double mutant D24RE88R and ethylenediamine-modified cSBL displayed similar anticancer activity (Ogawa et al., 2002).

Effect of Chemical Modification on Anticancer Activity of SBL

Chemical modification of SBL with a water-soluble carbodiimide, when nucleophiles such as ethylenediamine, taurine and glycine methylester were present, augmented the anticancer activity. The SBL obtained after chemical modification with ethylenediamine exhibited higher antitumor activity than SBL modified with glycine methylester, followed by SBL modified with taurine. RNase activity toward double-stranded RNA and RNA-oligo dA hybrids under physiologic ionic strength conditions was enhanced. On the other hand, RNase activity toward single-stranded RNA and poly U was reduced without affecting activity towards poly C. The base preference of the B2 base recognition site of modified SBL for guanine declined but the preference for adenine and cytosine was elevated. (Iwama et al., 2011).

Alpha-N-acetylgalactosamine–Binding Bullfrog Egg Lectin

A 21-kDa alpha-N-acetylgalactosamine–binding lectin isolated from *Rana catesbiana* eggs demonstrated hemagglutinating activity toward blood group A-erythrocytes but not blood group B- or 0-erythrocytes (Sakakibara et al., 1976).

Rana catesbeiana egg lectin specifically agglutinated both intact blood group A- adult human erythrocytes and umbilical cord human erythrocytes.

Similar to phytohemagglutinin P and protamine, *Rana catesbiana* egg agglutinin agglutinated mouse mammary carcinoma (FM3A) cells and exhibited cytotoxic activity toward the cells (Takeda et al., 1981).

14-kDa Homodimeric Soluble Beta-Galactoside- Binding S Type Lectin from Unfertilized Frog (*Rana catesbeiana*) Eggs

A 14-kDa homodimeric soluble beta-galactoside- binding S type lectin from unfertilized frog (*Rana catesbeiana*) eggs did not depend on Ca++ or SH-reagent for hemagglutination activity (Ozeki et al., 1991).

Distinct recognition mechanisms are involved in the cell adhesive activity of frog (*Rana catesbeiana*) S-type 14K lectin and a C-type lectin echinoidin from sea urchin plasma. Adhesion of human rhabdomyosarcoma cells (attached to and spread on lectin-coated plastic plates) brought about by the amphibian lectin was abolished by asialofetuin glycopeptide and lactose whereas echinoidin- induced cell adhesion was prevented by the peptide GRGDS which contains RGD sequence present in echinoidin. The results illustrate that some lectins may have cell adhesive activity instead of working through the carbohydrate recognition mechanism (Ozeki et al., 1991).

Bullfrog Skin Lectin

Lectins in the bullfrog skin exhibit analogous specific binding profiles in dorsal and ventral skin; During metamorphosis the larval glycoconjugate binding profiles are altered at the same time the skin changes histologically as well as physiologically. remodeling. The skin lectin may have a role in the regulation of water balance in the aquatic larvae and amphibian adult bullfrog (Kaltenbach et al., 2004).

Bullfrog Liver Lectin

Rana catesbeiana liver lectin agglutinated intact blood group A- adult human erythrocytes much more strongly than intact blood group A- umbilical cord human erythrocytes. *Xenopus laevis* skin lectin nonspecifically agglutinated adult and umbilical cord human erythrocytes. *Plecoglossus altivelis* egg lectin specifically agglutinated intact blood group B- adult and umbilical cord human erythrocytes, but weakly agglutinated blood group intact B- umbilical cord human erythrocytes (Nitta et al., 1988).

Japanese Frog (*Rana Japonica*) Egg Lectin

Japanese frog (*Rana japonica*) egg lectin is a 111- residue single-chain protein with an N- terminal pyroglutamyl residue and 4 disulfide bonds [Cys 19 -Cys72], [Cys34-Cys82], [Cys52-Cys97] and[Cys94-Cys111] resembling those of bull frog egg S-lectin, human angiogenin, and pancreatic ribonucleases (Kamiya et al., 1990)

Rana catesbeiana and *R. japonica* egg lectins which possess RNase activity and *R. catesbeiana* liver RNase demonstrate 65-83% sequence homology. The base specificities of their B1 and B2 sites are U > C and G >U >A, C, respectively. They exhibited greater stability than RNase A when subjected to pH-induced denaturation, guanidine hydrochloride treatment, and heat treatment. cells *R. catesbeiana* liver RNase did not agglutinate tumor cells, unlike *R. catesbeiana* and *R. japonica* egg lectins which agglutinated tumor cells (e.g., Ehrlich, S-180 and AH109A ascites carcinoma. *Rana catesbeiana* and *R. japonica* egg lectin induced AH109A cell agglutination was prevented by nucleotides. The agglutination sites are not the same as albeit related to active sites of the three frog proteins (Okabe et al., 1991).

RC-RNases prepared from *Rana catesbeiana* yolk granules and cytisolic fractions were the same. The ribosomal RNA was degraded in RC-RNase- treated HeLa cells but not in RNase A-treated HeLa cells (Liao et al., 1996).

Bullfrog Liver RNase

Bullfrog liver RNase was a 111-residue protein with three disulfide bridges in RNase A conserved but devoid of a [Cys65-Cys126] disulfide bridge in RNase A deleted, and instead a new C-terminal disulfide bridge. Bullfrog liver RNase contained all catalytic amino acid residues in bovine pancreatic RNase A conserved except for Asp121and there was 30.6% sequence homology between the two RNases. Bullfrog liver RNase showed 64.8% and 70.2% sequence homology to and *R. japonica* egg lectin and bullfrog egg lectin, respectively (Nitta et al., 1989).

Crystal Structures of RC-RNase Isozymes

RC-RNases are pyrimidine-guanine sequence specific ribonucleases from *Rana catesbeiana* oocytes with lectin and anticancer activities. RC-RNase isozyme 3 crystals are in the orthorhombic C222(1) space group, having the unit-cell parameters a = 66.66, b = 97.38, c = 85.74 A. RC-RNase isozyme 4 crystals are in the trigonal space group P3(1) or P3(2), having the unit-cell parameters a = b = 32.22, c = 92.12 A. RC-RNase isozyme 6 crystals complexed with cytidylyl 2'-5' guanosine are in the tetragonal space group

P4(1)2(1)2 or P4(3)2(1)2, having the unit-cell parameters a = b = 61.80, c = 65.96 A (Liu et al., 2011).

Solution Structure of a Cytotoxic Ribonuclease from Bullfrog Oocytes

The structure of the RNase consists of three alpha-helices and two triple-stranded anti-parallel beta-sheets and folds in a kidney-shape, very similar to the X-ray crystal structure of a homologous protein, onconase isolated from *Rana pipiens* The two inhibitors, 2'-deoxycytidylyl (3'-->5')-2'-deoxyguanosine (3',5'-dCpdG) and cytidylyl (2'-->5') guanosine (2',5'-CpG) bind to RC-RNase in a "retro-binding"manner, with the guanine base bonded to the B1 subsite. The His103 residue is located in the B state with the imidazole ring pointing away from the active site (Chang et al., 1998).

Xenopus Laevis Egg RNase

A 122 kDa RNase from eggs of *Xenopus laevis* two 60 kDa subunits was able to recognize pyrimidine bases specifically. The pH optimum of the RNase was 7.5 in Tris-HCl buffer. The enzyme activity was abolished by treatment at 80 degrees C for 5 min and pH 2 or 12 for 1 h. Since egg lectins with RNase activity obtained from Rana catesbeiana and R. japonica and bovine pancreatic RNase A show about 30% protein homology and these three proteins are 12-14 kDa heat-stable RNases(Nitta et al., 1993).

The RNase is a 111-amino acid protein with a cytotoxic action on several tumor cell lines. Its structure characterized by the presence of three alpha-helices and two triple-stranded antiparallel beta-sheets, exhibiting resemblance to the pyrimidine family ribonucleases. Two sets of resonances were observed for 11 amide protons and 8 alpha-protons located in the loop-1 region, an alpha 2 helix, and three beta-strands, (beta 1, beta 3 and beta 4), indicating nonlocalized multiple conformations for the RNase (Chen et al., 1996).

Xenopus Laevis Lectins

Xenopus lectin family (X-lectins) contain a number of members. Two of them are called XCL-1 and XCL-2, both of which are expressed in the embryo. XCL-1 is a 35-kDa calcium ion -dependent lectin from adult *Xenopus* serum. The expression of XCL-1 gene occurs in the adult stage and is not stimulated during tail regeneration. XCL-2 is a XCL-1-related protein. The expression of XCL-1 gene occurs in the middle/late tadpole stages gene expression was significantly increased in the regenerating tails and is stimulated during tail regeneration (Ishino et al., 2007).

The *Xenopus laevis* oocyte cortical granule lectin (XL35) plays a role in fertilization and embryonic development. The ascidian worm lamprey, trout, frog, mouse, and human produce proteins with sequence resemblance to XL35 and a common fibrinogen-like motif involved in binding carbohydrates. Release of the lectins from specialized vesicles in which they are stored occurs in response to microbial infection. Some of these lectins bind to bacterial oligosaccharides from pathogens. Hence they are important in defense against pathogens (Lee et al., 2004).

Salamander Lectins

Andrias davidianus skin lectin is a dimeric 17- kDa. protein with Ca^{2+} and Mg2+ ion-independent hemagglutinating activity toward native and trypsinized human B red blood cells which was stable between 4°C and 50°C. Glycoproteins but not monosaccharides inhibited its hemagglutinating activity. The lectin exhibited antifungal activity against *Saccharomyces cerevisiae*, and antibacterial activity against *Bacillus subtilis, Enterobacter aerogenes, Escherichia coli, Shewanella* sp. and *Staphylococcus aureus* (Qu et al., 2015).

The amino-acid sequence of *Pleurodeles waltl* oviduct lectin resembles C-type carbohydrate-recognition domains of other animal lectins and displays a greater extent of resemblance to animal lectins with defense function than lectins with roles in development and reproduction (Tiffoche et al., 1993).

DISCUSSION

Peptides isolated from the frog skin serve as anti-cancer, anti-viral, immunomodulatory, and anti-diabetic agents (Conlon et al., 2014). The frog ovary and liver are also the sources of bioactive proteins. Frog lectins represent mostly the work of Japanese investigators. The structures and activities of these lectins have been elucidated.

The antimicrobial and anticancer activities of the afore-mentioned frog egg lectins are in keeping with egg lectins of other animals. An L-rhamnose-binding egg lectin from the rock boring sea urchin *Echinometra lucunter* agglutinated Gram -positive bacterial cells, suggesting its ability to recognize pathogens (Carneiro et al., 2015). Snail (*Aplysia dactylomela*) egg lectin agglutinated *taphylococcus aureus* and suppressed biofilm formation (Carneiro et al., 2017). Zebrafish fish-egg lectins increased the phagocytic ability of macrophages toward Gram-positive and Gram- negative bacteria rendering the embryos resistant to the pathogenic bacterium *Aeromonas hydrophila* (Wang et al., 2016). Upregulation of turbot *(Scophthalmus maximus*) ovarian rhamnose-binding lectin in mucosal tissues was observed after exposure to *Streptococcus iniae* and *Vibrio anguillarum*, suggesting regulation of mucosal immunity by the lectin (Gao et al., 2018). *Pomacea scalaris* egg lectin agglutinated bacteria and altered the intestinal morphophysiology of potential predators thus protecting the embryos. (Ituarte et al., 2018). *Aplysia kurodai* lectin is involved in organogenesis in the developmental stage of *A. kurodai* (Motohashi et al., 2017).

An α-galactoside-binding catfish (*Silurus asotus*) egg lectin exerted antiproliferative but not cytotoxic activity toward Gb3-expressing Burkitt's lymphoma Raji cells. The lectin brought about cell cycle arrest in G0/1 phase, upregulated the CDK inhibitors p21 and p27, but downregulated the transcription of cell cycle-related factors such as cyclin D3, cyclin-dependent protein kinase (CDK)-4. and c-MYC The Ras-MEK-ERK pathway was stimulated (Sugawara et al., 2017).

The frog egg lectins are unique in that their characteristic ribonuclease activity has not been reported from the egg lectins of other animals although they may exhibit anticancer or antimicrobial activity in common. The egg lectins are potential therapeutic candidates.

Acknowledgments

We gratefully acknowledge the award of Health and Medical Research Fund (no. 12131221) from Food and Health Bureau, The Government of Hong Kong Special Administrative Region, and grants from National Natural Science Foundation of China (no. 81471927).

References

Akkouh O, Ng TB, Singh SS, Yin C, Dan X, Chan YS, Pan W, Cheung RC. Lectins with anti-HIV activity: a review. *Molecules.* 2015 Jan 6;20(1):648-68.

Arlinghaus FT, Eble JA. C-type lectin-like proteins from snake venoms. *Toxicon.* 2012;60(4):512-9.

Bordbar S, Anwar F, Saari N. High-value components and bioactives from sea cucumbers for functional foods--a review. *Mar Drugs.* 2011;9(10): 1761-805.

Carneiro RF, Teixeira CS, de Melo AA, de Almeida AS, Cavada BS, de Sousa OV, da Rocha BA, Nagano CS, Sampaio AH. L-Rhamnose-binding lectin from eggs of the *Echinometra lucunter*: Amino acid sequence and molecular modeling. *Int J Biol Macromol.* 2015;78:180-8.

Carneiro RF, Torres RC, Chaves RP, de Vasconcelos MA, de Sousa BL,Goveia AC, Arruda FV, Matos MN, Matthews-Cascon H, Freire VN,Teixeira EH, Nagano CS,

Sampaio AH. Purification, Biochemical Characterization, and Amino Acid Sequence of a Novel Type of Lectin from *Aplysia dactylomela* Eggs with antibacterial/Antibiofilm Potential. *Mar Biotechnol* (NY). 2017 Feb;19(1):49-64.

Chang CF, Chen C, Chen YC, Hom K, Huang RF, Huang TH. The solution structure of a cytotoxic ribonuclease from the oocytes of *Rana catesbeiana* (bullfrog). *J Mol Biol.* 1998;283(1):231-44.

Chen C, Hom K, Huang RF, Chou PJ, Liao YD, Huang T. The secondary structure of a pyrimidine- guanine sequence-specific ribonuclease possessing cytotoxic activity from the oocytes of *Rana catesbeiana. J Biomol NMR.* 1996;8(3):331-44.

Conlon JM, Mechkarska M, Lukic ML, Flatt PR. Potential therapeutic applications of multi- functional host-defense peptides from frog skin as anti-cancer, anti-viral, immunomodulatory, and anti-diabetic agents. *Peptides.*2014 Jul;57:67-77.

Csaba G. Lectins and Tetrahymena - A review. Acta Microbiol Immunol Hung. 2016 Sep;63(3):279-291. Danger JM, Guy J, Benyamina M, Jégou S, Leboulenger F, Coté J, Tonon MC, Pelletier G, Vaudry H. Localization and identification of neuropeptide Y (NPY)-like immunoreactivity in the frog brain. *Peptides.* 1985 Nov-Dec; 6(6):1225-36.

Gao C, Su B, Zhang D, Yang N, Song L, Fu Q, Zhou S, Tan F, Li C. l-rhamnose-binding lectins (RBLs) in turbot (*Scophthalmus maximus* L.): Characterization and expression profiling in mucosal tissues. *Fish Shellfish Immunol.* 2018 Sep;80:264-273.

Grubhoffer L, Kovár V, Rudenko N. Tick lectins: structural and functional properties. *Parasitology.* 2004;129 Suppl:S113-25.

Hemmi H, Kuno A, Hirabayashi J. NMR structure and dynamics of the C-terminal domain of R-type lectin from the earthworm *Lumbricus terrestris. FEBS J.* 2013 Jan;280(1):70-82.

Irie M, Nitta K, Nonaka T. Biochemistry of frog ribonucleases. *Cell Mol Life Sci.* 998;54(8):775-84.

Ishino T, Kunieda T, Natori S, Sekimizu K, Kubo T. Identification of novel members of the *Xenopus* Ca2+ -dependent lectin family and analysis of their gene expression during tail regeneration and development. *J Biochem.* 2007Apr;141(4):479-88.

Ituarte S, Brola TR, Fernández PE, Mu H, Qiu JW, Heras H,Dreon MS. A lectin of a non-invasive apple snail as an egg defense against predation alters the rat gut morphophysiology. *PLoS One.* 2018 Jun 1;13(6):e0198361.

Iwama M, Ogawa Y, Ohgi K, Tsuji T, Irie M. Enzymatic properties of sialic acid binding lectin from *Rana catesbeiana* modified with a water- soluble carbodiimide in the presence of various nucleophiles. *Biol Pharm Bull.* 2001;24 (12) :1366-9.

Iwama M, Ogawa Y, Sasaki N, Nitta K, Takayanagi Y, Ohgi K, Tsuji T, Irie M. Effect of modification of the carboxyl groups of the sialic acid binding lectin from bullfrog (*Rana catesbeiana*) oocyte on anti-tumor activity. *Biol Pharm Bull.* 2001;24(9):978-81.

Kaltenbach JC, Faszewski EE, Nytch KE, Potter CH, Shanthakumar N, Fakin A. Glycoconjugate localization in larval and adult skin of the bullfrog, *Rana catesbeiana*: a lectin histochemical study. *J Morphol.* 2004;261(2):184-95.

Kamiya Y, Oyama F, Oyama R, Sakakibara F, Nitta K, Kawauchi H, Takayanagi Y, Titani K Amino acid sequence of a lectin from Japanese frog (*Rana japonica*) eggs. *J Biochem.* 1990; 108 (1) :139-43.

Kariya Y, Tatsuta T, Sugawara S, Kariya Y, Nitta K, Hosono M. RNase activity of sialic acid-binding lectin from bullfrog eggs drives antitumor effect via the activation of p38 MAPK to caspase-3/7 signaling 7 pathway in human breast cancer cells. *Int J Oncol.* 2016;49(4):1334-42.

Kawsar SM, Matsumoto R, Fujii Y, Yasumitsu H, Uchiyama H, Hosono M, Nitta K, Hamako J, Matsui T, Kojima N, Ozeki Y. Glycan-binding profile and cell adhesion activity of American bullfrog (*Rana catesbeiana*) oocyte galectin-1. *Protein Pept Lett.* 2009;16(6):677-84.

Kobayashi Y, Kawagishi H. Fungal lectins: a growing family. *Methods Mol Biol.* 2014;1200: 15-38.

Konishi Y, Iwamuro S, Hasunuma I, Kobayashi T, Kikuyama S. Molecular cloning and multifunctional characterization of host defense peptides from the bullfrog Harderian gland with special reference to catesbeianalectin. *Zoolog Sci.* 2013;30(3):185-91.

Kumar VT, Holthausen D, Jacob J, George S. Host Defense Peptides from Asian Frogs as Potential Clinical Therapies. *Antibiotics* (Basel). 2015 Mar 30;4(2):136-59.

Kvennefors EC, Leggat W, Hoegh-Guldberg O, Degnan BM, Barnes AC. An ancient and variable mannose-binding lectin from the coral *Acropora millepora* binds both pathogens and symbionts. *Dev Comp Immunol.* 2008;32(12):1582-92.

Laursen SB, Nielsen OL. Mannan-binding lectin (MBL) in chickens: molecular and functional aspects. *Dev Comp Immunol.* 2000;24(2-3): 85-101.

Lázár G. Peptides in frog brain areas processing visual information. *Microsc Res Tech.* 2001 Aug 15;54(4):201-19.

Lee JK, Baum LG, Moremen K, Pierce M. The X-lectins: a new family with homology to the *Xenopus laevis* oocyte lectin XL-35. *Glycoconj J.* 2004;21(8-9):443-50.

Lewis MT, Hunt LT, Barker WC. Striking sequence similarity among sialic acid-binding lectin, pancreatic ribonucleases, and angiogenin: possible structural and functional relationships. *Protein Seq Data Anal.* 1989;2(2):101-5.

Liao YD. A pyrimidine-guanine sequence- specific ribonuclease from *Rana catesbeiana* (bullfrog) oocytes. *Nucleic Acids Res.* 1992; 20(6):1371-7.

Liao YD, Huang HC, Chan HJ, Kuo SJ. Large-scale preparation of a ribonuclease from *Rana catesbeiana* (bullfrog) oocytes and characterization of its specific cytotoxic activity against tumor cells. *Protein Expr Purif.* 1996;7(2):194-202.

Liu JH, Liao YD, Sun YJ. Crystallization and preliminary X-ray diffraction analysis of cytotoxic ribonucleases from bullfrog *Rana catesbeiana*. *Acta Crystallogr D Biol Crystallogr.* 2001;57(Pt 11): 1697 -9.

Mair CM, Ludwig K, Herrmann A, Sieben C. Receptor binding and pH stability -how influenza A virus hemagglutinin affects host-specific virus infection. *Biochim Biophys Acta.* 2014 Apr;1838(4):1153-68.

Mason CP, Tarr AW. Human lectins and their roles in viral infections. *Molecules.* 2015;20(2):2229-71.

Motohashi S, Jimbo M, Naito T, Suzuki T, Sakai R, Kamiya H. Isolation, Amino Acid Sequences, and Plausible Functions of the Galacturonic Acid-Binding Egg Lectin of the Sea Hare *Aplysia kurodai Mar Drugs*..2017 Jun 2;15(6). pii: E161.

Ng TB, Cheung RC, Ng CC, Fang EF, Wong JH. A review of fish lectins. *Curr Protein Pept Sci.* 2015;16(4):337-51.

Ng TB, Wong CM, Li WW, Yeung HW. Horseshoe crab (*Limulus polyphemus*) lectin but not garden snail (Helix pomatia) lectin elicits insulin-like activities *in vitro*. *Comp Biochem Physiol B.* 1985;81(2):503-4.

Ng TB, Wong JH. Fungal proteins with anti- proliferative and anticancer activities. *Protein Pept Lett.* 2013 Apr;20(4):433-8.

Nitta K, Ozaki K, Hagioita K, Tsukamoto Y, Terasaki Y, Suzuki N, Sano A, Hosono M, Kawauchi H, Takayanagi Y. Purification and some properties of ribonuclease from *Xenopus laevis* eggs. *Biol Pharm Bull.* 1993;16(4):353-6.

Nitta K, Ozaki K, Ishikawa M, Furusawa S, Hosono M, Kawauchi H, Sasaki K, Takayanagi. Y, Tsuiki S, Hakomori S. Inhibition of cell proliferation by *Rana catesbeiana* and *Rana japonica* lectins belonging to the ribonuclease superfamily. *Cancer Res.* 1994;54(4):920-7.

Nitta K, Oyama F, Oyama R, Sekiguchi K, Kawauchi H, Takayanagi Y, Hakomori S, Titani K. Ribonuclease activity of sialic acid- binding lectin from *Rana catesbeiana* eggs. *Glycobiology.* 1993;3(1):37-45.

Nitta K, Ozaki K, Tsukamoto Y, Furusawa S, Ohkubo Y, Takimoto H, Murata R, Hosono M, Hikichi N, Sasaki K, et al. Characterization of a *Rana catesbeian*a lectin-resistant mutant of leukemia P388 cells. *Cancer Res.* 1994; 54(4) : 54(4) :928-34.

Nitta K, Ozaki K, Tsukamoto Y, Hosono M, Ogawakonno Y, Kawauchi H, Takayanagi Y, Tsuiki S, Hakomori S. Catalytic lectin (leczyme) from bullfrog (*Rana catesbeiana*) eggs. *Int J Oncol.* 1996 9(1):19-23.

Nitta K, Terasaki Y, Kawauchi H, Takayanagi G. Comparative studies of the haemagglutination of adult and umbilical cord erythrocytes by animal lectins. *Comp Biochem Physiol B.* 1988;91 (4):657-61.

Nitta K, Takayanagi G, Kawauchi H, Hakomori S. Isolation and characterization of *Rana catesbeiana* lectin and demonstration of the lectin-binding glycoprotein of rodent and human tumor cell membranes. *Cancer Res.* 1987; 47(18) 4877-83.

Nitta R, Katayama N, Okabe Y, Iwama M, Watanabe H, Abe Y, Okazaki T, Ohgi K, Irie M. Primary structure of a ribonuclease from bullfrog (*Rana catesbeiana*) liver. *J Biochem.* 1989;106(5):729-35.

Ogawa Y, Iwama M, Ohgi K, Tsuji T, Irie M, Itagaki T, Kobayashi H, Inokuchi N. Effect of replacing the aspartic acid/glutamic acid residues of bullfrog sialic acid binding lectin with asparagine/glutamine and arginine on the inhibition of cell proliferation in murine leukemia P388 cells. *Biol Pharm Bull.* 2002;25(6):722-7.

Ogawa Y, Sugawara S, Tatsuta T, Hosono M, Nitta K, Fujii Y, Kobayashi H, Fujimura T, Taka H, Koide Y, Hasan I, Matsumoto R, Yasumitsu H, Kanaly RA, Kawsar SM, Ozeki Y. Sialyl- glycoconjugates in cholesterol-rich microdomains of P388 cells are the triggers for apoptosis induced by *Rana catesbeiana* oocyte ribonuclease *Glycoconj J.* 2014;31(2):171-84.

Okabe Y, Katayama N, Iwama M, Watanabe H, Ohgi K, Irie M, Nitta K, Kawauchi H, Takayanagi Y, Oyama F, et al., Comparative base specificity, stability, and lectin activity of two lectins from eggs of *Rana catesbeiana* and *R. japonica* and liver ribonuclease from *R. catesbeiana. J Biochem.* 1991; 109(5):786-90.

Ozeki Y, Matsui T, Nitta K, Kawauchi H, Takayanagi Y, Titani K. Purification and characterization of beta-galactoside binding lectin from frog (*Rana catesbeiana*) eggs. *Biochem Biophys Res Commun.* 1991;178(1):407-13.

Ozeki Y, Matsui T, Titani K. Cell adhesive activity of two animal lectins through different recognition mechanisms. *FEBS Lett.* 1991;289(2): 145-7.

Podgorskiĭ VS, Kovalenko ÉA, Karpova IS, Sashchuk EV, Get'man EI. Extracellular lectins from saprophytic strains of bacteria of the genus Bacillus (review)]. *Prikl Biokhim Mikrobiol.* 2014;50(3):256-63.

Qu M, Tong C, Kong L, Yan X, Chernikov OV, Lukyanov PA, Jin Q, Li W. Purification of a secreted lectin from *Andrias davidianus* skin and its antibacterial activity. *Comp Biochem Physiol C Toxicol Pharmacol.* 2015 Jan;167:140-6.

Rüdiger H, Gabius HJ. Plant lectins: occurrence, biochemistry, functions and applications. *Glycoconj J.* 2001 Aug;18(8):589-613.

Sakakibara F, Takayanagi G, Kawauchi H, Watanabe K, Hakomori S. An anti-A-like lectin of Rana catesbiana eggs showing unusual reactivity. *Biochim Biophys Acta.* 1976;444(2):386-95.

Sasaki K, Saitoh M, Takayanagi G. Antitumor activity of egg lectin from frog. I. Inhibitory effects of egg lectins from *Rana japonica* and *Rana catesbiana* on growth of Ehrlich solid tumor in mice. *Folia pharmacol.* japon. 73, 571-578 (1977).

Satoh T, Tatsuta T, Sugawara S, Hara A, Hosono M. Synergistic anti-tumor effect of bullfrog sialic acid-binding lectin and pemetrexed in malignant mesothelioma. *Oncotarget.* 2017;8(26): 42466 -42477.

Singh RS, Thakur SR, Bansal P. Algal lectins as promising biomolecules for biomedical research. *Crit Rev Microbiol.* 2015 Feb;41(1):77-88.

Singh SS, Wang H, Chan YS, Pan W, Dan X, Yin CM, Akkouh O, Ng TB. Lectins from edible mushrooms. *Molecules.* 2014 Dec 31;20(1):446-69.

Sugawara S, Im C, Kawano T, Tatsuta T, Koide Y, Yamamoto D, Ozeki Y, Nitta K, Hosono M. Catfish rhamnose-binding lectin induces G0/1 cell cycle arrest in Burkitt's lymphoma cells via membrane surface Gb3.*Glycoconj J.* 2017 Feb;34(1):127-138.

Takeda S, Kubota K, Endo Y, Matsuzawa T. Study of agglutination of mouse mammary carcinoma (FM3A) cell induced by egg agglutinin of *Rana catesbiana*. II. Phytohemagglutinin P and protamine. *Exp Cell Biol.* 1981;49(1):54-9.

Tateno H, Winter HC, Petryniak J, Goldstein IJ. Purification, characterization, molecular cloning, and expression of novel members of jacalin-related lectins from rhizomes of the true fern *Phlebodium aureum* (L) J.Smith (Polypodiaceae). *J Biol Chem.* 2003 Mar 28;278(13):10891-9.

Tatsuta T, Satoh T, Sugawara S, Hara A, Hosono M. Sialic acid-binding lectin from bullfrog eggs inhibits human malignant mesothelioma cell growth in vitro and in vivo. *PLoS One.* 2018;13(1): e0190653.

Tatsuta T, Hosono M, Miura Y, Sugawara S, Kariya Y, Hakomori S, Nitta K. Involvement of ER stress in apoptosis induced by sialic acid-binding lectin (leczyme) from bullfrog eggs. *Int J Oncol.* 2013;43(6):1799-808.

Tatsuta T, Hosono M, Ogawa Y, Inage K, Sugawara S, Nitta K. Downregulation of Hsp70 inhibits apoptosis induced by sialic acid-binding lectin (leczyme). *Oncol Rep.* 2014;31(1):13-8.

Tatsuta T, Hosono M, Sugawara S, Kariya Y, Ogawa Y, Hakomori S, Nitta K. Sialic acid-binding lectin (leczyme) induces caspase-dependent apoptosis-mediated mitochondrial perturbation in Jurkat cells. *Int J Oncol.* 2013;43(5):1402-12.

Tatsuta T, Hosono M, Takahashi K, Omoto T, Kariya Y, Sugawara S, Hakomori S, Nitta K. Sialic acid-binding lectin (leczyme) induces apoptosis to malignant mesothelioma and exerts synergistic antitumor effects with TRAIL. *Int J Oncol.* 2014;44(2):377-84.

Tatsuta T, Sugawara S, Takahashi K, Ogawa Y, Hosono M, Nitta K. Cancer-selective induction of apoptosis by leczyme. *Front Oncol.* 2014;4:139.

Tatsuta T, Sugawara S, Takahashi K, Ogawa Y, Hosono M, Nitta K. Leczyme: a new candidate drug for cancer therapy. *Biomed Res Int.* 2014;2014: 421415.

Tiffoche C, Chesnel A, Jego P, Le Pennec JP. Isolation and characterization of a cDNA clone encoding a *Pleurodeles* lectin. *Eur J Biochem.* 1993 May 1;213(3):901-7.

Titani K, Takio K, Kuwada M, Nitta K, Sakakibara F, Kawauchi H, Takayanagi G, Hakomori S. Amino acid sequence of sialic acid binding lectin from frog (*Rana catesbeiana*) eggs. *Biochemistry.* 1987;26 (8):2189-94.

Uchiyama H, Komazaki S, Oyama M, Matsui T, Ozeki Y. Distribution and localization of galectin purified from *Rana catesbeiana* oocytes. *Glycobiology.* 1997;7(8):1159-65.

Ukena K, Koda A, Yamamoto K, Kobayashi T, Iwakoshi- Ukena E, Minakata H,Kikuyama S, Tsutsui K. Novel neuropeptides related to frog growth hormone-releasing peptide:

isolation, sequence, and functional analysis. *Endocrinology*. 2003 Sep;144(9):3879-84.

Van Damme EJ. History of plant lectin research. *Methods Mol Biol.* 2014; 1200:3-13.

Vasta GR, Feng C, Bianchet MA, Bachvaroff TR, Tasumi S. Structural, functional, and evolutionary aspects of galectins in aquatic mollusks: From a sweet tooth to the Trojan horse. *Fish Shellfish Immunol*. 2015 Sep;46(1):94-106.

Vasta GR, Quesenberry M, Ahmed H, O'Leary N. C-type lectins and galectins mediate innate and adaptive immune functions: their roles in the complement activation pathway. *Dev Comp Immunol*. 1999 Jun-Jul;23(4-5):401-20.

Wang XW, Wang JX. Diversity and multiple functions of lectins in shrimp immunity. *Dev Comp Immunol*. 2013 Jan-Feb;39(1-2):27-38.

Wang Y, Bu L, Yang L, Li H, Zhang S. Identification and functional characterization of fish-egg lectin in zebrafish. *Fish Shellfish Immunol.* 2016 May;52:23-30.

Wang YW, Tan JM, Du CW, Luan N, Yan XW, Lai R, Lu QM. A Novel Trypsin Inhibitor-Like Cysteine-Rich Peptide from the Frog Lepidobatrachus laevis Containing Proteinase-Inhibiting Activity. *Nat Prod Bioprospect*. 2015 Aug;5(4):209-14.

Wright RK, Cooper EL. Protochordate immunity--II. Diverse hemolymph lectins in the solitary tunicate *Styela clava*. *Comp Biochem Physiol B.* 1984;79(2):269-77.

Xu X, Lai R. The chemistry and biological activities of peptides from amphibian skin secretions. *Chem Rev*. 2015 Feb 25;115(4):1760-846.

Yasumitsu H, Mochida K, Yasuda C, Isobe M, Kawsar SM, Fujii Y, Matsumoto R, Kanaly RA, Ozeki Y. Antiproliferative effects of galectin-1 from Rana catesbeiana eggs on human leukemia cells and its binding proteins in human cells. *In Vitro Cell Dev Biol Anim*. 2011;47(10):728-34.

Yagi F, Iwaya T, Haraguchi T, Goldstein IJ. The lectin from leaves of Japanese cycad, *Cycas revoluta* Thunb. (gymnosperm) is a member of the jacalin-related family. *Eur J Biochem*. 2002 Sep; 269(17):4335-41.

Yau T, Dan X, Ng CC, Ng TB. Lectins with potential for anti-cancer therapy. *Molecules*. 2015 20(3):3791-810.

Zhang YX, Lai R, Lee WH, Zhang Y. Frog albumin is expressed in skin and characterized as a novel potent trypsin inhibitor. *Protein Sci*. 2005 Sep;14(9):2469-77.

About the Editors

Dr. Jack Ho Wong

Pharmaceutical Formulation Scientist,

Vita Green Pharmaceutical (Hong Kong) Ltd, Hong Kong

Honorary Associate Professor,

The Third Affiliated Hospital of Shenzhen University, Shenzhen, China

Jack Ho Wong received his Bachelor's Honors and PhD degrees in Biochemistry and postdoctoral training from the Chinese University of Hong Kong (CUHK). Currently he is a pharmaceutical formulation scientist at Vita Green Pharmaceutical (Hong Kong) Ltd involved in formulation of vitamins and Chinese medicinal products. Formerly he was a research associate at the Faculty of Medicine, CUHK. He has an H index of 26 and has published 128 papers in internationally refereed reputable journals, and 51 invited book chapters on a diversity of biomolecules encompassing short interfering RNAs, chemokine receptor, melatonin, lactoferrin, lactoferricin, lactoferrampin, antifungal proteins, peptides and compounds with anti-*Candida albicans,* and anti-phytopathogen activities, antiviral proteins and peptides with anti-HIV reverse transcriptase, protease and integrase activities, antibacterial peptides with anti-*Helicobacter pylori* and anti-MRSA activities, antifreeze proteins, venom proteins, ribosome inactivating proteins, lectins, trypsin inhibitors, polysaccharide-peptides, enzymes, dopamine D4 receptor, synaptosomal-associated protein 25, cathelicidins, defensins, ribonucleases, chitosan, marine products, methyl gallate and other antioxidants, and Chinese medicinal products including those from *Dendrobium, Cordyceps, Ganoderma, Panax notoginseng, Momordica, Trichosanthes, Dioscorea,* and *Coriolus* species. He has researched extensively on the mechanisms of anticancer action of antineoplastic constituents of medicinal mushrooms, the Chinese drug Tianhuafen and the bitter melon. He has also written about essential oils of botanical origin.He is the principal investigator of several research grants and a coinvestigator of a number of research grants. He has served as a reviewer of manuscripts submitted to various international journals and research grant applications submitted to the National Science Foundation of China and also national funding agencies in Europe and Singapore. He is on the editorial board of journals including Toxins, Frontiers in Microbiology, Frontiers in Pharmacology and a number of other journals.

Email: jack.wong@vitagreen.com; jack1993@yahoo.com.

Dr. Tzi Bun Ng

Professor, Faculty of Medicine,
The Chinese University of Hong Kong, Hong Kong

Tzi Bun Ng received his Bachelor's degree from the University of Hong Kong. He was awarded a Commonwealth scholarship tenable in Canada and received his PhD degree in endocrinology and protein biochemistry from Memorial University of Newfoundland under the supervision of Professor David Richard Idler who was well known for his work on pituitary, adrenal and gonadal hormones. He subsequently pursued postdoctoral research in Hormone Research Laboratory, University of California in San Francisco under the supervision of Professor Choh Hao Li who contributed significantly to the knowledge about many of the pituitary hormones and insulin-like growth factor and has been nominated several times for the Nobel Prize in Medicine and Physiology. Tzi Bun Ng conducted sabbatical research in microbial biotechnology at Imperial College London under the supervision of Professor Brian Hartley and Professor Conrad Lichtenstein. He is a professor at the Chinese University of Hong Kong. He has an H index of 65 and has published over 720 papers in international journals and over 100 book chapters. He has served as a reviewer of manuscripts submitted to various international journals, research grants submitted to national funding agencies in European and SE Asian countries and South Africa, and as an assessor of candidates for the PhD degree and for promotion to various professorial ranks in universities in Hong Kong, South Africa, Singapore, Pakistan and India. He is on the editorial boards of Scientific Reports, Toxins, Frontiers in Microbiology, Frontiers in Pharmacology, Frontiers in Oncology, Molecules, Biomed Research International, Current Protein and Peptide Science and other journals. He has served as a reviewer of manuscripts submitted to various international journals and research grant applications submitted to national funding agencies in Europe and Singapore. His research interests encompass a multitude of biologically active biomolecules including milk proteins, pathogenesis related proteins, pineal hormones including melatonin, fungal proteins and enzymes, mushroom polysaccharides and polysaccharopeptides, and bioactive compounds from medicinal herbs.

Email: tzibunng@cuhk.edu.hk

Ryan Edward Tse

Managing Director of Genning Partners Ltd.
and Non-Executive Member of Vita Green Pharmaceutical
(Hong Kong) Ltd., Hong Kong

Ryan Edward Tse received his Bachelor's degree Magna Cum Laude from Georgetown University, USA and his Master's degree in Business Management from London Business School, London. He is Managing Director of Genning Partners Ltd and a non-executive member of Vita Green Pharmaceutical (Hong Kong) Ltd. (a manufacturer of traditional Chinese medicine, health products, vitamins and western food supplements) Prior to his appointment at Genning Partners Company Limited (a direct selling Chinese herbal supplement company in Hong Kong), he was the Strategic Planning Manager of Vita Green and was responsible for formulation of strategies and directions for management analysis. His research interests lie in milk proteins, soy protein, management of obesity, health benefits of beetroot, and spirulina, functional foods and health foods, and complementary and alternative medicine. He has products related to his research interests.

Email: ryan.tse@vitagreen.com.

Dr. Tak Fu Tse

Co-Founder and Non-Executive Chairman
of Vita Green Pharmaceutical (Hong Kong) Ltd,
Honorary Clinical Associate Professor, Li Ka Shing Faculty
of Medicine, The University of Hong Kong, Hong Kong
Division of Cardiology, Department of Medicine,
Queen Mary Hospital, Li Ka Shing Faculty of Medicine,
The University of Hong Kong, Hong Kong

Tak Fu Tse is the co-founder and non-executive Chairman of Vita Green Pharmaceutical (Hong Kong) Ltd. Dr. Tse obtained his MBBS degree from University of Hong Kong (HK), and also has qualification as a Registered Chinese Medicine Practitioner. He is the past president of the World Heart Federation and presently the Head of Cardiology of the 813 Medical Centre in Hong Kong. He has a variety of professional appointments including Honorary Clinical Associate Professor in Division of Cardiology, Department of Medicine, University of Hong Kong Queen Mary Hospital. He is currently a member of the Antidoping Committee of the Sports Federation & Olympic Committee of HK, China, and a member of Chinese Medical Council of HK, Consultant of HK Registered Chinese Medicine Practitioners Association, President of Society of

Practitioners of Chinese Herbal Medicine, President of HK Disaster Medical Association, Vice-Chairman of HK Heart Foundation, Council Member of HK College of Cardiology and Honorary Treasurer of HK College of Physicians. He also Honorary Associate Professor in Cardiology of the University of Hong Kong, Honorary Consultants of Hong Kong Sanatorium Hospital, Grantham Hospital and Haven of Hope Hospital and Honorary Professor of the Nanjing University of Chinese Medicine. He held the position of the head of the Cardiology Department of University Medical Unit of HKU for a long period of time. He served as the Chairman of World Congress of Cardiology in Beijing 2010. He was a recipient of the Award for Life-time Distinguished Service to the Hong Kong Institute of Allergy. His research interests lies in cardiovascular surgery and medicine, Chinese Herbal Medicine, functional foods and health foods. Dr. Tse has published papers on heart diseases, stroke and the anticancer efficacy of the medicinal mushroom Coriolus versicolor. He is a co-investigator of two major research grants awarded in 2018 and 2019 by The Innovation and Technology Commission, The Government of Hong Kong. One of the grants is on Western medicine and another one is on Chinese medicine.

Email: tf.tse@vitagreen.com

Dr. Helen Chan

Founder, CEO and Chairman of Vita Green Pharmaceutical (Hong Kong) Ltd., Hong Kong

Helen Chan is the founder, CEO and Chairman of Vita Green Pharmaceutical (Hong Kong) Ltd. Dr. Chan was awarded her MBBS degree by the University of Hong Kong, and is also a Registered Chinese Medicine Practitioner. She owns a private practice in pediatric medicine. She is International Distinguished Fellow of the ACAAI, a Fellow of Royal College of Physicians (Edinburgh), Royal College of Physicians (Glasgow), American College of Allergy, Asthma & Immunology, American Academy of Pediatrics, American Academy of Pediatrics, HK Academy of Medicine (Pediatrics), and a member of Chinese Medicine Council of HK. Her wide ranging research interests embrace anticancer proteins, milk proteins, soy proteins, cranberry proanthocyanidins and mannose for treatment of urinary tract infections, Ganoderma lucidum, Coriolus versicolor, ginseng, Cordycepin, and medicinal compounds and medications with therapeutic value for joint problems, malignant diseases, menstrual pain, hyperuricemia, obesity, anemia, allergy and oxidative stress; medications for promoting heart, lung, hair, eye and ear health and for maintaining normoglycemia and andrological status, and acaricidal and insecticidal compounds. Dr. Chan has published papers on the antineoplastic and immunomodulatory actions of the medicinal mushroom Coriolus versicolor, and health promoting activities of other medicinal products some of which are her own products. She is the editor of an allergy journal, and the principal investigator of two major research grants awarded in 2018 and 2019 by The Innovation and Technology Commission, The Government of Hong Kong. One of the grants is on Western medicine and another one is on Chinese medicine. She was the founder of the Hong Kong Institute of Allergy and the awardee of 2018 President's Medal of the Institute.

Email: helen.chan@vitagreen.com

INDEX

A

B

N

P

R

T

V

W

Polyunsaturated Fatty Acids (PUFAs): Food Sources, Health Effects and Significance in Biochemistry

Editor: Angel Catalá, Ph.D.

Series: Biochemistry Research Trends

Book Description: This book presents an overview of polyunsaturated fatty acids (pufas): food sources, health effects and significance in biochemistry. The topics analyzed cover a broad spectrum of polyunsaturated fatty acids and present new information in this area of research.

Hardcover ISBN: 978-1-53613-572-5
Retail Price: $195

Glycosylphosphatidylinositol-Anchored Proteins and Their Release from Cells: From Phenomenon to Meaning

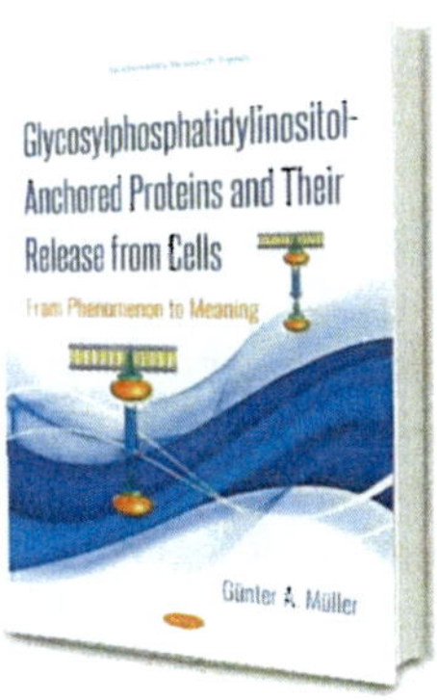

Author: Günter A. Müller

Series: Biochemistry Research Trends

Book Description: The book does not only represent a state-of-the-art compendium about the biology and (patho)physiology of GPI-AP, in general, and their cellular release including potential biomedical and biotechnological applications, in particular, but also introduces a novel concept for molecular life science research.

Hardcover ISBN: 978-1-53613-966-2
Retail Price: $230

Polyunsaturated Fatty Acids (PUFAs): Food Sources, Health Effects and Significance in Biochemistry

Editor: Angel Catalá, Ph.D.

Series: Biochemistry Research Trends

Book Description: This book presents an overview of polyunsaturated fatty acids (pufas): food sources, health effects and significance in biochemistry. The topics analyzed cover a broad spectrum of polyunsaturated fatty acids and present new information in this area of research.

Hardcover ISBN: 978-1-53613-572-5
Retail Price: $195

Glycosylphosphatidylinositol-Anchored Proteins and Their Release from Cells: From Phenomenon to Meaning

Author: Günter A. Müller

Series: Biochemistry Research Trends

Book Description: The book does not only represent a state-of-the-art compendium about the biology and (patho)physiology of GPI-AP, in general, and their cellular release including potential biomedical and biotechnological applications, in particular, but also introduces a novel concept for molecular life science research.

Hardcover ISBN: 978-1-53613-966-2
Retail Price: $230